Compressive Sensing
for Wireless Communication:
Challenges and Opportunities

RIVER PUBLISHERS SERIES IN COMMUNICATIONS

The "River Publishers Series in Communications" is a series of comprehensive academic and professional books which focus on communication and network systems. The series focuses on topics ranging from the theory and use of systems involving all terminals, computers, and information processors; wired and wireless networks; and network layouts, protocols, architectures, and implementations. Furthermore, developments toward new market demands in systems, products, and technologies such as personal communications services, multimedia systems, enterprise networks, and optical communications systems are also covered.

Books published in the series include research monographs, edited volumes, handbooks and textbooks. The books provide professionals, researchers, educators, and advanced students in the field with an invaluable insight into the latest research and developments.

Topics covered in the series include, but are by no means restricted to the following:

- Wireless Communications
- Networks
- Security
- Antennas & Propagation
- Microwaves
- Software Defined Radio

For a list of other books in this series, visit www.riverpublishers.com

Compressive Sensing for Wireless Communication: Challenges and Opportunities

Radha Sankararajan

SSN College of Engineering, India

Hemalatha Rajendran

SSN College of Engineering, India

Aasha Nandhini Sukumaran

SSN College of Engineering, India

River Publishers

Published, sold and distributed by:
River Publishers
Alsbjergvej 10
9260 Gistrup
Denmark

River Publishers
Lange Geer 44
2611 PW Delft
The Netherlands

Tel.: +45369953197
www.riverpublishers.com

ISBN: 978-87-93379-85-5 (Hardback)
      978-87-93379-86-2 (Ebook)

©2016 River Publishers

Content

Preface

Compressed sensing is a novel research area, which was introduced in 2006, and since then has already become a key concept in various areas of applied mathematics, computer science, and electrical engineering. It surprisingly predicts that high-dimensional signals, which allow a sparse representation by a suitable basis or, more generally, a frame, can be recovered from what was previously considered highly incomplete linear measurements by using efficient algorithms. This book is an extension of research work carried out by research scholars under the guidance of Dr. S. Radha. It is designed to serve as reference book for the elective subject on Compressive Sensing to the researchers, post graduate and undergraduate students.

In this book, an overview of the theories of sparse representation and compressive sampling is presented and several interesting imaging modalities based on these theories are examined. The use of linear and non-linear kernel sparse representation as well as compressive sensing in many computer vision problems including target tracking, background subtraction and object recognition are also explored. The use of CS for wireless networks, wireless sensor networks and cognitive radios are also discussed.

Chapter 2 deals with the concepts of compressed sensing and mathematics involved in it. This chapter explains in detail about sparse representation, sparsity and basis vectors. Different types of sparsifying transforms, different measurement matrices involved in CS and their properties are explained in detail. The applications of the compressed sensing such as object detection, target tracking, image reconstruction in holography, MRI etc. is also discussed.

Chapter 3 deals with the concept of existing greedy recovery algorithms and some of the non-iterative procedures for compressed data have been discussed with mathematical background. Special algorithms like focal under-determined system solution (FOCUSS) and multiple signal classification (MUSIC) algorithm have also been touched upon.

Chapter 4 deals with the issue in applying CS and sparse decompositions to speech and audio signals. It also describes the use of DCT and CS to obtain an efficient representation of audio signals, especially when they are sparse in the frequency domain. A new sensing matrix called orthogonal Symmetric Toeplitz Matrix (OSTM) generated with Binary, Ternary and PN sequence shows improved results compared to Random, Bernoulli, Hadamard and Fourier Matrices is also discussed in Chapter 4.

Chapter 5 provides a brief overview of some thrust CS based imaging applications to have energy efficient and complex representations of smaller magnitude. This chapter also deals with the methodologies and algorithms used for CS based image applications. Finally, a case study depicting the image transmission using compressed sensing in wireless multimedia sensor networks is also presented.

Chapter 6 discusses the application of compressed sensing to computer vision problems. Conventional object detection techniques and object tracking techniques such as point tracking, kernel tracking, silhouette tracking are explained in brief in this chapter. Compressive video processing with different transforms and compressed sensing based background subtraction are discussed in detail.

Chapter 7 explains the compressed sensing based wireless networking such as 3G wireless cellular networks, WiMAX, WiFi, Adhoc sensor networks etc. In this chapter, the different channel models like multipath channel estimation, random field estimation and multiple access techniques that uses compressed sensing concept has been discussed further.

Chapter 8 gives a brief introduction to various functions of cognitive radio, different approaches to spectrum sensing and then elaborates the use of compressed sampling for cognitive radios. It also gives a detailed insight on collaborative and distributed compressed sensing techniques followed by the research challenges in compressed sensing for cognitive radios.

Chapter 9 is concerned with WSN, its architecture, the associated wireless transmission technology, protocols utilized along with the operating system being used for WSN. It also provides details about the application of CS to WSN and the significance related to it.

Chapter 10 focuses on efficient strategies to improve the sampling and reconstruction process in CS.

Finally, Chapter 11 discusses the real time application of compressive sensing and their challenges in detail.

Acknowledgement

The authors are always thankful to the Almighty for guiding them in their perseverance and blessing them with achievements.

The authors wish to thank Mr. Shiv Nadar, Chairman, HCL Technologies, Founder, SSN Institutions, Ms. Kala Vijayakumar, President, SSN Institutions and Dr.S.Salivahanan, Principal, SSN College of Engineering for their whole hearted support given in this successful endeavor.

Dr. S. Radha would like to thank her parents Mr. R. Sankararajan and Ms. V. Vijayalakshmi for their constant motivation, inspiration and untiring support throughout the carrier: husband Mr. S. Kumaran and daughter K. Subhalakshmi, for their understanding, moral support and freedom to work.

Dr. R. Hemalatha would like to thank her parents Mr. M. Rajendran and Mrs. A. Nagarani for their inspiring guidance; husband Mr. D. Lakshmanakumar; son, L. H. Jeevithan and daughter L. H. Pradanyaa for their constant encouragement and moral support along with patience and understanding.

Ms. S. Aasha Nandhini would like to thank her parents Mr. M. P. Sukumaran and Mrs. M. Anantha Valli for their constant support and encouragement.

The authors wholeheartedly thank, appreciate and acknowledge Dr. K. Muthumeenakshi, Mrs. V. Angayarkanni, Ms. J. Florence Gnana Poovathy, and Ms. L. Umasaritha for their excellent and unforgettable help.

List of Figures

List of Tables

List of Algorithms

List of Abbreviations

3GPP	3G Partnership Project
AbS	Analysis-by-Synthesis
ACQUIRE	Active Query forwarding in sensor networks
ADC	Analog to Digital Converter
AGC	Automatic Gain Control
AIC	Analog to Information Converter
Aloha	Areal Locations of Hazardous Atmospheres
APSP	All-Pairs-shortest-paths
APTEEN	Adaptive periodic threshold-sensitive energy-efficient sensor network
ARMA	Autoregressive moving-average
AWGN	Additive white Gaussian noise
AWN	Additive White Gaussian
BCS	Block-based compressive sensing
BER	Bit error rate
BinDCT	Binary Discrete cosine transform
BMLD	Binaural masking level difference
BP	Basis Pursuit
BPDN	Basis Pursuit DeNoising
BPSK	Binary Phase Shift Keying
CBAH	Clustering-Based Aggregation Heuristic
CCD	Charge coupled device
CCE	Compressed Channel Estimation
CCK	Complementary Code Keying
CDF	Cohen-Daubechies-Feauveau wavelet
CDG	Compressive Data Gathering
CDM	Code division multiplexing
CDMA	Code Division Multiple Access
CH	Cluster head
CMOS	Complementary metal-oxide semiconductor

CoSaMP	Compressive Sampling Matching Pursuit
CPU	Central processing unit
CS	Compressed Sensing
CSEC	Compressive sensing erasure coding
CSMA	Carrier Sense Multiple Access
CT	Computerized tomography
DAM	Distributed Aggregate Management
DAQ	Data Acquisition System
DARPA	Defense Advanced Research Projects Agency
DBTMA	Dual busy tone multiple access
DC	Data-centric
DCF	Distributed coordination function
DCS	Distributed compressed sensing
DCT	Discrete Cosine transform
DFB	Directional Filter Bank
DFC	Data Fusion Centre
DFT	Discrete Fourier transform
DMD	Digital micro mirror device
DP	Directional Pursuit
DPM	Dynamic Power Management
DSA	Dynamic Spectrum Access
DSM	Dynamic Spectrum Management
DSSS	Direct sequence spread spectrum
DVB-T	Digital Video Broadcasting-Terrestrial
DVS	Dynamic Voltage Scaling
DWT	Discrete wavelet transform
EBAM	Energy-Based Activity Monitoring
EBW	Extended Baum-Welch
EMLAM	Expectation-Maximization Like Activity Monitoring
EMWV	Entropy metrics weighted average
EOMP	Enhanced Orthogonal Matching Pursuit
FAR	False Alarm Rate
FCC	Federal Communications Commission
FDMA	Frequency division multiplexing access
FFDs	Full function devices
FFS	Fast Fourier Sampling
FFT	Fast Fourier Transform
FHSS	frequency hopping spread spectrum
f-LTM	Forward light transport matrix

FOCUSS	Focal Underdetermined System Solution
FSK	Frequency shift keying
FSM	Finite state machine
GAF	Geographic Adaptive Fidelity
GBR	Gradient-Based Routing
GDLS	Gradient descend line search
GEAR	Geographic and Energy Aware Routing
GEDIR	Geographic Distance Routing
G-LRT	Generalized Likelihood Ratio Test
GMM	Gaussian Mixture Models
GP	Gradient Pursuit
GPS	Global Positioning System
GPU	Graphics processing units
GSCS	Group Sparse Compressed Sensing
GSM	Global System for Mobile
HDR	High dynamic range
HOG	Histogram of Oriented Gradients
IDEA	Incoherent detection and estimation algorithm
IF	Intermediate Frequency
IFFT	Inverse Fast Fourier Transform
IHT	Iterative Hard Thresholding
ILP	Integer Linear Program
i-LTM	inverse light transport matrix
IO – MUD	Individual Optimal Multiuser Detector
IRLS	Iteratively re-weighted least squares algorithm
ISM	Industrial, Scientific and Medical
ITU	International Telecommunication Union
JO	JO-MUD Joint Optimal Multiuser Detector
JPEG	Joint photographic experts group
JSCC	Joint source channel coding
JSM	Joint sparsity model
KF	Kalman Filter
KLT	Karhunen–Loève Transform
kNN	k-nearest neighbors
K-SVD	K-means Singular value decomposition
LAN	local-area network
LAs	Local Aggregators
LCD	Liquid crystal display
LDA	Linear Discriminative Analysis

LEACH	Low-energy adaptive clustering hierarchical
LLC	Logical link control
LML	Local Markov Loops
LNA	Low Noise Amplifier
LP	Laplacian Pyramid
LPC	Linear prediction coding
LSAF	Line Search A-functions
MAC	Medium Access Control
MAN	Metropolitan Area Networks
MANET	Mobile Ad Hoc Networks
MAR	Minimum-to-average ratio
MAs	Master Aggregators
MBCS	Model Based Compressive Sensing
MBWA	Mobile Broadband Wireless Access
MC-CDMA	Multicarrier code division multiple access
MC	Matrix Completion
MCFA	Minimum Cost Forwarding Algorithm
MCU	Microcontroller unit
MFR	Most Forward within Radius
MIMO	Multiple-input multiple-output
MITMOT	Mac and mImo Technologies for More Throughput
MMSE	Minimum Mean Square Error
MMV	Multiple measurement vector
MP	Matching Pursuit
MPE	Multi-Pulse Excitation
MRC	maximal ratio combining
MRI	Magnetic Resonance Imaging
MS	Maximum selection
MSE	Mean square error
MS-IDEA	Multi Sensor version of Incoherent Detection and Estimation
MSTP	Minimum Spanning Tree Projection
MSTs	Minimum-Spanning Trees
MU-MIMIO	Multiuser MIMO
MUSIC	Multiple Signal Classification
MWC	Modulated Wideband Converter
NiCd	Nickel-cadmium
NiMH	Nickel-metal hydride

NITRA	Non-Iterative Threshold based Reconstruction Algorithm
NiZn	Nickel-zinc
NL-means	Non-Local Means
NSCT	Non subsampled Contourlet Transform
NSP	Null space property
NUS	Non-Uniform Sampling
OCDMA	Optical Code Division Multiple Access
Ofcom	Office of Communications
OFDM	Orthogonal Frequency Division Multiplexing
OFDMA	Orthogonal Frequency Division Multiple Access
OMP	Orthogonal Matching Pursuit
OOMP	Ordered Orthogonal Matching Pursuit
OSTM	Orthogonal Symmetric Toeplitz Matrix
OTE	Object Tracking Error
PAN	Personal Area Network
PCM	Pulse Coding Modulation
PDU	Protocol data unit
PEGASIS	Power-Efficient Gathering in Sensor Information Systems
PESQ	Perceptual Evaluation of Speech Quality
PLL	Phase Locked Loop
PMMWI	Passive millimeter-wave imaging
PN	Pseudorandom
PRF	Pulse repetition frequency
PRI	Pulse repetition interval
PRNET	Packet Radio Networks
PSK	Phase shift keying
PSNR	Peak signal to noise ratio
QoS	Quality of Service
QPSK	Quadrature Phase Shift Keying
R3A	Reduced Runtime Recovery Algorithm
RAM	Random access memory
RF	Radio Frequency
RFDs	Reduced function devices
RIP	Restricted Isometry Property
ROC	Receiver Operating Characteristics
ROMP	Regularized Orthogonal Matching Pursuit
SAR	Synthetic aperture radar

SCDCS	Spatial Correlation-based Distributed Compressed Sensing
SCI	Sparsity Concentration Index
SDMA	Space-Division Multiple-Access
SDWV	Standard deviation weight average
SIMD	Single instruction, multiple data
SINR	Signal-to-interference and-noise
SMACS	Self-organizing medium access control for sensor networks
SMC	Spectral Management Center
SNR	Signal to Noise Ratio
SOMP	Simultaneous orthogonal matching pursuit
SOP	Self-Organizing Protocol
SP	Subspace Pursuit
SpAdOMP	Sparse Adaptive Orthogonal Matching Pursuit
SPC	Single Pixel Camera
SPIN	Sensor Protocols for Information via Negotiation
SPMT	Split Process Merge Technique
SR	Sparse representation
ST	Surfacelet Transform
STBC	Space-time block-coding
STCP	Sensor Transmission Control Protocol
StOMP	Stagewise Orthogonal Matching Pursuit
STTC	Space-time trellis coding
SURAN	Survivable Adaptive Radio Networks
SV	Sensed Value
SVM	Support vector machine
TDM	Time division multiplexing
TDMA	Time division multiplexing access
TEEN	Threshold-sensitive Energy Efficient Protocols
TG	Total Ground Truth
TRDR	Tracker Detection Rate
TTL	Time-to-live
TV	Total Variation
UMTS	Universal Mobile Telecommunications System
UWB	Ultra Wide Band
VCO	Voltage Controlled Oscillator
VGA	Virtual Grid Architecture routing

VQ	Vector Quantization
VSN	Visual Sensor Network
WBSN	Wireless body sensor network
WCDMA	Wideband Code Division Multiple Access
WMSN	Wireless Multimedia Sensor Network
WSN	Wireless Sensor Network
WWiSE	World-Wide Spectrum Efficiency

1

Introduction

1.1 Overview

Compressed sensing is a novel research area, which was introduced in 2006, and since then has already become a key concept in various areas of applied mathematics, computer science, and electrical engineering. It surprisingly predicts that high-dimensional signals, which allow a sparse representation by a suitable basis or, more generally, a frame, can be recovered from what was previously considered highly incomplete linear measurements by using efficient algorithms. This article shall serve as an introduction to and a survey about compressed sensing. The area of compressed sensing was initiated in 2006 by two groundbreaking papers, namely [1] by Donoho and [2] by Candes, Romberg, and Tao. Nowadays, after only 6 years, an abundance of theoretical aspects of compressed sensing is explored in more than 1000 articles. Moreover, this methodology is to date extensively utilized by applied mathematicians, computer scientists, and engineers for a variety of applications in astronomy, biology, medicine, radar, and seismology, to name a few. The key idea of compressed sensing is to recover a sparse signal from very few non-adaptive, linear measurements by convex optimization. Taking a different viewpoint, it concerns the exact recovery of a high-dimensional sparse vector after a dimension reduction step. From a yet another standpoint, we can regard the problem as computing a sparse coefficient vector for a signal with respect to an over complete system. The theoretical foundation of compressed sensing has links with and also explores methodologies from various other fields such as applied harmonic analysis, frame theory, geometric functional analysis, numerical linear algebra, optimization theory, and random matrix theory.

Compressed sensing has become an efficient compression algorithm over the last decade which has gained the attention of researchers of different fields wherever deep compression is required. Even today, many traditional

1

compression techniques remain significant due to their high compression ratio. In traditional compression algorithms, signals, images, and video frames are recovered from uniformly sampled values obtained from the captured original signal. This is nothing but Nyquist criterion which states that near optimal recovery of the original input is possible when the same is sampled at regular intervals which destined by the twice the highest frequency component present in the signal. Nyquist criterion has been strictly followed ever since its proposal until the last few years which was ridden by the introduction of compressed sensing or compressive sampling. Compressed sensing can be put in this way: It is possible to obtain the sampled signal directly while signal capturing with less sampling rate than sampling the signal with the Nyquist rate. Obtaining signal samples directly from the input, which are technically called measurements, obviously reduces the tediousness of the hardware used, decreases energy consumption, increases easy recovery, reduces calculation time, etc. Sometimes it is possible that while using Nyquist criterion, the resulting samples itself will be huge and cannot be processed as such leaving us the complexities as follows: It may be too costly to recover the signal, it may require immense time, devices which would support such high sampling rates might be necessary, etc. Thus, despite extraordinary growth in computational efficiency, the acquisition and processing huge volumes of data such as image, videos, etc., would become a tremendous challenge.

With huge collections of data to be processed, compression, which involves extraction of very little high-priority information from the input with a bearable error, becomes unavoidable. Compression in other hand can be done in various ways, specially using Nyquist criterion, but the efficiency of the compression process is of great importance. A commonly used compressed technique is transform coding which makes use of sparse approximation. Sparse approximation is the process of preserving only k eminent values out of n number of values in actual input, where $k << n$. Such situations can easily be handled by compressed sensing which converts actual signal values into measurements which are easily processed. This forms the basis of traditional transform coding schemes which utilizes the signal sparsity and compressibility as in JPEG, JPEG 2000, MPEG, etc. Compressed sensing also takes signal sparsity and compressibility into account but provides an efficient compression framework, potentially reducing the sampling and compression costs in sensing data to a greater extent. Compressed sensing works on the basis that, when the signal under consideration is sparse in a known basis, the number of measurements can be vastly reduced. In a nutshell, compressed sensing can be explained as follows: Instead of obtaining the signal first and then sampling the same at higher rate, the signal can be directly sensed in a compressed form.

Though compressed sensing has become popular and a highly prioritized procedure in various fields, its emergence dates back to late 1700s. In 1795, Prony proposed a method to estimate a signal with very small number of samples. This was not considered of greater importance, but in 1900s, many researchers found the possibility of reconstruction without abiding by the Nyquist criterion. It was still more generalized and much work started after 1900s. In early 2000s, Blu, Marzilliano, and Vetterli found that it is possible to recover certain signals which are governed by only K parameters, showing that these signals can be sampled and recovered with only 2K samples. More recently, Romberg, Tao, and Candes showed that signals can be perfectly recovered from a small set of linear samples. Their work gave rise to the term compressed sensing since any signal can be recovered from a very few measurements obtained from its sparsified form. This gave rise to the development of various algorithms for faster reconstruction and efficient compression methods in the field of compressed sensing. Now, it has become a field of importance where number of researchers work to give at most compression rate and efficient reconstruction.

This book portrays the various works based on compressed sensing, its advantages, and its application in different fields. The chapters in this book provide the details of sparsity, compressed sensing framework, various recovery algorithms, and applications of compressed sensing in image processing, audio processing, video processing, cognitive radio networks, wireless networks, wireless sensor networks, etc. Also, the real-time implementations of compressed sensing in many of the above-said fields are discussed. This chapter describes the motivation of compressed sensing in Section 1.2, Section 1.3 describes the traditional sampling, Section 1.4 provides in brief about the existing data acquisition system, and Section 1.5 gives an introduction of transform coding and drawbacks of transform coding. Section 1.6 discusses the sparsity and signal recovery concept of compressed sensing and the use of CS to different applications. Section 1.7 gives the outline of the chapters covered in this book.

1.2 Motivation

Due to redundancy in the data, it is clear that the signal can be reconstructed from few data without any loss. This leads to the success of lossy compression formats for sounds, images, and specialized technical data [2].

Consider a vector $x \in R^d$ that has at most s nonzero elements. That is,

$$\|x\|_o \underline{\underline{\text{def}}} |\{i : x_i \neq 0\}| \leq s$$

Clearly, we can compress x by representing it using s (index, value) pairs. Furthermore, this compression is lossless—we can reconstruct x exactly from the s (index, value) pairs. Now, let us take one step forward and assume that $x = U\alpha$ where α is a sparse vector, $\|\alpha\|_o \leq s$, and U is a fixed orthonormal matrix. That is, x has a sparse representation in another basis. It turns out that many natural vectors are (at least approximately) sparse in some representation. In fact, this assumption underlies many modern compression schemes. For example, the JPEG 2000 format for image compression relies on the fact that natural images are approximately sparse in a wavelet basis. Can we still compress x into roughly s numbers? Well, one simple way to do this is to multiply x by UT, which yields the sparse vector α, and then represent α by its s (index, value) pairs. However, this requires to first "sense" x, to store it, and then to multiply it by UT.

The phenomenon of ubiquitous compressibility raises very natural questions: What is the necessity to put much effort in acquiring all the data when most of it will be thrown away? Can we not just directly measure the part that will not end up being thrown away? The important information about the signals/images in effect is not acquiring that part of the data that would eventually just be "thrown away" by lossy compression [2].

An early breakthrough in signal processing was the Nyquist–Shannon sampling theorem. It states that if the signal's highest frequency is less than half of the sampling rate, then the signal can be reconstructed perfectly. The main idea is that with prior knowledge about constraints on the signal's frequencies, fewer samples are needed to reconstruct the signal. The Nyquist/Shannon sampling theory has been accepted as the doctrine for signal acquisition and processing ever since it was implied by the work of Nyquist in 1928 and proved by Shannon in 1949. The theorem says that to exactly reconstruct an *arbitrary* band-limited signal from its samples, the sampling rate needs to be at least twice the bandwidth [1].

Around 2004, Emmanuel Candès, Terence Tao, and David Donoho proved that given knowledge about a signal's sparsity, the signal may be reconstructed with fewer samples than the sampling theorem requires. This idea is the basis of compressed sensing. Compressed sensing is a technique that simultaneously acquires and compresses the data. The key result is that a random linear transformation can compress x without losing information. The number of measurements needed is order of $s \ log \ (d)$. That is, we roughly acquire only the important information about the signal. Another important application of compressed sensing is medical imaging, in which requiring less measurements translates to less radiation for the patient [3].

In many applications nowadays, including digital image and video cameras, the Nyquist rate is so high that the excessive numbers of samples make compression a necessity prior to storage or transmission. In other applications, including medical scanners, radars, and high speed analog to digital converters, increasing the sampling rate is very expensive, either due to the sensor itself being expensive or the measurement process being costly.

Moreover, the protocols are non-adaptive and do not require knowledge of the signal/image to be acquired in advance other than knowledge that the data will be compressible and do not attempt any "understanding" of the underlying object to guide an active or adaptive sensing strategy. The measurements made in the compressed sensing protocol are holographic thus, not simple pixel samples, and must be processed nonlinearly. In specific applications, this principle might enable dramatically reduced measurement time, sampling rates, or reduced use of analog-to-digital converter resources [2].

1.3 Traditional Sampling

In 1949, Shannon presented that any band-limited time-varying signal with "n" Hertz highest frequency component can be perfectly reconstructed by sampling the signal at regular intervals of at least $1/2n$ s. In traditional signal-processing techniques, the signal is uniformly sampled at Nyquist rate, prior to transmission, to generate "N" samples. These samples are then compressed to "M" samples, discarding N-M samples. At the receiver end, the signal is decompressed by retrieving "N" samples from "M" samples. The paradigm of Shannon's sampling theory is cumbersome when applied to the emerging wideband signal systems since high data rate A/D converters are computationally expensive and require more storage space. Initially, all the "N" samples are considered to extract and process "M" samples, which leads to storage, energy, and computational complexity. The alternative theory of compressive sensing [1, 2] by Candes et al. has made a significant contribution by giving sampling theory a new dimension by acquiring "M" measurements in the first place. Figure 1.1 represents the concept of traditional data sampling.

Conventional approaches to sampling signals or images follow Shannon's theory: The sampling rate must be at least twice the maximum frequency present in the signal. Whatever the field of application, most of the acquisition systems follows Nyquist Shannon sampling theorem. For instance, sampling

Figure 1.1   Traditional sampling.

intervals for image sampling are chosen based on the sufficient number of pixels for sampling smallest objects or object borders to secure their resolving, localization, or recognition. The sampling interval determines the effective image base band, which specifies the bandwidth of signal analog amplifiers or transmission systems. Since the area occupied in images by the smallest objects and object borders is usually a very small fraction of the total image area, such uniform sampling produces very many redundant samples. This property of signal/image spectra is called as "signal sparsity" in CS community. The capability of transforms to compress image energy in few spectral coefficients is called their energy compaction capability. It is used in transform signal coding for compressing signal discrete representation obtained with ordinary sampling, i.e., for substantial reduction of the quantity of data required for signal reproducing with a given acceptable quality. In transform signal coding, signals are replaced by their "band-limited" or "sparse" approximations, i.e., by their copies that contain only a limited number of nonzero transform components, fewer than the number of signal samples.

The Shannon–Nyquist sampling paradigm is a classical example for this purpose. Consider a signal y in the time domain. The Shannon–Nyquist theory provides the conditions for perfectly reconstructing the signal y from a set of discrete samples $\{Y_k\}_{k=1}^N$. For achieving this, it is necessary to acquire equally time-spaced samples at a rate exceeding twice the bandwidth of y. Thus, the Shannon–Nyquist sampling theory is valid assuming the following: (1) We know the signal bandwidth in advance and (2) sampling should be carried out guaranteeing equally spaced samples and at a sufficiently high sampling rate. These requirements are satisfied when the signal is represented in Fourier domain. The signals in time domain representation (y) and discrete representation (Y_k) are equivalent in the sense that we can perfectly recover one of them given the other. Thus, the discrete Fourier transform (DFT) X_k of the signal Y_k is as follows:

$$X_k = (1/\sqrt{N}) \sum_{j=1}^{N} Y_j \exp(-2\pi\,(j-1)(k-1)i/N)$$

Letting $X = \{X_k\}_{k=1}^{N} \in \mathbf{R}^N$ and $Y = \{Y_k\}_{k=1}^{N} \in \mathbf{R}^N$, the relation between the discrete representation of y and its Fourier transform can be written compactly as $X = FY$ with F being the unitary DFT matrix. The converse procedure of reconstructing the original signal from its Fourier representation can then be described by

$$Y = \overline{F}X \rightarrow y$$

where $\overline{F} = F^T$ is the inverse DFT matrix which equals the conjugate transpose of F. The original signal y has only a few significant entries in terms of magnitude when it is considered in the Fourier domain. Hence, X is called as a compressible vector or sparse vector in which the insignificant entries are discarded. This indicates that the signal energy does not spread uniformly over the spectrum, and hence, it can be adequately approximated by a reduced representation. The reduced representation of y could not be immediately obtained in the time domain, and we had to turn to an alternative domain in which it appears sparse or compressible. If we could tell the locations of the most significant entries of X in terms of magnitude, then we could obtain an almost identical representation of y using fewer samples in Y. This argument follows by considering a sparse version of X in which all insignificant entries are set to zero. Let us denote this vector by Z and let us look at the difference $\Delta Y = Y - \overline{Y} = \overline{F}(X - Z)$. This difference cannot be large, meaning that we can use Z to adequately approximate the original signal

$$Z \rightarrow \overline{Y} \approx Y \rightarrow y$$

Since the locations of significant entries in Z are known, only a fraction of the samples in Y are used to construct Z. Thus,

$$Y^m = \overline{F}_{m \times m} Z^m = \overline{F}_{m \times N} Z$$

where $m < N$ is the number of significant entries of Z, and $Y^m \in \mathbf{R}^m$, $Z^m \in \mathbf{R}^m$, $F_{m \times m} \in \mathbf{R}^{m \times m}$, $F_{m \times N} \in \mathbf{R}^{m \times N}$ are obtained by eliminating the columns/rows corresponding to insignificant entries in Z. The signal y can be reconstructed by sampling $m < N$ points at specific locations, a premise which translates into a sampling rate that may go far below the Nyquist rate.

1.4 Conventional Data Acquisition System

Data acquisition and data conversion systems interface between the real world of physical parameters which are analog and the artificial world of digital computation and control. With the current emphasis on digital systems, the interfacing function has become an important one. The devices which perform the interfacing function between analog to digital worlds are A/D and D/A converters which together are known as data converters. Computerized feedback control systems are used in many different industries today in order to achieve greater productivity in our modern industrial society. Industries which presently employ such automatic systems include steel making, food processing, paper production, oil refining, chemical manufacturing, textile production, and cement manufacturing. The devices which perform the inter-facing function between analog and digital worlds are analog-to-digital (A/D) and digital-to-analog (D/A) converters, which together are known as data converters. Some of the specific applications in which data converters are used include data telemetry systems, pulse code modulated communications, automatic test systems, computer display systems, video signal processing systems, data logging systems, and sampled data control systems. In addition, every laboratory digital multimeter or digital panel meter contains an A/D converter [2].

1.4.1 Data Acquisition System

Capabilities and the accuracy of data acquisition (DAQ) systems can be founded from analog input specifications. A basic specification is as follows: number of channels, sampling rate, resolution, and input range. The number of analog channel inputs will be indicated for both single-ended and differential inputs. The most common way used to transmit electrical signals over wires is the single-ended form. In general, one wire carries a voltage (the signal), while the other wire is connected to a reference voltage (ground). Instead, differential signal is a method of transmission of electrical signals with two complementary signals sent on a couple of wires. The main advantage of single-ended over differential signaling is that fewer wires are needed to transmit multiple signals. A disadvantage of single-ended signaling is that the return currents for all the signals can be shared with the same conductor, and this can sometimes cause interference ("cross talk") between the signals. Many DAQ systems are known as plug in boards used in scientific application to acquire data and transfer it directly to computer memory. The main components of DAQ system is the A/D board. It converts analog voltages from external signal sources (sensor) into a digital signal which can be read by the host computer [1].

1.4.2 Functional Components of DAQ

The main components of DAQ system is the analog input (A/D) boards; it converts analog voltages from external signal sources into a digital signal, which can be read by the host computer. Figure 1.2 shows the block diagram of components of DAQ.

Moreover, the functional diagram of a typical DAQ system can be described of the following main components:

  i. Input multiplexer
 ii. Input signal amplitude
iii. Sample and hold circuit.

1.4.3 Digital Image Acquisition

With respect to digital imaging, four major and several minor imaging techniques meet these requirements. The major techniques are as follows:

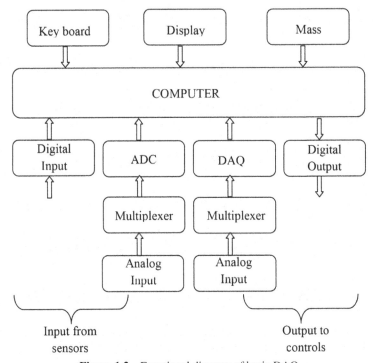

Figure 1.2 Functional diagram of basic DAQ.

X-ray imaging measures the absorption of shortwave electromagnetic waves, which is known to vary between different tissues. Magnetic resonance imaging measures the density and molecular binding of selected atoms (most notably hydrogen which is abundant in the human body), which varies with tissue type, molecular composition, and functional status. Ultrasound imaging captures reflections at the boundaries between and within tissues with different acoustic impedance. Nuclear imaging measures the distribution of radioactive tracer material administered to the subject through the blood flow. It measures function in the human body.

Other imaging techniques include EEG and MEG imaging, microscopy, and photography. All the techniques have in common that an approximate mapping is known between the diagnostic question, which was the reason for making the image and the measurement value that is depicted. This can be very helpful when selecting an analysis technique. If, for instance, bones need to be detected in an X-ray CT slice, a good first guess would be to select a thresholding technique with a high threshold because it is known that X-ray attenuation in bone is higher than that in soft tissues and fluids. Many of the imaging techniques come in two varieties: Projection images show a projection of the 3D human body onto a 2D plane and slice images show a distribution of the measurement value in a 2D slice through the human body. Slice images may be stacked to form a volume. Digitized images consist of a finite number of image elements. Elements of a 2D picture are called pixels (picture elements), and elements of stacked 2D slices are called voxels (volume elements).

1.5 Transform Coding

One of the most popular techniques for signal compression is known as transform coding, and typically relies on finding a basis or frame that provides *sparse* or compressible representations for signals in a class of interest. By a sparse representation, for a signal of length n, it can be represented with $k << n$ nonzero coefficients; by a compressible representation, the signal is well approximated by a signal with only k nonzero coefficients. Both sparse and compressible signals can be represented with high fidelity by preserving only the values and locations of the largest coefficients of the signal. This process is called sparse approximation and forms the foundation of transform coding schemes that exploit signal sparsity and compressibility, including the JPEG, JPEG 2000, MPEG, and MP3 standards [2]. The basic idea of transform coding is to transform the image from the spatial domain to another domain where coefficients have lower entropy so they can be coded more efficiently. This approach is used in both JPEG and MPEG because it yields much higher compression rates than other methods. The image is subdivided into n × n blocks prior to calculating the transform to improve speed and also to localize any errors that may occur.

The basic principle of transform coding is to map the pixel values into a set of linear transform coefficients, which are subsequently quantized and encoded. By applying an inverse transformation on the decoded transform coefficients, it is possible to reconstruct the image with some loss. It must be noted that the loss is not due to the process of transformation and inverse transformation, but due to quantization alone. Since the details of an image and hence its spatial frequency content vary from one local region to the other, it leads to a better coding efficiency if we apply the transformation on local areas of the image, rather than applying global transformation on the entire image. Such local transformations require manageable size of the hardware, which can be replicated for parallel processing. For transform coding, the first and foremost step is to subdivide the image into non-overlapping blocks of fixed size. Without loss of generality, we can consider a square image of size $N \times N$ pixels and divide it into n^2 number of blocks, each of size $(N/n) \times (N/n)$, where $n < N$ and is a factor of N. Figure 1.3 shows the block diagram of transform coding. This consists of transmitter and receiver. The main components in transmitter section are segment into blocks, forward transform, quantization, and coder. The receiver sections consist of decoder, inverse transform, and combine blocks.

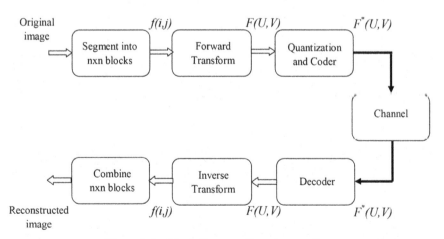

Figure 1.3 Block diagram of transform coding.

Quantization process of transform coding depends on three factors such as desired average bit rate, statistics of various elements of transformed sequence, and effects of distortion in the transform coefficients on the reconstructed sequence. Quantized values need to be encoded using some binary encoding technique (e.g., run length coding, Huffman coding).

1.5.1 Need for Transform Coding

The purpose of transformation is to convert the data into a form where compression is easier. This transformation will transform the pixels which are correlated into a representation where they are decorrelated. The new values are usually smaller on average than the original values. The net effect is to reduce the redundancy of representation. For lossy compression, the transform coefficients can now be quantized according to their statistical properties, producing a much compressed representation of the original image data. The basic criteria for the selection of a particular transform coding are decorrelation, energy concentration, transform should provide energy compaction, visually pleasant basis functions, pseudorandom noise, m-sequences, lapped transforms, quantization errors make basis functions visible, low complexity of computation, separability in 2D and simple quantization of transform coefficients. A transformation must necessarily fulfill the following properties:

i. The coefficients in the transformed space should be decorrelated.
ii. Only a limited number of transform coefficients should carry most of the signal energy (in other words, the transformation should possess energy compaction capabilities), and most of the coefficients should carry insignificant energy. Only then, the quantization process can coarsely quantize those coefficients to achieve compression, without much of perceptible degradation. A number of transformation techniques, such as discrete Fourier transforms (DFT), discrete cosine transforms (DCT), discrete wavelet transforms (DWT), KL transforms (KLT), discrete Haar transforms, and discrete Hadamard transforms, are available.

1.5.2 Drawbacks of Transform Coding

DFT have some drawbacks such as storage and manipulation of complex quantities and creation of spurious spectral components due to the assumed periodicity of image blocks. The drawbacks of haar transform include its energy compaction is fair and it is not suitable for compression algorithms. The disadvantages of KL transform (KLT) are it depends on signal statistics,

KLT is not separable for image blocks, and it is rarely used in practice because of the following limitations: (i) The transformation matrix for a block of image is derived from the covariance matrix, which needs to be computed for every block. This makes the transformation data dependent and involves non-trivial computations. (ii) Perfect decorrelation in transform domain is not possible, since rarely, the image blocks can be modeled as a random field. (iii) No fast computational algorithms are available for its implementation. Transform matrix cannot be factored into sparse matrices [4].

Despite excellent energy compaction capabilities, mean-square reconstruction error performance closely matching that of KLT, and availability of fast computational approaches, DCT offers a few limitations which restrict its use in very low bit rate applications. The limitations are listed below: (i) Truncation of higher spectral coefficients results in blurring of the images, especially wherever the details are high. (ii) Coarse quantization of some of the low spectral coefficients introduces graininess in the smooth portions of the images. (iii) Serious blocking artifacts are introduced at the block boundaries, since each block is independently encoded, often with a different encoding strategy and the extent of quantization. The drawbacks of Walsh–Hadamard transform include Walsh–Hadamard transform requires additions and subtractions and use of high-speed signal processing has reduced its use due to improvements with DCT. These drawbacks are addressed with the help of CS theory. In this chapter, we will see how CS theory is applied to various domains. CS can be successfully applied to audio, image, and computer vision problems in which the need for reducing the dimension of the signal is high.

1.6 Compressed Sensing

This section discusses the concept behind compressed sensing and application of CS to various fields.

1.6.1 Sparsity and Signal Recovery

CS is based on the principle that, through optimization, the sparsity of a signal can be exploited to recover it from far fewer samples than required by the Shannon–Nyquist sampling theorem. There are two conditions under which recovery is possible. The first one is sparsity which requires the signal to be sparse in some domain. The second one is incoherence which is applied through the isometric property which is sufficient for sparse signals.

1.6.2 CS Recovery Algorithms

Right at the heart of the compressive sensing theory is the ability for recovery algorithms to provide accurate signal estimations in an efficient manner. Recovery algorithms generally fall into two categories: those based on convex optimization and greedy algorithms. One of the basic algorithms from each category is considered. Convex optimization is a minimization problem subject to a number of constraints where the functions involved are convex. A greedy algorithm iteratively makes decisions based on some locally optimal solution. One of the simplest greedy algorithms suitable for the sparse signal approximation problem is orthogonal matching pursuit. Some of the earliest orthogonal matching pursuit algorithms can be found in [5, 6], although notable more recent work can be found in [7] and [8]. This method can often perform faster than basis pursuit due to its simplicity [9].

1.6.3 Compressed Sensing for Audio

Using spectral analysis and the properties of the DCT, we can treat audio signals as sparse signals in the frequency domain. This is especially true for sounds representing tones. On the other hand, CS has been traditionally used to acquire and compress certain sparse images. We propose the use of DCT and CS to obtain an efficient representation of audio signals, especially when they are sparse in the frequency domain. By using the DCT as signal preprocessor in order to obtain a sparse representation in the frequency domain, we show that the subsequent application of CS represents our signals with less information than the well-known sampling theorem. This means that our results could be the basis for a new compression method for audio and speech signals. Despite the above-mentioned work, there still exists a huge gap between the CS theory and applications to audio signals [10]. In particular, it is still unknown how to construct a sparse audio signal, especially when CS relies on two principles: sparsity (which pertains to the signal of interest) and incoherence (which pertains to the sensing modality) [10]. For the problem of making a sparse representation of an audio signal, we introduce the DCT which is at present the most widely used transform for image and video compression systems. Its popularity is due mainly to the fact that it achieves a good data compaction, because it concentrates the information content in relatively few coefficients [10]. This means that we can obtain a compressed version of an audio signal by first obtaining a sparse representation in the frequency domain, and later processing the result with a CS algorithm. The DCT speech signal representation has the ability to pack input data into as few coefficients as possible. This allows the quantizer to discard coefficients with relatively small amplitudes without introducing audio distortion in the reconstructed signal. Although the compressive sampling technique is used primarily for compression sample images, we achieve reasonable results due to the preprocessing of the audio signal. This means that our hypothesis is satisfied in the sense that our proposed technique can achieve a significant reduction in the number of samples required to represent certain audio signals and therefore a decrease in the required number of bytes for encoding.

1.6.4 Compressed Sensing for Image

In conventional imaging systems, natural images are often first sampled into the digital format at a higher rate and then compressed through the JPEG or the JPEG 2000 codec for efficient storage purpose. However, this approach is not applicable for low-power, low-resolution imaging devices (e.g., those used in a sensor network) due to their limited computation capabilities. Over the past few years, a new framework called as compressive sampling (CS) has been developed for simultaneous sampling and compression. It builds upon the groundbreaking work by Candes et al. [2] and Donoho [1], who showed that under certain conditions, a signal can be precisely reconstructed from only a small set of measurements. The CS principle provides the potential of dramatic reduction in sampling rates, power consumption, and computation complexity in digital data acquisitions. Due to its great practical potentials, it has stirred great excitements in both academia and industries in the past few years [10]. However, most of existing works in CS remain at the theoretical study. In particular, they are not suitable for real-time sensing of natural image as the sampling process requires accessing the entire target at once [11]. In addition, the reconstruction algorithms are generally very expensive [11].

1.6.5 Compressed Sensing for Video

Research into the use of CS in video applications has only started very recently. The first use of CS in video processing is proposed in [12]. Their approach is based on the single-pixel camera [13]. The camera architecture employs a digital micro mirror array to perform optical calculations of linear projections of an image onto pseudorandom binary patterns. It directly acquires random projections. They have assumed that the image changes slowly enough across a sequence of snapshots which constitutes one frame.

They acquired the video sequence using a total of M measurements, which are either 2D or 3D random measurements. For 2D frame-by-frame reconstruction, 2D wavelets are used as the sparsity-inducing basis. For 3D joint reconstruction, 3D wavelets are used. The matching pursuit reconstruction algorithm [14] is used for reconstruction. Another implementation of CS video coding is proposed in [15]. In this implementation, each video frame classified as a reference or non-reference frame. A reference frame (or key frame) is sampled in the conventional manner, while non-reference frames are sampled by CS techniques. The sampled reference frame is divided into non-overlapping blocks whereby discrete cosine transform (DCT) is applied. A compressed sensing test is applied to the DCT coefficients of each block to identify the sparse blocks in the non-reference frame. This test basically involves comparing the number of significant DCT coefficients against a threshold. If the number of significant coefficients is small, then the block concerned is a candidate for CS to be applied. The sparse blocks are compressively sampled using an i.i.d. Gaussian measurement matrix and an inverse DCT sensing matrix. The remaining blocks are sampled in the traditional way.

1.6.6 Compressed Sensing for Computer Vision

One of the main motivations for developing new computer vision applications is the recent introduction of compressive sensing. With the advent of compressive sensing, a large number of new methods have been developed for image analysis in computer vision. This particular work derives mathematical formulations from the recently developed compressive sensing, sparse representation, and matrix completion for related applications in image processing and computer vision. While image acquisition and preprocessing play an important role in acquiring raw input data, image analysis, image restoration, and image enhancement are three important aspects of a computer vision rendering system. Image analysis system which consists of feature extraction, segmentation, and classification/recognition forms the first important step of understanding the raw image data. The analyzed data is useful in making decisions in general applications such as video surveillance for event and activity detection, organizing information for content-based data retrieval, for computer human interaction, etc [16].

Embedded smart camera networks represent an emerging direction of next-generation surveillance systems. A big challenge to implement computer vision applications on embedded cameras is the limit of memory and computational capacity. Since background subtraction algorithms play a fundamental yet significant role of most computer vision applications, their memory requirements and computational efficiency should be taken into account in the design. In this paper, we propose an efficient hierarchical lightweight background subtraction approach by combining the pixel-level and the block-level background subtraction modules into a single framework so that it is capable of dealing with dynamic background scenes. Block compressed sensing theory is for the block-level module design to save memory and improves computational efficiency. Moreover, considering the continuity of foreground objects, a novel integral filter is designed for the pixel-level module to eliminate perturbations efficiently. Experimental results on various videos demonstrate superior performance of the proposed algorithm. The proposed lightweight algorithm only requires about 6.5 bytes per pixel and is applicable for embedded smart cameras. Furthermore, as each block is processed independently, it can be implemented in parallel.

Robust and efficient foreground extraction is a crucial topic in many computer vision applications. In this paper, we propose an accurate and computationally efficient background subtraction method. The key idea is to reduce the data dimensionality of image frame based on compressive sensing and in the meanwhile apply sparse representation to build the current background by a set of preceding background images. According to greedy iterative optimization, the background image and background subtracted image can be recovered by using a few compressive measurements. The proposed method is validated through multiple challenging video sequences. Experimental results demonstrate the fact that the performance of our approach is comparable to those of existing classical background subtraction techniques.

Foreground extraction is often the first step of many visual surveillance applications such as object tracking, recognition, and anomaly detection. Background subtraction [16] is the most frequently used method to detect and extract objects automatically in video sequences. The basic principle can be formulated as a technique that builds the background model and compares this model with the current image frame in order to distinguish foreground, that is, moving objects from static or slow moving background. Many pixel-based methods have been investigated in the past decades. Among these, Gaussian mixture model (GMM) [16] is a representative method for robustly modeling complicated backgrounds with slow illumination changes and small repetitive movements. This method models the distribution of the values observed over time at each pixel by a weighted mixture of Gaussians. Several modified methods concerning the number of Gaussian components, learning rate, and parameters update [17–19] have been proposed. However, the main drawback of these approaches is being computationally intensive. In [20], a nonparametric kernel density estimation (KDE) has been proposed to model the background, but this method consumes too much memory. Besides, in [21], each pixel is represented by a codebook which is able to capture background motion over a long period of time with a limited amount of memory. The codebooks can evolve with illumination variations and moving backgrounds. Nevertheless, the codebook update will not allow the creation of new code words once codebooks have been learned from a typically training sequence. Recently, a universal background subtraction method called ViBe has been presented in [22], where each pixel is modeled with a set of real observed pixel values. It is reported that this method outperforms current mainstream methods in terms of both computation speed and detection accuracy. Most of pixel-level background subtraction techniques have considerable computation costs; thus, some other methods are proposed to improve the computation efficiency.

1.6.7 Compressed Sensing for Cognitive Radio Networks

In the emerging paradigm of open spectrum access, cognitive radios dynamically sense the radio spectrum environment and must rapidly tune their transmitter parameters to efficiently utilize the available spectrum. The unprecedented radio agility envisioned, calls for fast and accurate spectrum sensing over a wide bandwidth, which challenges traditional spectral estimation methods typically operating at or above Nyquist rates. Capitalizing on the sparseness of the signal spectrum in open-access networks, this paper develops compressed sensing techniques tailored for the coarse sensing task of spectrum hole identification. Sub-Nyquist rate samples are utilized to detect and classify frequency bands via a wavelet-based edge detector. Because spectrum location estimation takes priority over fine-scale signal reconstruction, the proposed novel sensing algorithms are robust to noise and can afford reduced sampling rates [23].

Compressive sensing-based technique is used for speedy and accurate spectrum sensing in cognitive radio technology-based standards and systems [24]. IEEE 802.22 is the first standard to use the concept of cognitive radio, providing an air interface for wireless communication in the TV spectrum band. Although no spectrum sensing method is explicitly defined in the standard, it has to be fast and precise. Fast Fourier sampling (FFS)—an algorithm based on CS—is used to detect wireless signals as proposed in [24]. According to the algorithm, only "m" (where m Â« n) most energetic frequencies of the spectrum are detected and the whole spectrum is approximated from these samples using non-uniform inverse fast Fourier transform (IFFT). Using fewer samples, FFS results in faster sensing, enabling more spectra to be sensed in the same time window.

1.6.8 Compressed Sensing for Wireless Networks

A wireless network refers to a telecommunication network that interconnects between nodes that are implemented without the use of wires. Wireless networks have experienced unprecedented growth over the past few decades, and they are expected to continue to evolve in the future. Seamless mobility and coverage ensure that various types of wireless connections can be made anytime, anywhere. Wireless networks use electromagnetic waves, such as radio waves, for carrying the information. Therefore, their performance is greatly affected by the randomly fluctuating wireless channels. To develop an understanding of channels, the radio frequency band is studied first, then the existing wireless channel models used for different network scenarios, and finally the interference channel. There exist many wireless standards, and we describe them according to the order of coverage area, starting with cellular wireless networks. Compressive sensing is a new signal-processing paradigm that aims to encode sparse signals by using far lower sampling rates than those in the traditional Nyquist approach. It helps acquire, store, fuse, and process large data sets efficiently and accurately. This method, which links data acquisition, compression, dimensionality reduction, and optimization, has attracted significant attention from researchers and engineers in various areas. This comprehensive reference develops a unified view on how to incorporate efficiently the idea of compressive sensing over assorted wireless network scenarios, interweaving concepts from signal processing, optimization, information theory, communications, and networking to address the issues in question from an engineering perspective. It enables students, researchers, and communication engineers to develop a working knowledge of compressive sensing, including background on the basics of compressive sensing theory, an understanding of its benefits and limitations, and the skills needed to take advantage of compressive sensing in wireless networks [25]. To exploit CS in wireless communication, many applications using CS are given in detail. However, CS has more places to be adopted and emphasized.

1.6.9 Compressed Sensing for Wireless Sensor Networks

CS finds its applications in data gathering for large wireless sensor networks (WSNs), consisting of thousands of sensors deployed for tasks such as infrastructure or environment monitoring. This approach of using compressive data gathering (CDG) helps in overcoming the challenges of high communication costs and uneven energy consumption by sending "m" weighted sums of all sensor readings to a sink which recovers data from these measurements [26]. Although this increases the number of signals sent by the initial "m" sensors, the overall reduction in transmissions and energy consumption is significant since m Â« n (where n is the total number of sensors in large-scale WSN). This also results in load balancing which in turn enhances lifetime of the network. Lou et al. propose a scheme that can detect abnormal readings from sensors by utilizing the fact that abnormalities are sparse in time domain. CS is also used, in a decentralized manner, to recover sparse signals in energy-efficient large-scale WSNs [27]. Various phenomena monitored by large-scale WSNs usually occur at scattered localized positions, hence can be represented by sparse signals in the spatial domain. Exploiting this sparsity through CS results in accurate detection of the phenomenon. According to the proposed scheme [27], most of the sensors are in sleeping mode, whereas only a few are active. The active sensors sense their own information and also come up with optimum sensor values for their sleeping one-hop neighbors through "consensus optimization"—an iterative exchange of information with other active one-hop neighbors. Using a sleeping WSN strategy not only makes the network energy-efficient but also ensures detection of a physical phenomenon with high accuracy. The task of localization and mapping of the environment as quickly as possible, for reliable navigation of robots in WSNs, has utilized compressive sensing technique for its benefit [28]. A mobile robot, working in an indoor WSN for event detection application, may need to know its own position in order to locate where an incident has happened. The conventional approach of using navigation system to build a map requires estimation of features of the whole surrounding. This result in data coming to mobile robot from all sensors in the network, most of which is highly correlated—computational load may increase substantially, especially in case of large-scale sensor networks. CS enables the making of high-quality maps without directly sensing large areas. The correlation among signals renders them compressible. The nodes exploit sparse representation of parameters of interest in order to build localized maps using compressive cooperative mapping framework, which gives superior performance over traditional techniques [24].

1.7 Book Outline

In this book, an overview of the theories of sparse representation and compressive sampling is presented and several interesting imaging modalities based on these theories are examined. The use of linear and nonlinear kernel sparse representation as well as compressive sensing in many computer vision problems including target tracking, background subtraction, and object recognition is also explored. The use of CS for wireless networks, wireless sensor networks, and cognitive radios is also discussed.

Chapter 2 deals with the concepts of compressed sensing and mathematics involved in it. This chapter explains in detail about sparse representation, sparsity, and basis vectors. Different types of sparsifying transforms such as discrete cosine transform, discrete wavelet transform, Karhunen Loeve transform, Fourier transform, surfacelet transform, curvelet transform, and contourlet transform are explained in detail. Different measurement matrices involved in CS such as Gaussian matrix, Bernouille matrix, binomial matrix, hybrid matrix, and combination matrix are also discussed. The properties such as restricted isometry property and incoherence are also given an emphasis. This chapter also discusses the requirement of minimum number of measurements, stable recovery, and signal recovery in noise. The sparse recovery algorithms are introduced in this chapter. The applications of the compressed sensing such as object detection, target tracking, image reconstruction in holography, and MRI are also discussed.

Chapter 3 deals with the concept of reconstruction of compressed data using CS recovery algorithms. This chapter gives an insight about the various conditions, constrains, and properties such as null-space conditions and restricted isometry property that are mandatory for perfect recovery. It also explains the sensing matrix construction. The existing greedy recovery algorithms such as orthogonal matching pursuit (OMP), directional pursuit (DP), gradient pursuit (GP), subspace pursuit (SP), stagewise orthogonal matching pursuit (StOMP), compressive sampling matching pursuit (CoSaMP), and iterative hard thresholding (IHT) have been discussed with mathematical background. Special algorithms such as focal underdetermined system solution (FOCUSS) and multiple signal classification (MUSIC) algorithm have also been touched upon. Some of the greedy algorithms have been modeled to form model-based algorithms. Two such algorithms, namely model-based CoSaMP and model-based IHT, are also explained in this chapter. The rear end of the chapter discusses some of the non-iterative procedures including non-iterative threshold-based reconstruction algorithm (NITRA), reduced runtime recovery algorithm (R3A), etc.

Chapter 4 deals with the issue in applying CS and sparse decompositions to speech and audio signals. A multisensor version of incoherent detection and estimation algorithm (IDEA) called as MS-IDEA is developed in the data fusion centre (DFC) to detect the presence of a signal. This allows for intrasignal compression without intrasensor communication. CS can be successfully applied to both musical and speech signals, but the speech signals are more demanding in terms of the number of observations. Speech signals have more complex nature, and therefore, they are less sparse in the frequency domain, compared to the pure musical tones. Since sinusoidal signals satisfy the sparsity property, it can be concluded that the musical tones are convenient for the CS application. Chapter 4 also describes the use of DCT and CS to obtain an efficient representation of audio signals, especially when they are sparse in the frequency domain. Single-channel and multi-channel sinusoidal audio coding using CS is also discussed. In CS framework, reconstruction of a signal relies on the knowledge of the sparse basis and measurement matrix used for sensing. A new sensing matrix called orthogonal symmetric Toeplitz matrix (OSTM) generated with binary, ternary and PN sequence that shows improved results compared to random, Bernoulli, Hadamard, and Fourier matrices is also discussed in Chapter 4. The text-independent speaker identification that can be performed based solely on a speaker's voice is also discussed.

Chapter 5 provides a brief overview of some thrust CS-based imaging applications to have energy-efficient and complex representations of smaller magnitude. This chapter also deals with the methodologies and algorithms used for CS-based image applications. Camera architectures uniquely designed for direct acquisition of random projections of the signal without first collecting the pixels/voxels are available for taking compressive measurements. It also provides insight into the compressive imaging architectures such as single-pixel camera and lensless imaging. The applications of CS in specific imaging applications such as magnetic resonance imaging, synthetic aperture radar imaging, passive millimeter wave imaging, and light transport system are also provided. A case study depicting the image transmission using compressed sensing in wireless multimedia sensor networks is also presented.

Chapter 6 discusses the application of compressed sensing to computer vision problems. Conventional object detection techniques such as optical flow, temporal differencing, and background subtraction; and object-tracking techniques such as point tracking, kernel tracking, and silhouette tracking are explained in brief in this chapter. Compressive video processing with DCT,

DWT, and hybrid DWT–DCT approach is used for obtaining the sparse vector. Compressed sensing-based background subtraction is discussed in detail. The sparsity background constraint and adaptation of threshold are also given an emphasis. Compressive sensing-based background subtraction deals with object detection, object tracking, object recognition, and multi-view tracking applications.

Chapter 7 explains the compressed sensing-based wireless networking. It starts with the general discussions about wireless networks and their categories such as 3G wireless cellular networks, WiMAX, WiFi, and ad hoc sensor networks. It is followed by explanations regarding advanced wireless technologies such as orthogonal frequency division multiplexing (OFDM) and multiple antenna systems. This is extended to explain the part of compressed sensing in wireless networks. The different channel models such as multi-path channel estimation, random field estimation, and other channel estimation models that use compressed sensing concept have been discussed further. Then comes the multiple access techniques using compressed sensing. This is subdivided into multiuser detection and multiuser access. Comparison between multiuse detection and compressed sensing is portrayed. Multiuser access explains the uplink and downlink procedures carried out for data transfer.

Chapter 8 deals with cognitive radios. Cognitive radios are unlicensed users which try to coexist with the licensed users using various approaches. They observe the licensed spectrum for spectrum hole availability through a process known as spectrum sensing. It is a technically challenging functionality of cognitive radio especially in wideband spectrum monitoring as they require very high sampling rates. Compressive sensing is found to be an optimal solution to solve this problem. Cognitive radio spectrum sensing via compressed sampling techniques is becoming popular in the recent times. However, compressed sampling techniques do not reduce complexity to a large extent, but solves the higher sampling rate problem. There are many approaches proposed for compressed sampling-based sensing in the literature. This chapter gives a brief introduction to various functions of cognitive radio and different approaches to spectrum sensing, and then elaborates the use of compressed sampling for cognitive radios. It also gives a detailed insight on collaborative and distributed compressed sensing techniques followed by the research challenges in compressed sensing for cognitive radios.

Chapter 9 is concerned with WSN, its architecture, the associated wireless transmission technology, and protocols utilized along with the operating system being used for WSN. It also provides details about the application of CS to WSN and the significance related to it.

Chapter 10 focuses on efficient strategies to improve the sampling and reconstruction process in CS. Conventional CS process requires the same number of measurements from all blocks irrespective of the image content and thereby a larger total number of measurements for an image. Alternatively, a sampling methodology using two measurement matrices is proposed to vary the number of measurements per block, based on the image content, and is utilized in both CS and NUS measurement strategies. Recovery of sparse signals from a reduced number of measurements is an NP-hard problem. l_1 minimization and greedy algorithms are the two major recovery approaches available. The l_1 minimization provides uniform guarantees (recovering all sparse signals with as few measurements as possible). However, its implementation is time consuming and complex. Greedy approaches are fast. But, they lag in stability and uniform guarantee. Hence, an enhanced OMP algorithm is proposed in this chapter, which enables better recovery rate under reduced measurements.

Chapter 11 deals with magnetic resonance imaging (MRI). MRI is an important application of compressive sensing which accelerates MR acquisitions by reducing the amount of the required data. Real-time Compressive Sensing based MRI Reconstruction Using GPU Computing and Split Bregman Methods are discussed. GPU computation significantly accelerated CS MRI reconstruction of all but the smallest of the tested image sizes. Most part of the theoretical performance of the GPU can be realized over the combination of Matlab and Jacket processing package by requiring minimal code development. Surveillance video processing is able to detect anomalies and moving objects in a scene automatically and quickly. The video is acquired by compressive measurements, and the measurements are used to reconstruct the video by a low rank and sparse decomposition of matrix. CS has proposed some novel solutions in many practical applications. Focusing on the pixel-level multi-source image-fusion problem in wireless sensor networks, an algorithm of CS image fusion is proposed on the basis of multi-resolution analysis. The advantages of CS-based image fusion are also discussed. Magnetic resonance imaging (MRI) is an essential medical imaging tool with an inherently slow data acquisition process. Applying CS to MRI offers potentially significant scan time reduction, with benefits for patients and in the area of healthcare economics. A focal plane array–compressive sensing (FPA-CS) is presented with a new imaging architecture for parallel compressive measurement acquisition that can provide quality videos at high spatial and temporal resolutions in SWIR. CS can be used for the improvement of image reconstruction in holography by increasing the number of voxels that can be inferred from a single hologram. CS is also used for image retrieval from under sampled measurements in millimeter-wave holography and off-axis frequency-shifting holography. CS has applications in the detection and estimation of wireless signals, source coding, multi-access channels, data collection in sensor networks, and network monitoring. Finally, the architecture for compressive imaging without a lens is also discussed.

References

[1] Donoho, D. L. (2006). "Compressed sensing," IEEE Trans. Inform. Theory, 52, 1289–1306, July.

[2] Candes, E., Romberg, J., and Tao, T. (2006). "Robust uncertainty principles: Exact signal reconstruction from highly complete frequency information," IEEE Trans. Inform. Theory, 52, 489–509, Feb.

[3] Shalev-Shwartz, Shai, and Shai Ben-David. Understanding machine learning: From theory to algorithms. Cambridge University Press, 2014.

[4] Moreno-Alvarado, R. G., and Mauricio Martinez-Garcia. "DCT-compressive sampling of frequency-sparse audio signals." In Proceedings of the World Congress on Engineering, vol. 2, pp. 6–8. 2011.

[5] Pati, R. R., and Krishnaprasad, P. (1993). "Orthogonal matching pursuit: recursive function approximation with applications to wavelet decomposition," Proceedings of the 27th Asilomar Conference on Signals, Systems and Computers, pp. 40–44.

[6] Davis, G., Mallat, S., and Avellaneda, M. (1997). "Adaptive greedy approximations," Constr. Approx. 13(1), 57–98, Mar.

[7] Tropp, J. (2004). "Greed is good: Algorithmic results for sparse approximation," IEEE Trans. Inf. Theory. 50(10), 2231–2242.

[8] Tropp, J., and Gilbert, A. (2007). "Signal recovery from random measurements via orthogonal matching pursuit," IEEE Trans. Inf. Theory. 53(12), 4655–4666, 2007.

[9] Davies, R., et al. (2013). "The effect of recovery algorithms on compressive sensing background subtraction." Sensor Data Fusion: Trends, Solutions, Applications (SDF), 2013 Workshop on IEEE.

[10] Gao, Guangchun, Kai Xiong, and Lina Shang. "Bit-plane image coding scheme based on compressed sensing." Appl. Math 6, no. 3 (2012): 721–727.

[11] Gan, L. (2007). "Block compressed sensing of natural images." Digital Signal Processing, 2007 15th International Conference on IEEE.

[12] M. B. Wakin, J. N. Laska, M. F. Duarte, D. Baron, S. Sarvotham, D. Takhar, K. F. Kelly, and R. G. Baraniuk, "Compressive imaging for video representation and coding," in Proceedings of Picture Coding Symposium, Beijing, China, 24–26 April 2006.

[13] D. Takhar, J. N. Laska, M. B. Wakin, M. F. Duarte, D. Baron, S. Sarvotham, K. F. Kelly, and R. G. Baraniuk, "A new camera architecture based on optical domain compression," in Proceedings of SPIE Symposium on Electronic Imaging: Computational Imaging, vol. 6065, 2006.

[14] S. Mallat and Z. Zhang, "Matching pursuit with time-frequency dictionaries," IEEE Transactions on Signal Processing, vol. 41, no. 2, pp. 3397–3415, Dec. 1993.

[15] V. Stankovic, L. Stankovic, and S. Chencg, "Compressive video sampling," in Proceedings of 16th European Signal Processing Conference, Lausanne, Switzerland, Aug. 2008.

[16] Kulkarni, N. (2011). Compressive sensing for computer vision and image processing. (Diss. Arizona State University).

[17] Y. Wang, J. W. Tian, and Y. H. Tan, "Effective Gaussian mixture learning and shadow suppression for video foreground segmentation," in MIPPR 2007: Automatic Target Recognition and Image Analysis; and Multispectral Image Acquisition, 67861D, vol. 6786 of Proceedings of SPIE, Wuhan, China, November 2007.

[18] Z. Zivkovic, "Improved adaptive Gaussian mixture model for background subtraction," in Proceedings of the 17th International Conference on Pattern Recognition (ICPR '04), vol. 2, pp. 28–31, Washington, DC, USA, 2004.

[19] J. Cheng, J. Yang, Y. Zhou, and Y. Cui, "Flexible background mixture models for foreground segmentation," Image and Vision Computing, vol. 24, no. 5, pp. 473–482, 2006.

[20] A. Elgammal, D. Harwood, and L. Davis, "Non-parametric model for background subtraction," in Computer Vision—ECCV 2000: 6th European Conference on Computer Vision Dublin, Ireland, June 26–July 1, 2000 Proceedings, Part II, vol. 1843 of Lecture Notes in Computer Science, pp. 751–767, Springer, Berlin, Germany, 2000.

[21] K. Kim, T. H. Chalidabhongse, D. Harwood, and L. Davis, "Real-time foreground-background segmentation using codebook model," Real-Time Imaging, vol. 11, no. 3, pp. 172–185, 2005.

[22] O. Barnich and M. Van Droogenbroeck, "ViBe: a universal background subtraction algorithm for video sequences," IEEE Transactions on Image Processing, vol. 20, no. 6, pp. 1709–1724, 2011.

[23] Tian, Z., and Giannakis, G. B. (2007). "Compressed Sensing for Wideband Cognitive Radios," in Acoustics, Speech and Signal Processing, 2007. ICASSP 2007. International Conference on IEEE, 4, IV-1357–IV-1360, 15–20 April.

[24] Saad, Q., et al. (2013). "Compressive sensing: From theory to applications, a survey." J. ommun. Network. 15.5, 443–456.

[25] Zhu, H., Li, H., and Yin, W. (2013). Compressive sensing for wireless networks. (Cambridge: Cambridge University Press).

[26] Luo, F., Wu, J. S., and Chen, C. (2009). "Compressive data gathering for largescale wireless sensor networks," in Proceedings of the 15th annual international conference on Mobile computing and networking, pp. 145–156, ACM.

[27] Ling, Q., and Tian, Z. (2010). "Decentralized Sparse Signal Recovery for Compressive Sleeping Wireless Sensor Networks," IEEE Transactions on Signal Processing, 58(7), 3816–3827.

[28] Fu, S., Kuai, X., Zheng, R., Yang, G., and Hou, Z. (2010). "Compressive sensing approach based mapping and localization for mobile robot in an indoor wireless sensor network," in Networking, Sensing and Control (ICNSC), 2010 International Conference on IEEE, pp. 122–127.

2

Compressed Sensing: Sparsity and Signal Recovery

2.1 Introduction

The Nyquist/Shannon sampling theory was suggested by Nyquist in 1928 [1] and proved by Shannon in 1949 [2]. The theorem asserts the requirement of sampling at least twice faster than the signal bandwidth in order to obviate loss of information while capturing a signal. Signal processing deals with efficient algorithms for acquiring, processing, and extracting information from signals or data for which accurate models for the signals of interest are required. There are different models such as generative models, deterministic classes, and probabilistic Bayesian models which are used for distinguishing classes of interesting or probable signals from uninteresting or improbable signals with prior knowledge.

The sampling method produces a large volume of data with considerable redundant information wherein location of relevant information is often a difficult process. Processing the information is fast and simple in a sparse representation where only a few coefficients are used for disclosure of signif-icant information. Such representations can be constructed by decomposing signals over elementary waveforms chosen in a family called a dictionary. It is difficult to find an ideal sparse transform which can be adapted for all signals. The discovery of wavelet orthogonal bases and local time–frequency dictionaries has led to the development of new transforms. Signals are said to be sparse when they can be represented as a linear combination of a few elements from a known basis or dictionary. In sparse signal models, the signal contains relatively limited information when compared to the dimension of the signal.

It is necessary to adapt sparse representations to signal properties and to derive efficient processing operators. An orthogonal basis is a dictionary of a minimum size that can yield a sparse representation if designed for focusing on the signal energy over a set of a few vectors. A bigger dictionary helps to build shorter and more precise sentences. Similarly, dictionaries of vectors that are larger than bases are needed to build sparse representations of complex signals. But the choice of an optimal dictionary is difficult and requires more complex algorithms. Sparse representations in redundant dictionaries can improve pattern recognition, compression, and noise reduction, as also the resolution of new inverse problems. This includes super resolution, source separation, and compressive sensing.

Researchers have found that many important signals have a property called sparseness, which proves the ability of the information of a signal with relatively small number of samples. This emerging technique, called compressive sensing or compressive sampling (CS), can be used to exploit signal sparseness and allow signals to be under sampled without any loss of information [3].

2.2 Compressed Sensing

Compressed sensing (CS) is an emerging and promising technique that has attracted significant attention in the field of applied mathematics, electrical engineering, and computer science. CS ensures reconstruction of a signal using less volume of data, typically much less than the signal's dimension before it is compressed. CS framework has the ability to simultaneously sense and compress the finite dimensional vector that relies on linear dimensionality reduction. CS asserts that sparse high-dimensional signals can be recovered from random measurements by using efficient CS recovery algorithms. It is a very useful concept when dealing with limited and redundant data. Most of the signals are sparse; i.e., they have few nonzero coefficients and many zero coefficients. CS exploits the sparse nature of signals and thus allows the signal to be sampled at a sampling rate significantly lower than the Nyquist rate. It seems to violate the sampling theorem, as it depends on the sparsity of the signal and not on the highest frequency.

The conventional data acquisition system involves acquisition and compression stages. In the acquisition stage, adequate data are collected, while, in the compression stage, only useful information is retained by using a dedicated compression algorithm that involves the elimination of redundant and insignificant data. In image and video applications, the Nyquist rate is high resulting in many samples, making compression a necessity before storage or transmission. In the case of other imaging systems such as medical scanners and radars, increasing the sampling rate is very expensive due to either expensive nature of the sensors or the measurement process. For example, JPEG 2000 [4] can compress a megapixel image taken from a camera into just a few thousand wavelet coefficients by applying wavelet basis, reducing its size from a few megabytes to a few kilobytes, leading to economical storage and fast transfer. Despite storage of only a few coefficients, initially all the samples are considered for the process which increases the storage complexity and computational complexity. These complexity issues can be addressed with the help of the CS process which considers only a small number of measurements and a sophisticated yet fast algorithm for recovering the original image. The key idea of compressed sensing is to recover a sparse signal from very few non-adaptive, linear measurements through use of convex optimization techniques. Figure 2.1 shows the conventional transform coding technique used for obtaining the transformed signal.

An underdetermined system of linear equations has more unknowns than equations and generally an infinite number of solutions too. It is necessary to choose a solution to such a system by imposing additional constraints or conditions as appropriate. There is a constraint of sparsity, in CS allowing only solutions which have a small number of nonzero coefficients. It is also known that all underdetermined systems of linear equations do not have a sparse solution. However, if there is a unique sparse solution for the underdetermined system, the compressed sensing framework allows the recovery of that solution. Consider a signal that has a sparse representation, then the basic idea of compressed sensing is that if one takes non-adaptive samples in the form of projections of the signal onto a set of test vectors, and if the test vectors are incoherent with the basis vectors comprising the basis

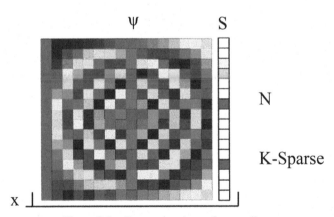

Figure 2.1 Conventional transform coding.

in which the signal is sparse, exact recovery of the signal representation from a relatively small number of such samples is possible, roughly proportional to the number of components in the sparse representation [1, 2]. Compressed sensing also remains stable in the presence of random noise; i.e., the recovery degrades gracefully as the noise level is increased [3, 5].

2.2.1 Compressed Sensing Process

The entire process of CS consists of three parts, which are depicted in Figure 2.2

The main process is the measurement process in which a sampling matrix is applied on the sparse signal to obtain the measurements. These measurements are far smaller than the actual number of samples in the signal. These measurements will be transmitted for reconstruction using the CS recovery algorithm. There are many CS recovery algorithms designed for reconstructing the sparse vector such as l_1 minimization algorithm, greedy algorithms, iterative algorithms, and model-based algorithms. In general, all signals are not inherently sparse, but many can be transformed into some basis, thus making them sparse. For example, a signal can be made sparse by applying a Fourier transform, in which only a few of the Fourier coefficients would be nonzero. Many real-world signals such as images or audio signals are sparse with respect to some basis. Sparse signals lie in a lower dimensional space, which means that they can be represented by few linear measurements. CS reduces the number of sampling points which directly correspond to the amount of data acquired from the signal. In addition, CS operates under the assumption that it is possible to directly acquire just the important information about the signals while discarding the unimportant information. This enables the creation of net-centric and stand-alone applications with many fewer resources [3]. Figure 2.3 shows the CS measurement process.

CS does not attempt measurement of the dominant coefficients of a transformed signal but, instead, measures the weighted combinations of all transform coefficients. The weights can be predetermined for each signal without the need for adaptive calculation as only a fixed set of weights are used for all signals under a given class.

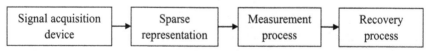

Figure 2.2   Block diagram of compressed sensing process.

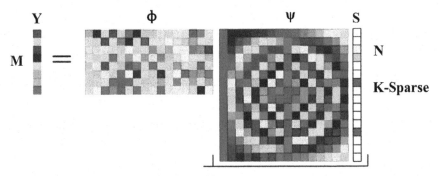

Figure 2.3 Measurement process of compressed sensing [3].

2.2.2 What Is the Need for Compressed Sensing?

In transform coding, initially all the samples are considered for obtaining the coefficients before discarding those with the least information. This process is a wasteful as well as a complex process as it considers all the samples to do the processing. There are different transforms with each having its own advantages and disadvantages. For example, discrete Fourier transform (DFT) has some drawbacks such as the need for large storage, manipulation of complex quantities, and creation of spurious spectral components due to the assumed periodicity of image blocks. Energy compaction is fair in Haar transform not suitable for compression algorithms. The Karhunen–Loève Transform (KLT) depends on signal statistics and has a few limitations: (i) The transformation matrix for a block of image is derived from the covariance matrix, which needs computation for every block. This makes the transformation data dependent and involves non-trivial computations. (ii) Perfect decorrelation in transform domain is not possible, since rarely, the image blocks can be modeled as a random field. (iii) No fast computational algorithms are available for its implementation. Transform matrix cannot be factored into sparse matrices [3].

Discrete cosine transform (DCT) has excellent energy compaction capabilities, mean-square reconstruction error performance closely matching that of KLT and fast computational approaches. DCT suffers from a few limitations which restrict its use in very low bit rate applications. The limitations are listed below: (i) Truncation of higher spectral coefficients results in blurring of the images, especially wherever the details are comprehensive. (ii) Coarse quantization of some of the low spectral coefficients introduces graininess in the smooth portions of the images. (iii) Serious blocking artifacts are introduced at the block boundaries, since each block is independently encoded, often with a different encoding strategy and the extent of quantization. Despite the computational simplicity of Walsh–Hadamard transform, the compression performance is smaller than the Fourier transform. Hence, coding CS is recommended for overcoming the drawbacks of transform coding. CS can reduce the computational, storage, and energy complexity by acquiring a few measurements rather than all samples required in the case of transform coding.

2.2.3 Adaptations of CS Theory

Over the past two decades, there have been many different adaptations of the CS theory, being perfectly adaptable methods, despite differences. The three major variations of CS algorithms found in literature are described in this chapter. They are distributed and multi-sensor compressive sensing (DCS), Bayesian compressive sensing (BCS), and model-based compressive sensing (MBCS). Each of the CS methods is explained in detail.

DCS is a branch of compressive sensing that enables new distributed coding algorithms for multi-signal ensembles. This method exploits correlation structures within and between the signals. DCS theory rests on a new concept called the joint sparsity of a signal ensemble. There are many algorithms for joint recovery of multiple signals from incoherent projections which characterize the number of distributed compression of sources with memory; this has remained a challenging problem for quite some time [5].

BCS is a Bayesian framework for solving the inverse problem of compressive sensing which remains extremely challenged. The basic BCS algorithm adopts the relevance vector machine and can marginalize the noise variance with improved robustness. Besides providing a Bayesian solution, the Bayesian analysis of CS, more importantly, provides a new framework that allows a user the facility of addressing a variety of issues that have not been addressed earlier in other compressive sensing algorithms. These issues include a stopping criterion for determining when a sufficient number of CS measurements have been performed, as well as adaptive design of the projection matrix and simultaneous inverse of multiple related CS measurements [6].

MBCS theory parallels the conventional theory and provides concrete guidelines on the modus operandi of creating model-based recovery algorithms with provable performance guarantees, which are extremely beneficial for the application of compressive sensing in industry-quality software. A highlight of this technique is the introduction of a new class of compressible signals and a new sufficient condition for robust model compressible signal recovery. The reduction in the degrees of freedom of a compressible signal by permitting only certain configurations of the large and small coefficients allows signal models to provide two immediate benefits to CS. The first benefit is that MBCS enables a user to reduce, in some cases significantly, the number of measurements "M" required to recover a signal. The second benefit is that, during signal recovery, MBCS enables a user to better differentiate true signal information from recovery artifacts, which leads to a more robust recovery [7].

2.2.4 Mathematical Background

Let us consider "x" to be an unknown vector in "R" where "R" could be a signal or image. It is necessary to acquire the data contained in vector x through sampling and then reconstruct it for obtaining the complete vector. Traditionally, this should require "N" samples, where "N" is decided on the basis of the Nyquist rate. If it is known beforehand that x is compressible by some transform domain such as wavelet or cosine, then the data x can be acquired by measuring "M" general linear functions rather than N. If the collection of linear functions is well-chosen, and a small degree of reconstruction error is allowed, the size of "M" can be dramatically smaller than the size "N" usually considered necessary. Thus, certain natural classes of signals or images with "N" data points need only "M" non-adaptive samples for faithful recovery, as opposed to the usual "N" samples. The CS concept is explained in detail below.

Consider a real-valued, finite-length, one-dimensional, discrete-time signal "x", which can be viewed as an $N \times 1$ column vector represented in terms of an orthonormal basis of $N \times 1$ vectors denoted as "ψ". Using the $N \times N$ basis matrix with the vectors $\{\psi i\}$ as columns, a signal "x" can be expressed as

$$x = \sum_{i=1}^{N} r_i \psi_i \qquad (2.1)$$

where "r" is the $N \times 1$ sparse vector. "x" and "r" are equivalent representations of the signal, with "x" in the time or space domain and "r" in the "ψ" domain. The signal "x" is said to be K-sparse when it is a linear combination of only K basis vectors; i.e., only K of the coefficients in (2.1) are nonzero and $(N - K)$ are zero. The signal "x" is compressible if it has few large coefficients and many small coefficients.

In the conventional data acquisition system, transform coding plays a vital role where initially, the full N-sample signal "x" is acquired and the transform coefficients are computed later for locating and encoding the "K" largest values and discard rest of the coefficients. Unfortunately, this sample-then-compress framework suffers from three drawbacks. First, the initial number of samples N may be large even when the desired K is small. Second, the set of all N transform coefficients must be computed despite discarding of all but K of them. Third, the locations of the large coefficients must be encoded, thus introducing an overhead. CS addresses these drawbacks by directly acquiring a compressed signal representation without going through the intermediate stage of acquiring N samples. Consider a general linear measurement process that computes $M < N$ inner products between "x" and a collection of vectors Φ. The measurement vector is represented as follows:

$$y = \Phi x \qquad (2.2)$$

Substitute the value of x in y, we get

$$y = \Phi \psi r = \Theta r \qquad (2.3)$$

where $\Theta = \Phi \psi$ an $M \times N$ matrix. The measurement process is not adaptive, meaning that Φ is fixed and does not depend on the signal "x".

Designing an efficient measurement matrix is essential for obtaining the measurement considered necessary, and for perfect recovery, an efficient recovery algorithm is required. The measurement matrix Φ must allow the reconstruction of the signal "x" from $M < N$ measurements. The signal reconstruction algorithm must take the "M" measurements in the vector "y", the random measurement matrix, and the basis and reconstruct the signal "x" or its sparse coefficient vector "r" [8].

2.2.5 Sparse Filtering and Dynamic Compressed Sensing

The basic CS framework is mainly concerned with parameter estimation, or time-invariant signals. However, currently, an effort is yet being made for developing efficient CS techniques that would be able to perform in high-dimensional non-dynamic settings. The fundamentals of CS build upon convex optimization perspectives, and as such, it is conventionally assumed that the measurements are available in a batch form. Thus, the theory of CS cannot be used for complex signals. Furthermore, the treatment of process dynamics, which are normally governed by probabilistic transition kernels, is not a straightforward task as far as optimization approaches are concerned. In light of the above, a much more practical approach for treating dynamic sparse signals would be somehow based on state filtering methodologies.

The Kalman filter (KF) algorithm is elegant and simple. It is the *linear* optimal minimum mean-square error (MMSE) estimator irrespective of noise statistics. Despite its advantages, it is rarely used in its standard formulation which is primarily designed for linear time-varying models. Recently, the KF structure has seen modifications and its capabilities are extended for use in many engineering and scientific fields. The resulting KF-based methods are used extensively for nonlinear filtering, constrained state estimation, distributed estimation, learning in neural networks, and fault-tolerant filtering.

The KF-based methodologies for dynamic CS can be divided into two broad classes namely hybrid and self-reliant. The hybrid method [9, 10] refers to KF-based approaches involving the utilization of peripheral optimization schemes for handling sparseness and support variations, whereas the self-reliant [11] class refers to methods that are entirely independent of any such scheme. The self-reliant KF method in [11] has the advantage of easy implementation. It avoids intervening in the KF process, thereby maintaining the filtering statistics as adequate as possible. The key idea behind it is to apply the KF in constrained filtering settings using the so-called pseudo-measurement technique. It may, however, exhibit an inferior performance when improperly tuned or when insufficient number of iterations had been carried out.

2.3 Signal Representation

Sparse and redundant signal representations are used in vision, signal, and image processing due to the fact that signals and images of interest can be sparse or compressible in some dictionary. The dictionary can be generated on the basis of a mathematical model of the data. Alternatively, it can be learned directly from the data. Using a dictionary helps better representation compared to using a wavelet or Fourier dictionary, thereby resulting in improved results in applications such as restoration and classification.

More precisely, suppose $x \in R^M$ is K-sparse in a basis (or a dictionary) ψ, so that $x = \psi x_0$, with $\|x_0\|_0 = K << M$. In the case when x is compressible in ψ, it can be well approximated by the best K-term representation. Consider a random $M \times N$ measurement matrix Φ with $M < N$ and assume that M measurements, which make up a vector y, are made as shown in Equation (2.3). According to CS theory, when θ satisfies the restricted isometry property

(RIP) [10], one can reconstruct x via its coefficients x_0 by solving the l^1 minimization problem [9, 11] as given below:

$$\hat{x}_0 = \arg \min_{x_0 \in R^N} \|x_0\|_1 \text{ subject to } y = \psi x. \qquad (2.4)$$

A matrix θ is said to satisfy the RIP of order K with constants $\delta_K \in (0, 1)$ if

$$(1 - \delta_K) \|v\|_2^2 \le \|\theta v\|_2^2 \le (1 + \delta_K) \|v\|_2^2$$

For any v such that $\|v\|_0 \le K$. One popular class of measurement matrices satisfying an RIP is the one consisting of i.i.d. Gaussian entries. It is a well-known fact that, when Φ is an $M \times N$ Gaussian matrix where $M > O(K \log M)$ and ψ is a sparsifying basis, $\pounds$ satisfies the RIP with high probability. One can also use greedy pursuits and iterative soft or hard thresholding algorithms for recovering signals from compressive measurements.

2.3.1 Sparsity

The term "sparsity" states that a vector has at most K non zero coefficients, at most. This is measured by the 'l_0 "norm", which is mathematically denoted as $\|.\|_0$.

A vector $x = (x_i)_{i=1}^N \in R^N$ is called K-sparse if

$$\|x\|_0 = \# \{i : x_i \ne 0\} \le K \qquad (2.5)$$

The set of all K-sparse vectors denoted by Σ_K is a highly nonlinear set. The set Σ_K be a K-sparse signal, if it belongs to the linear subspace consisting of all vectors with the same support set. In the following definition, the K-sparse approximation is given where the decay rate of the lp error of the best K-term approximation of a vector is analyzed [12].

Let $1 \le p < \infty$ and $r > 0$. A vector $x = (x_i)_{i=1}^N \in R^N$ is called p-compressible with constant C and rate r, if

$$\sigma_K(x)_p := \frac{\min}{\tilde{x} \in \Sigma_K} \|x - \tilde{x}\|_p \le C.K^r \text{ for any } K \in \{1, ..., N\}. \qquad (2.6)$$

2.4 Basis Vectors

CS states that many signals can be represented by using only a few non zero coefficients in a suitable basis or dictionary. The most popular conventional technique for signal compression is known as transform coding that finds a basis for providing sparse or compressible representations for signals. CS states that the number of measurements required for reconstruction can be significantly reduced when the signal is sparse in a known basis unlike the Nyquist–Shannon sampling theorem which states that a certain minimum number of samples are required for ensuring perfect reconstruction of a signal.

A set $\{\Phi_i\}_{i=1}^{N}$ is called a basis for R^N if the vectors are linearly independent. For any $x \in R^N$, there exist coefficients $\{c_i\}_{i=1}^{N}$ such that

$$x = \sum_{i=1}^{N} c_i \Phi_i \tag{2.7}$$

Let Φ represents the $N \times N$ matrix with columns given by Φ_i and let c represents the length-N vector with entries c_i, then the relation can be expressed as

$$x = \Phi c \tag{2.8}$$

The orthonormal basis is defined as a set of vectors $\{\Phi_i\}_{i=1}^{N}$ satisfying

$$\langle \Phi_i, \Phi_j \rangle = \begin{cases} 1, & i = j \\ 0, & i \neq j \end{cases} \tag{2.9}$$

In orthonormal basis, the coefficients c are computed as

$$c_i = \langle x, \Phi_i \rangle \tag{2.10}$$

or represented as

$$c = \Phi^T x \tag{2.11}$$

in matrix notation. This equation can be verified since the orthonormality of the columns of Φ means that $\Phi^T \Phi = I$, where I denotes the $N \times N$ identity matrix [12].

2.4.1 Fourier Transform

The Fourier transform [13] decomposes a signal into frequencies and is a complex-valued function of frequency, whose absolute value represents the amount of that frequency present in the original function, and whose complex argument is the phase offset of the basic sinusoid in that frequency. The term "Fourier transform" refers to both the frequency domain representation and the mathematical operation that associates the frequency domain representation to a function of time. The inverse Fourier transformation (also called Fourier synthesis) of a frequency domain representation combines the contributions of all the different frequencies to recover the original function of time.

Linear operations performed in one domain have corresponding operations in the others, which are sometimes easier to perform. For example, differentiation in the time domain corresponds to multiplication by the frequency. Hence, some differential equations are easier to analyze in the frequency domain. In addition, convolution in the time domain corresponds to ordinary multiplication in the frequency domain. Any linear time-invariant system, as e.g., a filter applied to a signal, can be expressed in a relatively simple manner in frequency domain. After performing the desired operations, transformation of the result can be made back to the time domain. The Fourier transform and the inverse Fourier transform are explained for both 1D and 2D signals.

The 1D Fourier transform pair (forward and inverse transform) is represented as

$$F(u) = \int_{\infty}^{\infty} f(x)e^{j2\pi ux}\mathrm{d}x \tag{2.12}$$

$$f(x) = \int_{\infty}^{\infty} F(u)e^{j2\pi ux}\mathrm{d}u \tag{2.13}$$

The 2D Fourier transform pair (forward and inverse transform) is given as

$$F(u,v) = \int_{\infty}^{\infty}\int_{\infty}^{\infty} f(x,y)e^{j2\pi(ux+vy)}\mathrm{d}x\,\mathrm{d}y \tag{2.14}$$

$$f(x,y) = \int_{\infty}^{\infty}\int_{\infty}^{\infty} F(u,v)e^{j2\pi(ux+vy)}\mathrm{d}u\,\mathrm{d}v \tag{2.15}$$

where u and v are spatial frequencies. The advantage of Fourier transform is that hardware acceleration is available on CPU, but it yields relatively poor compression performance.

2.4.2 Discrete Cosine Transform

The DCT transforms a signal from a spatial representation into a frequency representation. When an image is represented in frequency domain, most of the energy is concentrated in the lower frequencies, enabling discarding of the high-frequency components. Hence, the volume of data needed to describe the image can be reduced without affecting the image quality. Compared to DFT, the DCT results in less blocking artifacts due to the even symmetric extension properties of DCT. DCT uses real computations, whereas DFT uses complex computations. This makes DCT hardware simpler, as compared to that of DFT. These advantages have made DCT-based image compression a standard in still-image and multimedia coding standards. The type-2 DCT transforms a block of image of size $n \times n$ having pixel intensities f(x,y) into a transform array of coefficients $F(u,v)$ as shown in Equation (2.16). The inverse DCT for 2D signal image block $f(x, y)$ of size $n \times n$ is defined as (2.17) follows:

$$F(u, v) = \sqrt{\frac{4}{n^2}} C(u)C(v) \sum_{x=0}^{n} \sum_{y=0}^{n} f(x, y) \, \cos\left(\frac{\pi(2x + 1)u}{2n}\right)$$

$$\cos\left(\frac{\pi(2y + 1)v}{2n}\right) \tag{2.16}$$

$$f(x, y) = \frac{2}{n} c_u c_v \sum_{u=0}^{n} \sum_{v=0}^{n} F(u, v) \, \cos\left[\frac{\pi(2x + 1)u}{2n}\right] \cos\left[\frac{\pi(2y + 1)v}{2n}\right] \tag{2.17}$$

$$\text{where } u, v = 0, 1, ..., n - 1 \text{ and } c_u, c_v = \begin{cases} \frac{1}{\sqrt{2}}, & u, v = 0 \\ 1, & u, v \neq 0 \end{cases} \tag{2.18}$$

where u, v represent the spatial frequencies in the direction of x and y, respectively. The transformed array $F(u,v)$ obtained through Equation (2.16) is also of the size $n \times n$, same as that of the original image block. $u = v = 0$ corresponds to the DC component, whereas the remaining ones are the AC components corresponding to higher spatial frequencies as u and v increase. In DCT, the transformation requires $O(n^4)$ computations, whereas with the help of fast Fourier transform (FFT), the computations can be reduced to $O(2n^2 \log n)$. Speedy computational approaches and the use of real arithmetic have made DCT popular for image compression applications. Most of the transformed coefficients have very small values, and only a few coefficients have higher magnitudes. This shows the energy compaction capabilities of DCT. Selection of block sizes in DCT is an important consideration as the images should be subdivided to ensure reduction in the redundancies between the adjacent sub-images to an acceptable level, while the dimension of the sub-images should be an integer power of 2. The adjacent block redundancies can be reduced by increasing the block size. The mean-square reconstruction error can also be reduced using truncated and quantized coefficients, involving more computations. Most popular block sizes used in image compression are 8×8 and 16×16. Figure 2.4 shows the discrete cosine basis for 8×8 image block.

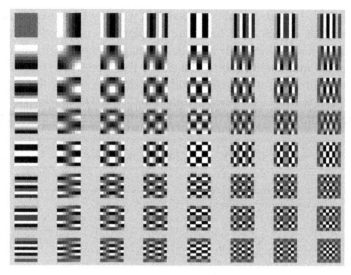

Figure 2.4 Discrete cosine transform basis [14].

DCT has become an international standard for transform coding as it is computationally efficient, is easy to implement in hardware, and has higher compression performance.

2.4.3 Discrete Wavelet Transform

In numerical and functional analyses, a discrete wavelet transform (DWT) is any wavelet transform for which the wavelets are discretely sampled. The advantage of DWT is the temporal resolution as it captures both frequency and location information. The wavelet transform uses functions that are localized in both the real and Fourier space unlike Fourier transform which decomposes the signal into sines and cosines. The different types of wavelet transforms are categorized on the basis of the wavelet orthogonality. The orthogonal wavelets are used for DWT development, while non-orthogonal wavelets are used for continuous wavelet transform development. These two transforms have the following properties:

1. The DWT returns a data vector of the same length as of the input. Usually, even in this vector, many data are almost zero. This corresponds to the fact that it decomposes into a set of wavelets (functions) that are orthogonal to its translations and scaling. Therefore, the signal is decomposed to the same or a lower number of the wavelet coefficient spectrum as the number

of signal data points. Such a wavelet spectrum is very good for signal processing and compression, for example, as no redundant information is here.

2. The continuous wavelet transform in contrary returns an array one dimension larger than the input data. For a 1D data, an image of the time–frequency plane is obtained. The signal frequencies evolution can be seen during the duration of the signal and compare the spectrum with other signals spectra.

For 2D signal, the DWT is shown in Figure 2.5 and the equations for DWT can be given as follows.

Given an image block $f(x, y)$ of size $n \times n$, the forward and inverse DWT are defined as [15]

$$W_\varphi(j_0, k_1, k_2) = \frac{1}{\sqrt{n^2}} \sum_{x=0}^{n-1} \sum_{y=0}^{n-1} f(x, y)_{j_0, k_1, k_2}(x, y) \qquad (2.19)$$

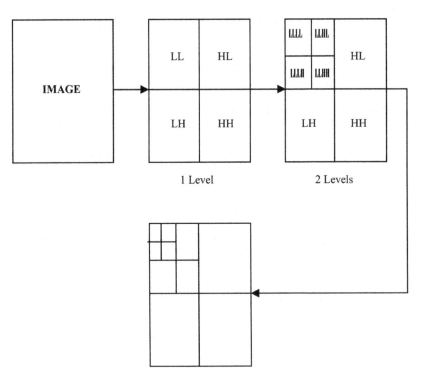

Figure 2.5 Discrete wavelet transform [15].

$$W_\psi^q(j_1, k_1, k_2) = \frac{1}{\sqrt{n^2}} \sum_{x=0}^{n-1} \sum_{y=0}^{n-1} f(x, y) \psi_{j_1,k_1,k_2}^q(x, y) \tag{2.20}$$

$$
f(i_1, i_2) = \left\{ \frac{1}{\sqrt{n^2}} \sum_x \sum_y W_\varphi(j_0, k_1, k_2)_{j_0,k_1,k_2}(x, y) \right.
$$

$$
\left. + \frac{1}{\sqrt{n^2}} \sum_{q=H_1,V_1,D_1} \sum_{j_1=0}^{m} \sum_{k_1} \sum_{k_2} W_\psi^q(j_1, k_1, k_2) \psi_{j_1,k_1,k_2}^q(x, y) \right\}
$$

$$\tag{2.21}$$

where $W_\phi(j_0, k_1, k_2)$, $W_\psi^q(j_1, k_1, k_2)$ denote the approximation coefficients and detail coefficients, respectively; $\varphi_{j_0,k_1,k_2}(x, y)$, $\psi_{j_1,k_1,k_2}^q(x, y)$ denote the scaling and wavelet functions, respectively; $j_0 = 0, j_1 \geq j_0$ represents the scale; and $q = \{H_1, V_1, D_1\}$ indicates the index of LH, HL, HH bands. The LL band contains the approximation coefficients, and the other three bands contain the detail coefficients. The LH, HL, and HH bands are sparse, that is, converted into a sparse vector. DWT has multi-resolution capability, thus achieving better quality [15].

2.4.4 Curvelet Transform

Curvelet transform [16–19] is the same as wavelet which is a multi-scale transform, but can provide sparse representation to smooth and edge parts of image providing good effect in energy concentration. Curvelet transform is anisotropy which has strong directivity with ability to more information for image. Curvelet transform appears to be the same as ridgelet transform in several orientations, positions, and scales, but the basic scale is fixed for ridgelet transform, hence curvelet transform anisotropy. The curvelet transform is a higher dimensional generalization of the wavelet transform designed to represent images at different scales and different angles. Curvelets enjoy two unique mathematical properties, namely:

- Curved singularities can be well approximated with very few coefficients and in a non-adaptive manner—hence the name "curvelets."
- Curvelets remain coherent waveforms under the action of the wave equation in a smooth medium.

Curvelets constitute a non-adaptive technique for multi-scale object representation. Curvelets are an extension of the wavelet concept, and hence, they

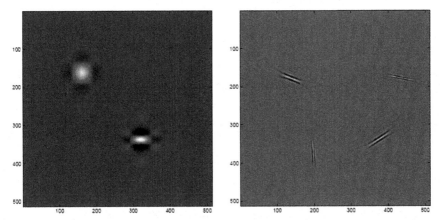

Figure 2.6 The elements of wavelets (*left*) and curvelets on various scales, directions, and translations in the spatial domain (*right*). Note that the tensor–product 2D wavelets are not strictly isotropic but have directional selectivity [16].

are becoming popular image processing and scientific computing techniques. Wavelets generalize Fourier transform by using a basis that represents both location and spatial frequency. For 2D as shown in Figure 2.6 or 3D signals, directional wavelet transforms use the basis functions that are localized in orientation. Curvelets define the higher resolution curvelets to be more elongated than the lower resolution curvelets. The use of some sort of directional wavelet transform whose wavelets have the same aspect ratio at every scale is suggested considering the absence of this property in the natural images. For an ideal image, curvelets provide a sparser representation compared to other wavelet transforms. For a Fourier transform, the squared error decreases only as $O\left(1\big/\sqrt{N}\right)$. For a wide variety of wavelet transforms, including both directional and non-directional variants, the squared error decreases as $O(1/N)$. The extra assumption underlying the curvelet transform enables achievement of to $O((\log N)^3/N^2)$ [20]. Efficient numerical algorithms exist for computing the curvelet transform of discrete data. The computational cost of a curvelet transform is approximately 10–20 times that of an FFT and has the same dependence of $O(N^2 \log N)$ for an image of size $N \times N$. The second-generation curvelet transform [21–23] has proved to be a very efficient tool for many different applications in image processing, seismic data exploration, fluid mechanics, and solving partial different equations (PDEs). Mathematically, the strength of the curvelet approach is their ability to formulate strong theorems in approximation and operator theory.

Despite the discrete curvelet transform being very efficient in representing curve-like edges, the current curvelet systems have two main drawbacks: (1) They are not optimal for sparse approximation of curve features beyond C^2 singularities and (2) the discrete curvelet transform is highly redundant. The currently available implementations of the discrete curvelet transform aims at reduction in the redundancy efficiently. However, the theoretical results show the inappropriateness of the discrete curvelet transform for image compression [24]. Curvelet constructions require a rotation operation and correspond to a partition of the 2D frequency plane based on polar coordinates. This property makes the curvelet idea simple in the continuous case but causes problems in the implementation for discrete images. In particular, approaching critical sampling seems difficult in discretized constructions of curvelets.

2.4.5 Contourlet Transform

Contourlets, as proposed by Do and Vetterli [25], form a discrete filter bank structure that can deal effectively with piecewise smooth images with smooth contours. This discrete transform can be connected to curvelet-like structures in the continuous domain. Hence, the contourlet transform can be seen as a discrete form of a particular curvelet transform. The advantage of contourlet transform lies in easy implementation of critical sampling using this transform. There exists an orthogonal version of the contourlet transform that is faster than current discrete curvelet algorithms. The contourlet transform has a number of useful features and qualities, but it also has its flaws too. One of the more notable variations of the contourlet transform was developed and proposed by da Cunha, Zhou, and Do in 2006. The non-subsampled contourlet transform (NSCT) was developed mainly due to the contourlet transform being not shift invariant. The reason for this lies in the up-sampling and down-sampling present in both the Laplacian pyramid and the directional filter banks. The method used in this variation was inspired by the non-subsampled wavelet transform or the stationary wavelet transform which were computed with the à trous algorithm. The applications of the contourlet transform include synthetic aperture radar despeckling, image enhancement, and texture classification [26]. Figure 2.7 shows the non-subsampled contourlet transform.

The Contourlet transform uses a double filter bank structure for getting smooth contours of images. The Laplacian pyramid (LP) and directional filter bank (DFB) are used in this double filter bank for capturing the point discontinuities and linear structures, respectively. The LP decomposition produces one band-pass image in a multidimensional signal processing that

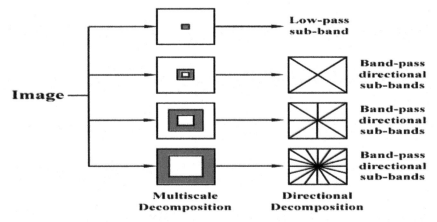

Figure 2.7 Contourlet transform [26].

can avoid frequency scrambling. The DFB is fit only for high frequency since it leaks the low frequency of signals in its directional sub-bands. Hence, DFB is combined with LP to remove the low frequency. Therefore, the image signals are passed through LP sub-bands initially to get band-pass signals and the through DFB to capture the directional information of image. The contourlet transform approximates the original image by using basic contour, so it is also called discrete contourlet transform [26].

2.4.6 Surfacelet Transform

Surfacelets [20, 27] are 3D extensions of the 2D contourlets that are obtained by a higher-dimensional directional filter bank and a multi-scale pyramid. They can be used for efficient capture and representation of surface-like singularities in multidimensional volumetric data involving biomedical imaging, seismic imaging, video processing, and computer vision. Surfacelets and the 3D curvelets aim at the same frequency partitioning, but the two transforms achieve this goal with different approaches as in the 2D case. The surfacelet transform is less redundant than the 3D curvelet transform with certain loss of directional features.

Surfacelet transform (ST), proposed by Lu and Do, can efficiently capture the surface intrinsic geometrical structure within N-dimensional signals. It offers directional sub-bands with decomposition level L_s by combining the multi-scale pyramid with the 3-dimensional directional filter banks (3D-DFB). Thus, surfacelet with more directions may help reducing the blocky artifacts caused by orthogonal wavelets. The input signal first goes through the 3D hourglass filter which is a three-channel undecimated filter bank. The output is then fed into a 2D filter bank, which operates on the planes. The tree structured filter bank produces output sub-bands, and each output is then fed into another 2D filter bank operating on the planes. The advantage of ST is that its angular resolution can be refined by invoking more levels of decomposition.

2.4.7 Karhunen–Loève Theorem

In the theory of stochastic processes, the Karhunen–Loève theorem, also known as the Kosambi–Karhunen–Loève theorem, is a representation of a stochastic process as an infinite linear combination of orthogonal functions, analogous to a Fourier series representation of a function on a bounded interval [28]. The importance of the Karhunen–Loève theorem is that it yields the best basis with minimum the total mean-squared error. The coefficients of the Karhunen–Loève theorem are random variables, and the expansion basis depends on the process. In fact, the orthogonal basis functions used in this representation are determined by the covariance function of the process.

In the case of a centered stochastic process $\{X_t\}_{t \in [a,b]}$ (centered means $E[X_t] = 0$ for all $t \in [a, b]$) satisfying a technical continuity condition, X_t admits a decomposition where Z_k are pairwise uncorrelated random variables and the functions e_k are continuous real-valued functions on $[a, b]$ that are pairwise orthogonal in $L^2([a, b])$. It is therefore sometimes said that the expansion is bi-orthogonal since the random coefficients Z_k are orthogonal in the probability space, while the deterministic functions e_k are orthogonal in the time domain. The general case of a process X_t that is not centered can be brought back to the case of a centered process by considering $X_t - E[X_t]$ which is a centered process.

Moreover, the random variables Z_k, when the process is Gaussian, are Gaussian and stochastically independent. This result generalizes the Karhunen–Loève transform. An important example of a centered real stochastic process on [0, 1] is the Wiener process; the Karhunen–Loève theorem can be used for providing a canonical orthogonal representation for it. Let A be a matrix whose rows are formed from the eigenvectors of the covariance matrix C. The first row of A is the eigenvector corresponding to the largest eigenvalue, and the last row is the eigenvector corresponding to the smallest Eigen value.

The forward transform is defined as follows:

$$Y = A(\underline{x} - \underline{m}) \tag{2.22}$$

It is called the Karhunen–Loève transform [29].

Inverse Transform
To reconstruct the original vectors $\underline{x}$ from its corresponding y

$$A^{-1} = A^T \tag{2.23}$$

$$\underline{x} = A^T Y + \underline{m} \tag{2.24}$$

A matrix A_k formed from the k eigenvectors corresponds to the k largest eigenvalues, yielding a transformation matrix of size k $\times$ n. The Y vectors would then be k dimensional. The construction of the original vector $\hat{x}$ is

$$\underline{x} = AK^T \underline{Y} + \underline{m}x \tag{2.25}$$

KLT has few drawbacks making it unsuitable for practical applications. The limitations are as follows:

- Its basis functions depend on the covariance matrix of the image, requiring recomputation and transmission for every image.
- Perfect decorrelation is not possible, since images can rarely be modeled as realizations of ergodic fields.
- There are no fast computational algorithms for its implementation.

2.5 Restricted Isometry Property

In CS, the restricted isometry property (RIP) is a powerful condition on measurement operators to ensure the robust recovery of sparse vectors from noisy, undersampled measurements via computationally tractable algorithms. A matrix Φ of size $M \times N$ is said to satisfy the RIP with RIP constant $R(K, M, N; \Phi)$ if, for every

$$x \in x^N(K) := \left\{ x \in R^N : \|x\|_0 \le K \right\}, \tag{2.26}$$

$$R(K, M, N; \Phi) := \min_{c \ge 0} c \quad \text{subject to } (1-c)\|x\|_2^2 \le \|\Phi x\|_2^2 \le (1+c)\|x\|_2^2 \tag{2.27}$$

The RIP constants measure the extent of the matrix Φ acting like an isometry when restricted to K columns; it describes the most significant distortions of the l^2 norm of any K-sparse vector. Typically, $R(K, M, N; \Phi)$ is measured for matrices with unit l^2 norm columns, and in this special case, $R(1, M, N) = 0$. Specifically, the RIP constant $R(K, M, N; \Phi)$ is the maximum distance from 1 of all the eigenvalues of the $\binom{N}{K}$ sub-matrices, $\Phi_K^T \Phi_K$, derived from Φ, where

K_1 is an index set of cardinality K which restricts Φ to those columns indexed by K_1. The RIP is predominantly used to establish theoretical performance guarantees when either the measurement vector y is corrupted with noise or the vector x is not strictly K-sparse. It is essential to prove that an algorithm is stable to noisy measurements since measurements are rarely free from noise.

For many CS encoder/decoder pairs, it has been shown that if the RIP constants for the encoder remain bounded as M and N increase with $M/N \rightarrow \delta \in (0, 1)$, the decoder can be guaranteed to recover the sparsest x for K up to a critical threshold, which can be expressed as a fraction of M, $\rho(\delta).M$. Typically, each encoder/decoder pair has an exclusive $\rho(\delta)$. Since the magnitude of the encoder/decoder pairs is not known, it becomes difficult for a practitioner to know how aggressively they may undersample, or which decoder has stronger performance guarantees. They adopt a proportional growth asymptotic, for quantifying the sparsity/undersampling trade-off wherein they consider sequences of triples (K, M, N) where all elements grow large in a coordinated way, $M \sim \delta N$ and $K \sim \rho M$ for some constants $\delta, \rho > 0$. This defines a two-dimensional phase space (δ, ρ) in [0,1] for asymptotic analysis [12].

2.6 Coherence

CS argues that the mutual coherence of the measurement probes is related to the reconstruction performance in imaging sparse scenes. While the spark, NSP, and RIP all provide guarantees for the recovery of K-sparse signals, establishing the satisfaction by a general matrix Φ of any of these properties would typically require a combinatorial search over all $\binom{N}{K}$ sub-matrices. The easily computable properties are used for providing recovery guarantees. The coherence of a matrix is one such property. The coherence of a matrix Φ, $\mu(\Phi)$, is the largest absolute inner product between any two columns Φi, Φj of Φ:

$$\mu(\Phi) = \max_{1 \leq i < j \leq n} \frac{|\langle \Phi_i, \Phi_j \rangle|}{\|\Phi_i\|_2 \|\Phi_j\|_2} \tag{2.28}$$

It is possible to show that the coherence of a matrix is always in the range $\mu(\Phi) \in \left[\sqrt{\frac{n-m}{m(n-1)}}, 1\right]$; the lower bound is known as the Welch bound. Note that when n $>>$ m, the lower bound is approximately $\mu(\Phi) \geq 1/\sqrt{m}$. The concept of coherence can also be extended to certain structured sparsity models and specific classes of analog signals [12].

2.7 Stable Recovery

The sparse signal recovery in the standard requires the absence of meeting a priori for sensing matrix Φ which may not be met in practical applications where various errors and fluctuations exist in the sensing instruments. The task of standard CS is to recover the original signal x^0 via an efficient CS recovery algorithm given the sensing matrix Φ, acquired sample y, and upper bound ε for the measurement noise. The restricted isometry property (RIP) is an important tool for stable recovery, which is defined as follows. The K restricted isometry constant (RIC) of a matrix Φ, denoted by $\delta_k(\Phi)$, as the smallest number such that

$$(1 - \delta_k(\Phi)) \|v\|_2^2 \leq \|\Phi v\|_2^2 \leq (1 + \delta_k(\Phi)) \|v\|_2^2 \qquad (2.29)$$

holds for all K-sparse vectors v, Φ is said to satisfy the K-RIP with constant $\delta_k(\Phi)$ if $\delta_k(\Phi) < 1$ [12] (Figure 2.8).

2.8 Number of Measurements

The number of measurements required for reconstruction depends on the sparsity level K. When the choice of M is too small compared to the critical number, there are known information–theoretic barriers to the accurate reconstruction of Arias-Castro et al. The number of measurements is unique for different CS recovery algorithms. At the same time, the measurement process is wasteful, when the chosen M is much larger as there are known

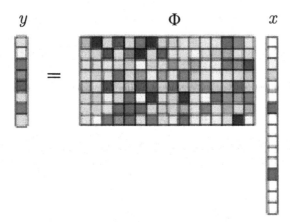

Figure 2.8 Stable recovery process [12].

algorithms that can reliably recover with approximately measurements. To deal with the selection of M, a sparsity estimate may be used in two different ways, depending on whether measurements are collected sequentially, or in a single batch. In the sequential case, an estimate of sparsity can be computed from preliminary measurements, and then, the estimated value determines how many additional measurements should be collected to recover the full signal. Alternatively, if all of the measurements must be taken in one batch, the sparsity estimate can be used to certify whether or not enough measurements were actually taken [12].

2.9 Sensing Matrix

Sensing matrix is used for obtaining the measurements from the signal, and two important characteristics of the matrix are sparsity and RIP. The RIP property includes the restricted isometry property of order K (RIP-K), and the restricted null-space property of order K (NSP-K), where K is a presumed upper bound on the sparsity level of the true signal. Since many recovery guarantees are closely tied to RIP-K and NSP-K, a growing body of work has been devoted to understanding the satisfaction or otherwise of a given matrix [12]. When K is treated as given, this problem is already computationally difficult. The sensing mechanisms collect information about a signal $x(t)$ by linear functional recordings,

$$yk = <x,\ ak> k = 1, ..., M. (1) \tag{2.30}$$

That is, the object is simply correlated to the waveforms $ak(t)$. This is a standard setup. When the sensing waveforms are Dirac delta functions (spikes), for example, y is a vector of the sampled values of x in the time or space domain. When the sensing waveforms are sinusoids, y is a vector of Fourier coefficients; this is the sensing modality used in magnetic resonance imaging (MRI). The process is mathematically represented as follows:

$$y = \Phi x; \tag{2.31}$$

where Φ is an $M \times N$ matrix whose rows are the sensing waveforms ak; and $y \in RM$. The matrix Φ represents a dimensionality reduction; i.e., it maps RN, where N is generally large, into RM, where M is typically much smaller than N. In the standard CS framework, it is assumed that the measurements are non-adaptive, meaning that the rows of Φ are fixed in advance and do not depend on the previously acquired measurements. There are two main theoretical questions in CS. First, how should we design the sensing matrix Φ

to ensure that it preserves the information in the signal x? Second, how can we recover the original signal x from measurements y? In the case where the data are sparse or compressible, we can see the ability to that design matrices Φ with $M << N$ that ensure the ability to recover the original signal accurately and efficiently using a variety of practical algorithms. Restrictions are imposed on Φ like satisfying the NSP, RIP, and/or some desired coherence for recovering a unique K-sparse vector (a vector with at most $K < N$ nonzero entries). The measurement matrix is generated based on the number of measurements required for reconstruction.

Sensing matrices are required for satisfaction of mutual coherence when the sensing matrix is chosen freely, the best choice consist of random matrices such as Gaussian IID matrices. It is still an open question whether deterministic matrices can be carefully constructed to have similar properties with respect to CS problems. Moreover, most applications do not allow for a free choice of the sensing matrix and enforce a particularly structured matrix. Exemplary situations are the application of data separation, in which the sensing matrix has to consist of two or more orthonormal bases or frames or high-resolution radar, for which the sensing matrix has to bear a particular time–frequency structure, uniform random ortho-projectors, or Bernoulli matrices.

2.9.1 Null-Space Conditions

The null space of Φ is denoted as

$$N(\Phi) = \{z : \Phi z = 0\} \tag{2.32}$$

To recover all sparse signals x from the measurements Φx, it is necessary for any pair of distinct vectors x, $x' \in \sum_K$, the condition must satisfy $\Phi x = \Phi x'$, or else, it would be impossible to distinguish x from x' based solely on the measurements y. More formally, by observing that if $\Phi x = \Phi x'$, then $\Phi(x - x') = 0$ with $x - x' \in \sum_{2k}$, then Φ uniquely represents all $x \in \sum_k$ if and only if $N(\Phi)$ contains no vectors in $\sum_{2k}$. While there are many equivalent ways of characterizing this property, one of the most common is known as the spark.

2.9.2 Restricted Isometry Property

The NSP provides the necessary and appropriate conditions for establishing guarantees that do not account for noise. When the measurements are contaminated with noise or have been corrupted by some error such as quantization, it is important to consider somewhat stronger conditions. Candes and Tao introduced the following isometry condition on matrices Φ and established its important role in CS. A matrix Φ satisfies the RIP of order K if there exists a $\delta_K \in (0, 1)$ such that $(1\,\delta_K)\,\|x\|_2^2 \leq \|\Phi x\|_2^2 \leq (1+\delta_K)\,\|x\|_2^2$ holds for all $x \in \sum_K$. A matrix Φ that satisfies the RIP of order 2K approximately preserves the distance between any pair of K-sparse vectors. This will clearly have fundamental implications concerning robustness to noise. Moreover, the potential applications of such stable embeddings range far beyond the acquisition for the sole purpose of signal recovery.

2.9.3 Gaussian Matrix

It is important to construct an efficient matrix that satisfies the necessary properties in the context of CS. Consider an example of $M \times N$ Vandermonde matrix V constructed from m distinct scalars having a spark $(V) = M + 1$. Unfortunately, these matrices are poorly conditioned for large values of N, making the recovery unstable. Similarly, there are known matrices Φ of size $M \times N$ that achieve the coherence lower bound $\mu(\Phi) = 1 / \sqrt{M}$, such as the Gabor frame generated from the Alltop sequence and more general equiangular tight frames. These constructions restrict the number of measurements needed for recovery of a K-sparse signal to be $M = O\left(K^2 \log N\right)$. It is also possible to deterministically construct matrices of size $M \times N$ that satisfy the RIP of order K, but requires M to be relatively large.

For practical implementations, these results would lead to an unacceptably large requirement on M. Using random matrices to construct Φ has a number of additional benefits. First, one can show that the measurements are democratic, for random constructions indicating the possibility of recovering a signal using any sufficiently large subset of the measurement. Thus, by using random Φ, one can be robust to the loss or corruption of a small fraction of the measurements. Second, and perhaps more significantly, it is necessary in practice to set x as sparse with respect to some basis Ψ. In this case, it is necessary that the product $\Phi\Psi$ satisfies the RIP.

For example, if Φ is chosen according to a Gaussian distribution and Ψ is an orthonormal basis, one can easily show that $\Phi\Psi$ will also have a Gaussian distribution. Hence, M being sufficiently high $\Phi\Psi$ will satisfy the RIP with high probability, just as before. Although less obvious, similar results hold for sub-Gaussian distributions as well. This property, sometimes referred to as universality, constitutes a significant advantage of using random matrices to construct Φ [31].

2.9.4 Toeplitz and Circulant Matrix

Toeplitz CS matrices have some additional benefits compared to completely independent (i.i.d.) random CS matrices. First, Toeplitz matrices are more efficient for generate and storage. A $M \times N$ (random) partial Toeplitz matrix requires only the generation and storage of $M + N$ independent realizations of a random variable, while a fully random matrix of the same size requires the generation and storage of MN random quantities. In addition, the use of Toeplitz matrices in CS applications leads to a general reduction in computational complexity. Performing a matrix–vector multiplication between a fully random $M \times N$ matrix and an $M \times 1$ vector requires MN operations. In contrast, multiplication by a Toeplitz matrix can be performed in the frequency domain, due to the convolution nature of Toeplitz matrices. Using fast Fourier transforms, the complexity of the multiplication can be reduced to O(N log N) operations, resulting in a significant speedup of the mixed-norm optimizations that are essential for several commonly utilized CS reconstruction procedures such as GPSR and SpaRSA. Depending on the computational resources available, this speedup can literally be the difference between intractable and solvable problems.

Bajwa et al. [32] prove that random Toeplitz matrices satisfy the RIP by using bounds on the coherence of this matrix. Random Toeplitz matrices can therefore be used as measurement matrices for CS. However, the number of measurements required for accurate signal recovery with random Toeplitz matrices grows as the square of sparsity K. In order to reduce the required number of measurements, Rauhut [33] proves that the necessary number of measurements for accurate signal recovery with Toeplitz or circulant matrices grows linearly as the sparsity K. However, they do not use the restricted isometry constants of a random Toeplitz or circulant matrix as [32, 34, 35], and so a good estimation of the restricted isometry constants is still open. Toeplitz and circulant matrices arise from the convolutional structure inherent in linear system identification problems. Compared with Gaussian random matrices, Toeplitz and circulant matrices are advantageous using fewer independent random variables. Furthermore, multiplication with Toeplitz or circulant matrices can be efficiently implemented by fast Fourier transform (FFT), resulting in faster projection and reconstruction algorithms. However, Toeplitz and circulant matrices cannot support block-based processing in CS applications, such as multi-channel and multi-dimensional filtering. In order to support block-based processing in CS applications, the matrices combine block matrices [27] with Toeplitz or circulant matrices, called Toeplitz or circulant block matrices. Consider (truncated circulant) Toeplitz block matrices of the form

$$
\Phi = \begin{pmatrix}
\Psi_K & \Psi_{K-1} & \cdots & \Psi_2 & \Psi_1 \\
\Phi_1 & \Phi_K & \cdots & \Phi_3 & \Phi_2 \\
\vdots & \vdots & \ddots & \ddots & \vdots \\
\Phi_{l-1} & \Phi_{l-2} & \cdots & \cdots & \Phi_l
\end{pmatrix} \in \Re^{n \times N},
$$

where $l < K$ and the block $\Phi_i \in \Re^{d \times e}$ are themselves (truncated circulant) Toeplitz matrices:

$$
\Phi_i = \begin{pmatrix}
\varphi_p^i & \varphi_{p-1}^i & \cdots & \varphi_2^i & \varphi_1^i \\
\varphi_1^i & \varphi_p^i & \cdots & \varphi_3^i & \varphi_2^i \\
\vdots & \vdots & \ddots & \ddots & \vdots \\
\varphi_{q-1}^i & \varphi_{q-2}^i & \cdots & \cdots & \varphi_q^i
\end{pmatrix} \in \Re^{p \times p}
$$

whose elements $\varphi_p^i, \varphi_{p-1}^i, \ldots, \varphi_1^i$ are drawn independently from certain probability distributions with $q < p$. Let

$$
A = \begin{pmatrix}
A_{1,1} & A_{1,2} & \cdots & A_{1,L-1} & A_{1,L} \\
A_{2,1} & A_{2,2} & \cdots & A_{2,L-1} & A_{2,L} \\
\vdots & \vdots & \vdots & \vdots & \vdots \\
A_{D,1} & A_{D,2} & \cdots & A_{D,L-1} & A_{D,L}
\end{pmatrix}
$$

be the matrix composed by sub-matrices $A_{i,j}(i = 1, 2, \ldots, D, j = 1, 2, \ldots, L)$, and $A_{i,j}$ is a circulant matrix.

2.9.5 Binomial Sampling Matrix

Binary sampling matrices are RIP-fulfilling matrices with 0, 1 elements prior to column normalization. A study in the subset of such matrices was made in the field of optical code division multiple access (OCDMA) with the name of OOC [25]; since in the optical communication, only positive values can be transmitted, each user is assigned a binary vector (signature) with a fixed weight (number of 1's) where the inner product of different vectors is small compared to the weight (in contrast to what OOC stands for, the signatures are not orthogonal).

2.9.6 Structured Random Matrix

While Gaussian and Bernoulli matrices ensure sparse recovery via minimization with the optimal bound (2.22) on the number of measurements, they have only limited utility in applications for several reasons. Often the design of the measurement matrix is subject to physical or other constraints of the application, or it is actually given to us without having the freedom to design anything, and therefore, the matrix following a Gaussian or Bernoulli distribution is not justifiable on many occasions. Moreover, Gaussian or other unstructured matrices have the disadvantage that no fast matrix multiplication is available, which may speed up recovery algorithms significantly, so that large-scale problems are not practicable with Gaussian or Bernoulli matrices. Even storing an unstructured matrix may be difficult. From a computational and an application-oriented viewpoint, it is desirable to have measurement matrices with structure. Since it is hard to rigorously prove good recovery conditions for deterministic matrices, the randomness is never allowed to come into play. This leads to the study of structured random matrices. The larger part of these notes are devoted to the recovery of randomly sampled functions that have a sparse expansion in terms of an orthonormal system $\{\psi_j, j = 1, \ldots, N\}$ with uniformly bounded $L^\infty - \mathrm{norm}, \sup_j \in |N| \|\psi_j\|_\infty = \sup_{j \in |N|} \sup_x |\phi_j(x)| \leq K$ [36].

The corresponding measurement matrix has entries $(\psi_j(t_\ell))_{\ell, j}$ where t_1 are random sampling points. So the structure is determined by the function system ψ_j, while the randomness comes from the sampling locations. The random partial Fourier matrix, which consists of randomly chosen rows of the discrete Fourier matrix, can be seen as a special case of this setup and has been studied already in the very first papers on compressive sensing [36]. It is important to note that in this case, the fast Fourier transform (FFT) algorithm can be used to compute a fast application of a partial Fourier matrix in $O(N \log(N))$ operations to be compared with the usual $O(MN)$ operations for a matrix vector multiply with an $M \times N$ matrix. Commonly, $M > Cs \log(N)$ in CS, so that an $O(N \log(N))$ matrix multiply implies a substantial complexity gain. The second type of structured random matrices is partial random circulant and Toeplitz matrices. They arise in applications where convolutions are involved. Since circulant and Toeplitz matrices can be applied efficiently using again the FFT, they are also of interest for computationally efficient sparse recovery.

2.9.7 Kronecker Product Matrix

It is possible to design measurement matrices that are Kronecker products. Such matrices correspond to measurement processes that operate individually on portions of the multidimensional signal. For simplicity, it is assumed that each portion consists of a single d-section of the multidimensional signal, despite the impossibility of other configurations. The resulting measurement matrix can be expressed as $\overline{\Phi} = \Phi_1 \otimes \cdots \otimes \Phi_D$ Consider the example of distributed sensing of signal ensembles was separate measurements obtained, in the sense that each measurement depends on only one of the signals. More formally, for each signal $x_j, 1 \leq j \leq J$ separate measurements are obtained $y_j = \Phi_j x_{\cdot,j}$ with an individual measurement matrix being applied to each 1-section. The structure of such measurements can be succinctly captured by Kronecker products. For compact representation of the signal and measurement ensembles, 0 denotes a matrix of appropriate size with all entries equal to 0. Then, $Y = \overline{\Phi}\,\overline{x}$ shows that the measurement matrix that arises from distributed sensing has a characteristic block-diagonal structure when the entries of the sparse vector are grouped by signal. If a matrix $\Phi_j = \Phi'$ is used at each sensor to obtain its individual measurements, then the joint measurement matrix can be expressed as $\overline{\Phi} = I_J \otimes \Phi'$ where I_J denotes $J \times J$, the identity matrix [37].

2.9.8 Combination Matrix

This matrix is obtained by taking the Kronecker product of the Gaussian matrix and the Toeplitz matrix [38]. The Gaussian matrix is known for universal usage, as it can be paired with any sparse matrix. It also yields better results in terms of PSNR. The Toeplitz matrix is highly preferred considering the number of elements used in the measurement matrix is small. The Toeplitz matrix is a square matrix, and the elements are constant along its diagonal. The Kronecker product of matrix "A" of size $m_1 \times n_1$ and matrix "B" of size $p \times q$ yields a matrix of size $m_1 p \times n_1 q$ that is given as

$$A \otimes B = \begin{bmatrix} a_{11}B & \cdots & a_{1n,}B \\ \vdots & \ddots & \vdots \\ a_{m,1}B & \cdots & a_{m,n,}B \end{bmatrix}$$

where A is a Gaussian matrix and B is a Toeplitz matrix. For example, let A be a 2×2 Gaussian matrix represented as

$$A = \begin{bmatrix} g_{11} & g_{12} \\ g_{21} & g_{22} \end{bmatrix}$$

and B be a 2×2 Toeplitz matrix represented as

$$B = \begin{bmatrix} t_{11} & t_{12} \\ t_{21} & t_{11} \end{bmatrix}$$

The combination matrix of size 4×4 is generated by taking the Kronecker product of A and B represented as

$$C = A \otimes B = \begin{bmatrix} g_{11}t_{11} & g_{11}t_{12} & g_{12}t_{11} & g_{12}t_{12} \\ g_{11}t_{21} & g_{11}t_{11} & g_{12}t_{21} & g_{12}t_{11} \\ g_{21}t_{11} & g_{21}t_{12} & g_{22}t_{11} & g_{22}t_{12} \\ g_{21}t_{21} & g_{21}t_{11} & g_{22}t_{21} & g_{22}t_{11} \end{bmatrix}$$

2.9.9 Hybrid Matrix

The hybrid measurement matrix of size $M_1 \times n^2$ is generated by combining the Toeplitz matrix of size $M_1 \times n^2/2$ and the binary matrix of size $M_1 \times n^2/2$ [38]. The Toeplitz matrix is generated with -1 and $+1$ entries, and the binary matrix is generated with 0 and 1 values. For example, let A be a 2×2 Toeplitz matrix represented as

$$A = \begin{bmatrix} t_{11} & t_{12} \\ t_{21} & t_{11} \end{bmatrix}$$

and B be a 2×2 Binary matrix represented as

$$B = \begin{bmatrix} b_{11} & b_{12} \\ b_{21} & b_{22} \end{bmatrix}$$

then the hybrid matrix of size 2×4 is given as

$$H = \begin{bmatrix} t_{11} & t_{12} & b_{11} & b_{12} \\ t_{21} & t_{11} & b_{21} & b_{22} \end{bmatrix}$$

The hybrid matrix is memory efficient, as it requires fewer elements to generate the matrix when compared with the Gaussian matrix. The measurement matrix is applied to the sparse vector for obtaining the measurement vector y that is transmitted to the receiver side, where it is then reconstructed using CS recovery algorithms. Different CS recovery algorithms are discussed in detail in Chapter 3.

2.10 Sparse Recovery Algorithms

The design of sparse recovery algorithms are guided by various criteria. Some important ones are listed as follows [39].

- Minimal number of measurements: Sparse recovery algorithms must require approximately the same number of measurements (up to a small constant) required for the stable embedding of K-sparse signals.
- Robustness to measurement noise and model mismatch: Sparse recovery algorithms must be stable vis-a-vis (pronounced as visavee) to perturbations of the input signal, as well as noise added to the measurements; both types of errors arise naturally in practical systems.
- Speed: Sparse recovery algorithms must strive toward expending minimal computational resources, keeping in mind that a lot of applications in CS deal with very high-dimensional signals.
- Performance guarantees: Choose to design algorithms that possess instance optimal or probabilistic guarantees. One can also choose to focus on algorithm performance for the recovery of exactly K-sparse signals x, or consider performance for the recovery of general signals xs. Alternately, it can also consider algorithms that are accompanied by performance guarantees in either the noise-free or noisy settings.

Broadly speaking, recovery methods tend to fall under three categories: convex optimization-based approaches, greedy methods, and combinatorial techniques. The convex optimization methods include l_1 minimization, greedy algorithms include orthogonal matching pursuit, stagewise orthogonal matching pursuit, regularized orthogonal matching pursuit, compressively samples orthogonal matching pursuit, model-based algorithms, hard iterative algorithm, and soft iterative algorithms.

While convex optimization techniques are powerful methods for computing sparse representations, there is also a variety of greedy/iterative methods for solving such problems. Greedy algorithms makes use of iterations to obtain an approximate of the signal coefficients and support, either by iteratively identifying the support of the signal until a convergence criterion is met, or alternatively by obtaining an improved estimate of the sparse signal at each iteration that attempts accounting for the mismatch to the measured data. Some greedy algorithms show results similar to a convex optimization algorithm. In fact, some of the more sophisticated greedy algorithms are remarkably similar to those used for l^1 minimization. However, the techniques required to prove performance guarantees are substantially different.

In addition to l^1 minimization and greedy algorithms, combinatorial algorithms are also used in practice. These algorithms have been developed by the theoretical computer science community and target the sparse signal recovery problem. For example, to identify defective products in an industrial setting, or a subset of diseased tissue samples in a medical context, the vector x indicates anomalous elements; i.e., $x_i = 0$ for the K anomalous elements and $x_i = 0$ otherwise. Our goal is to design a collection of tests that allow identification of the support (and possibly the values of the nonzeros) of x while also minimizing the number of tests performed. In the simplest practical setting, these tests are represented by a binary matrix Φ whose entries a_{ij} are equal to 1 if and only if the jth item is used in the ith test. If the output of the test is linear with respect to the inputs, then the problem of recovering the vector x is essentially the same as the standard sparse recovery problem in CS. Another application area in which combinatorial algorithms have proven useful is computation on data streams.

2.10.1 Signal Recovery in Noise

The ability to perfectly reconstruct a sparse signal from noise-free measurements leads to accuracy in results. However, in practical scenarios, the measurements are likely to be contaminated by some form of noise. For instance, in order to process data in a computer, one must be able to represent it using a finite number of bits, and hence, the measurements will typically be subject to quantization error. Moreover, systems which are implemented in physical hardware will be subject to a variety of different types of noise depending on the setting. Another important noise source is present on the signal itself. In many settings, the signal x to be estimated is contaminated by some form of random noise. The implications of this type of noise on the achievable sampling rates have been recently analyzed. Here, the focus is on the measurement of noise, which has received much more attention in literature. Perhaps, somewhat surprisingly, one can demonstrate the possibility of stable recovery of sparse signals under a variety of common noise models. As might be expected, both the RIP and coherence are useful in establishing performance guarantees in noise.

Given noisy compressive measurements $y = \Phi x + e$ of a signal x, a core problem in compressive sensing (CS) is to recover a sparse signal x from a set of measurements y. Considerable efforts have been directed toward developing algorithms that perform fast, accurate, and stable reconstruction of x from y. All the methods discussed in this section optimize a convex function (usually the $\ell 1$-norm) over a convex (possibly unbounded) set. This implies guaranteed convergence to the global optimum. In other words, given that the sampling matrix Φ satisfies the conditions specified in "signal recovery via $\ell 1$ minimization," convex optimization methods can recover the underlying signal x. In addition, convex relaxation methods also guarantee stable recovery by reformulating the recovery problem as the SOCP, or the unconstrained formulation [40].

While convex optimization techniques are powerful methods for computing sparse representations, there are also a variety of greedy/iterative methods for solving such problems. Greedy algorithms rely on iterative approximation of the signal coefficients and support, either by iteratively identifying the support of the signal until a convergence criterion is met or alternatively by obtaining an improved estimate of the sparse signal at each iteration by accounting for the mismatch to the measured data. Some greedy methods can actually be shown to have performance guarantees that match those obtained for convex optimization approaches. In fact, some of the more sophisticated greedy algorithms are remarkably similar to those used for $\ell 1$ minimization described previously. However, the techniques required to prove performance guarantees are substantially different. There also exist iterative techniques for sparse recovery based on message passing schemes for sparse graphical models. In fact, some greedy algorithms can be directly interpreted as message passing methods.

Although the ultimate aim is to recover a sparse signal from a small number of linear measurements in both of these settings, there are some important differences between such settings and the compressive sensing setting studied in this course. First, it is natural in these settings to assume that the designer of the reconstruction algorithm also has full control over Φ and is thus free to choose Φ in a manner that reduces the amount of computation required to perform recovery. For example, it is often useful to design Φ to ensure it has very few nonzeros; i.e., the sensing matrix itself is also sparse. In general, most methods involve careful construction of the sensing matrix Φ, which is in contrast to the optimization and greedy methods that work with any matrix satisfying a generic condition such as the restricted isometry property. This additional degree of freedom can lead to significantly faster algorithms.

Second, the fact of the computational complexity of all the convex methods and greedy algorithms always being at least linear in N may be noted since we must at least incur the computational cost of reading out all N entries of x for recovering x. This may be acceptable in many typical compressive sensing applications, but becomes impractical when N is extremely large, as in the network monitoring example. In this context, one may seek to develop algorithms whose complexity is linear only in the length of the representation of the signal, i.e., its sparsity K. In this case, the algorithm does not return a complete reconstruction of x but instead returns only its K largest elements (and their indices) [12].

2.11 Applications of Compressed Sensing

Compressed sensing has already had a notable impact on several applications. One example is medical imaging [12], where it has enabled speedups by a factor of seven in pediatric MRI while preserving diagnostic quality. Moreover, the broad applicability of this framework has inspired research that extends. The benefit of compressive sensing lies in the ability to drastically reduce the computation time of electromagnetic simulations through processing of fewer data points.

More particularly, video-based object tracking is widely investigated with CS methods. The volume of data provided by video cameras in real time is enormous. In order to cope with this increased data flow, CS techniques are employed for background subtraction over a part of the video frame. Whereas traditional background subtraction techniques require that the full image is available, the CS-based background subtraction utilizes a reduced image size. The first CS-based background subtraction algorithm [41] performs background subtraction on compressive measurements of a scene, while retaining the ability to reconstruct the foreground. However, in this algorithm, the measurement matrix is fixed. In [42], a technique is proposed that adaptively adjusts the number of compressive measurements. This leads to an adaptive scheme to outperform the basic CS-based background subtraction algorithm [41].

In target tracking with video data, the object template has a sparse representation. For instance, in [43], the target is modeled as a sparse representation of multiple predefined templates. The convex relaxation-based tracking algorithm needs copying with the underlying complexity requiring the use of different $l1$ minimization techniques, e.g., the orthogonal matching pursuit (OMP) [44] or the $l1$-regularized least squares [45].

Compressive sensing or sampling has many applications combined with computationally intense problems. The field of compressive sensing is related to applications such as signal processing and computational mathematics, underdetermined linear systems, group testing, heavy hitters, sparse coding, multiplexing, sparse sampling, and finite rate of innovation. CS theory mainly finds applications in fields such as signal processing and compression, solution of inverse problems, design of radiating systems, radar and through-the-wall imaging, and antenna characterization. Imaging techniques having a strong affinity with compressive sensing include coded aperture and computational photography. Implementations of compressive sensing in hardware platform at different technology readiness levels are available.

Conventional CS reconstruction uses sparse signals (usually sampled at a rate less than the Nyquist sampling rate) for reconstruction through constrained l_1 minimization. One of the earliest applications of such an approach was in reflection seismology which used sparse reflected signals from band-limited data for tracking changes between sub-surface layers. When the LASSO model came into prominence in the 1990s as a statistical method for selection of sparse models, this method was further used in computational harmonic analysis for sparse signal representation from over-complete dictionaries. Some of the other applications include incoherent sampling of radar pulses. Boyd in one of his works has applied the LASSO model—for selection of sparse models—toward analog to digital converters (the current ones use a sampling rate higher than the Nyquist rate along with the quantized Shannon representation). This would involve a parallel architecture in which the polarity of the analog signal changes at a high rate followed by digitizing the integral at the end of each time interval for obtaining the converted digital signal.

Compressed sensing is used in a mobile phone camera sensor. The approach allows a reduction in image acquisition energy per image by as much as a factor of 15 at the cost of complex decompression algorithms. The computation may require an off-device implementation. Compressed sensing is used in single-pixel cameras from Rice University [46]. Bell Laboratories employed the technique in a lensless single-pixel camera that takes stills using repeated snapshots of randomly chosen apertures from a grid. Image quality improves with the number of snapshots and generally requires a small fraction of the data of conventional imaging, while eliminating lens/focus-related aberrations.

Compressed sensing can be used for the improvement of image reconstruction in holography by increasing the number of voxels that can be inferred from a single hologram. It is also used for image retrieval from under sampled measurements in optical and millimeter-wave holography. Compressed sensing is being used in facial recognition applications and also to shorten magnetic resonance imaging scanning sessions on conventional hardware.

Compressed sensing addresses the issue of high scan time by enabling faster acquisition through measurement of fewer Fourier coefficients. This produces a high-quality image with relatively lower scan time. Another application (also discussed ahead) is for CT reconstruction with fewer X-ray projections. Compressed sensing, in this case, removes the high spatial gradient parts—mainly image noise and artifacts. This holds tremendous potential as one can obtain high-resolution CT images at low radiation doses.

Compressed sensing has also been used in the application of network tomography for obtaining accurate results. Network delay estimation and network congestion detection can both be modeled as underdetermined systems of linear equations where the coefficient matrix is the network routing matrix. Moreover, in the Internet, network routing matrices usually satisfy the criterion for using compressed sensing. Commercial shortwave-infrared cameras based upon compressed sensing are available. These cameras have light sensitivity from 0.9 μm to 1.7 μm, which are wavelengths invisible to the human eye. In the field of radio astronomy, compressed sensing has been proposed for deconvolving an interferometric image. In fact, the Högbom CLEAN algorithm that has been in use for the deconvolution of radio images since 1974 is similar to compressed sensing's matching pursuit algorithm.

2.12 Summary

Compressed sensing is an exciting, rapidly growing field that has attracted considerable attention in signal processing, statistics, and computer science, as well as the broader scientific community [12]. Since its initial development, just a few years ago, thousands of papers have appeared in this area, and hundreds of conferences, workshops, and special sessions have been dedicated to this growing research field. In this chapter, we have reviewed some of the basics of the theory underlying CS and the drawbacks of transform coding. The mathematical background is also given an emphasis. The transforms for sparse representation are also reviewed along with different matrices that are used in the CS process. This chapter also explains the conditions for a good measurement matrix and perfect recovery. Recent researches on the CS measurement matrix are also given importance. Finally, different applications of CS are also discussed in this chapter.

References

[1] Nyquist, H. (1928). "Certain topics in telegraph transmission theory," IEEE Trans. Am. Inst. Elect. Eng. 47, (2), 617–644, Apr.

[2] Shannon, C. (1949). "Communication in the presence of noise," Proc. Inst. Radio Eng. 37, (1), 10–21.

[3] Donoho, D. L. (2006). "Compressed sensing," IEEE Trans. Inform. Theory 52, 1289 1306.

[4] http://jpeg.org/jpeg2000

[5] Baron, D., Duarte, M. F., Wakin, M. B., Sarvotham, S., and Baraniuk. R. G. (2009). "Distributed compressive sensing." arXiv preprint arXiv:0901.3403.

[6] Shihao, J., Xue, Y., and Carin, L. (2008). "Bayesian compressive sensing." IEEE Trans. Signal Proces. 56.6: 2346–2356.

[7] Baraniuk, R. G., Cevher, V., Duarte, M. F., and Hegde. C. (2010). "Model-based compressive sensing." IEEE Trans. In. Theory, 56, (4), 1982–2001.

[8] Baraniuk, R. G. (2007). "Compressive sensing." IEEE Signal Process. Mag. 24.4.

[9] Asif, M. S., Charles, A., Romberg, J., and Rozell, C. (2011). Estimation and dynamic updating of time varying signals with sparse variations. In: Proceedings of the international conference on acoustics, speech sig process (ICASSP), pp. 3908–3911.

[10] Charles, A., Asif, M. S., Romberg, J., and Rozell, C. (2011). Sparsity penalties in dynamical system estimation. In: Proceedings from the conference on information sciences and systems, pp. 1–6.

[11] Carmi, A., Gurfil, P., and Kanevsky, D. (2010). Methods for sparse signal recovery using Kalman filtering with embedded pseudo-measurement norms and quasi-norms. IEEE Trans. Signal Process. 58 (4), 2405–2409.

[12] Eldar, Y. C., and Kutyniok, G. (Eds.) (2012). Compressed sensing: theory and applications. (Cambridge: Cambridge University Press).

[13] https://en.wikipedia.org/wiki/Fourier_transform

[14] Watson, Andrew B. "Image compression using the discrete cosine Transform." *Mathematica journal* 4, no. 1 (1994): 81.

[15] Dua S. Acharya R. Ng EYK. Computational analysis of the human eye with applications. World Scientific; 2011.

[16] http://www.curvelet.org/

[17] Donoho, D., Elad, M., and Temlyahov, V. (2006). Stable recovery of sparse overcomplete representations in the presence of noise. IEEE Trans Inform Theory, 52, (1), 6–18.

[18] Candè, E. J., and Donoho, D. L. (1999), Curvelets Available: [Online] Available: http://www-stat.stanford.edu/~donoho/Reports/1999/curvelets.pdf

[19] Candè, E. J., and Donoho, D. L., Cohen, A., Rabut, C. and Schumaker, L. L. (1999). "Curvelets—A surprisingly effective nonadaptive representation for objects with edges", Curve and Surface Fitting: Saint-Malo 1999, Vanderbilt Univ. Press.

[20] Ma, Jianwei, and Gerlind Plonka. "The curvelet transform." Signal Processing Magazine, IEEE 27.2 (2010): 118–133.

[21] https://en.wikipedia.org/wiki/Curvelet

[22] Candµes, E., and Donoho, D. (2003). Continuous curvelet transform: I. Resolution of the wavefront set, Appl. Comput. Harmon. Anal.19, 162–197.

[23] Candµes, E., and Donoho, D. (2004). New tight frames of curvelets and optimal representations of objects with piecewise singularities, Comm. Pure Appl. Math. 57, 219–266.

[24] Do, M. N., and Vetterli, M. (2002). "Contourlets: a directional multiresolution image representation." Image Processing. 2002. Proceedings. 2002 International Conference on. Vol. 1. IEEE.

[25] https://en.wikipedia.org/wiki/Contourlet

[26] Lu, Y., and Do, M. N. (2007). "Multidimensional directional filter banks and surfacelets," IEEE Trans. Image Processing 16 (4), 918–931.

[27] Jain, A. K. (1976). "A fast Karhunen-Loeve transform for a class of random processes." NASA STI/Recon Technical Report A 76, 42860.

[28] http://www.commsp.ee.ic.ac.uk/~tania/teaching/DIP%202014/KLT.pdf

[29] Blanchard, Jeffrey D., Coralia Cartis, and Jared Tanner. "Compressed sensing: How sharp is the restricted isometry property?." SIAM review 53, no. 1 (2011): 105–125.

[30] Chen, S., Billings, S.A., and Luo, W. (1989). Orthogonal least squares methods and their application to non-linear system identification. Int. J. Contro. 50, 1873–1896.

[31] Bajwa, W. U., Haupt, J. D., Raz, G. M., Wright, S. J., Nowak, R. D. (2007). "Toeplitz-Structured Compressed Sensing Matrices", IEEE/SP 14th Workshop on Statistical Signal Processing (SSP), 26–29 Aug., pp. 294, 298.

[32] Holger. R. (2009). "Circulant and Toeplitz matrices in compressed sensing." arXiv preprint arXiv:0902.4394.

[33] Lei, Y., Barbot, J. P., Gang Z., and Hong, S. (2010). "Toeplitz-structured Chaotic Sensing Matrix for Compressive Sensing", 7th International

Symposium on Communication Systems Networks and Digital Signal Processing (CSNDSP), 21–23 July, pp. 229–233.

[34] Wotao, Y., Morgan, S., Yang, J., and Zhang, Y. (2010). "Practical Compressive Sensing with Toeplitz and Circulant matrices", Proceedings on SPIE7744, Visual Communications and Image processing, 7744K.

[35] Rauhut, H. (2010). "Compressive sensing and structured random matrices." Theor. Found. Num. Methods Sparse Recover. 9, 1–92.

[36] Duarte, Marco F., and Richard G. Baraniuk. "Kronecker product matrices for compressive sensing." In *2010 IEEE International Conference on Acoustics, Speech and Signal Processing*, pp. 3650–3653. IEEE, 2010.

[37] Aasha Nandhini, S., et al. (2015). Video compressed sensing framework for wireless multimedia sensor networks using a combination of multiple matrices. Elsevier's Comput. Electric. Eng. 44, 51–66. http://dx.doi.org/10.1016/j.compeleceng.2015.02.008

[38] Foucart, S. (2012). "Sparse recovery algorithms: sufficient conditions in terms of restricted isometry constants." Approximation Theory XIII: San Antonio 2010. Springer, New York. 65–77.

[39] https://cnx.org/contents/9wtroLnw@5.12:-G49NN0v@3/Sparse-recovery-algorithms.

[40] Cevher, V., Sankaranarayanan, A., Duarte, M. F., Reddy, D., Baraniuk, R. G., and Chellappa, R. (2008). "Compressive sensing for background subtraction." In Computer Vision–ECCV 2008, pp. 155–168. Springer Berlin Heidelberg, 2008

[41] Garrett, W., Reddy, D., and Chellappa, R. (2012). "Adaptive rate compressive sensing for background subtraction." IEEE International Conference on in Acoustics, Speech and Signal Processing (ICASSP), pp. 1477–1480.

[42] Carmi, A. Y., Mihaylova, L. S., and Godsill, S. J. (2013). "Compressed Sensing & Sparse Filtering" (Berlin: Springer).

[43] Tropp, J., and Gilbert, A. (2007). Signal Recovery from Partial Information via Orthogonal Matching Pursuit. IEEE Trans. Inform. Theor. 53(12), 4655–4666.

[44] Yin, W., et al. (2008). "Bregman iterative algorithms for\ell_1-minimization with applications to compressed sensing." SIAM J. Imag. Sci. 1.1, 143–168.

[45] Duarte, M. F., and Mark, A. Davenport. "Single-pixel camera."

3

Recovery Algorithms

3.1 Introduction

Various algorithmic approaches to the problem of signal recovery from CS measurements are discussed in this chapter. Algorithms of various types have been used in applications such as sparse approximation, which exploit sparsity in other contexts and can be brought to bear on the CS recovery problem. Some of these algorithms are overviewed below.

3.2 Conditions for Perfect Recovery

3.2.1 Sensing Matrices

The standard finite-dimensional CS model is considered as a default framework for the purpose of making the discussion more concrete. More particularly, given a signal $x \in R^n$, measurement systems that acquire "m" linear measurements are considered. This process can be mathematically represented as

$$y = Ax \tag{3.1}$$

where A is an $N \times M$ matrix and $y \in \mathrm{R}^m$.

The matrix A represents a dimensionality reduction; i.e., it maps R^N, where N is generally large, into R^M, where M is typically much smaller than N. Assumption has been made that measurements are non-adaptive in the standard CS framework. This means that the rows of A are fixed in advance and do not depend on previously obtained measurements [1]. In certain settings, adaptive measurement schemes can lead to significant performance gains. Although the standard CS framework assumes that x is a finite-length vector with a time- or space-valued index, in practice, it will often be interested to design measurement systems for acquiring continuously indexed signals such as continuous-time signals or images. It is sometimes possible to extend this model to continuously indexed signals using an intermediate discrete representation. For now, x is considered as a finite set of samples.

There are two main events to be accomplished in CS. First, the sensing matrix A must be designed to ensure that it preserves the information in the signal x; second, the original signal x should be recovered from the measurements y. In the case where the data are sparse or compressible, matrices A can be designed to ensure the accurate recovery of the original signal and as also efficient use of a variety of practical algorithms.

This section first addresses the question of how to design the sensing matrix A. Rather than directly proposing a design procedure, a number of desirable properties are considered that A might have. Some important examples of matrix constructions that satisfy these properties are also provided.

3.2.1.1 Null-space conditions

A natural place to begin is by considering the null space of A that can be denoted by

$$N(A) = \{Z \: : \: Az = 0\} \; [2]$$

In linear algebra, null space is also called as the kernel. This is depicted in Figure 3.1.

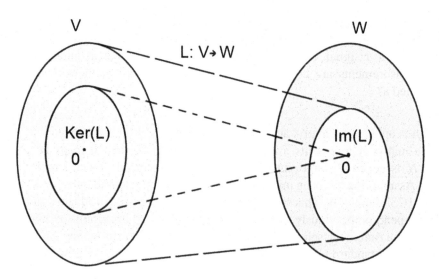

Figure 3.1   Depiction of null space or kernel.

In linear algebra and functional analysis, the kernel of a linear map $L : V \rightarrow W$ between two vector spaces V and W is the set of all elements $\mathbf{v}$ of V for which $L(\mathbf{v}) = \mathbf{0}$, where $\mathbf{0}$ denotes the zero vector in W. That is, in set-builder notation,

$$ker(L) = \{v \in V | L(v) = 0\}$$

Considering the problem at hand, if *all* sparse signals x are to be recovered from the measurements Ax, it is clear that, for any pair of vectors, $x, x' \in \Sigma_k$, it is a must that $AX \neq AX'$, is imperative since; otherwise, it would be impossible to distinguish x from x' based solely on the measurements y. By observing that if $A(x - x') = 0$ with $x - x' \in \Sigma_{2k}$, when $Ax = Ax'$, it can be seen that A uniquely represents all $x \in \Sigma_k$ if and only if $N(A)$ contains no vectors in Σ_{2k}. Null-space property (NSP) is given by the following definition.

Definition 3.1 *A matrix A satisfies the* null-space property *(NSP) of order k if there exists a constant C > 0 such that*

$$\|h_\Lambda\|_2 \leq C \frac{\|h_\Lambda c\|_1}{\sqrt{k}} \tag{3.2}$$

holds for all $h \in N(A)$ and for all Λ such that $|\Lambda| \leq k$.

NSP quantifies the notion that vectors in the null space of A should not be too concentrated on a small subset of indices. For example, if a vector h is exactly k sparse, then there exists a Λ such that $\|h_\Lambda c\|_1 = 0$ and hence (3.2) implies that $h_\Lambda = 0$ as well. Thus, if a matrix A satisfies the NSP, then the only k-sparse vector in $N(A)$ is $h = 0$. In order to fully illustrate the implications of the NSP in the context of sparse recovery, the measurement of the performance of sparse recovery algorithms when dealing with general non-sparse x is discussed. Let $\Delta : R^m \rightarrow R^n$ represent the specific recovery method. The guarantees in perfect recovery are of the form

$$\|\Delta(Ax) - x\|_2 \leq C \frac{\sigma k(x)_1}{\sqrt{k}} \tag{3.3}$$

for all x, where $\sigma_{k(x)_1}$ is as defined as

$$\sigma_k(x)_p = \min_{x \in \sum k} \|x - \hat{x}\|_p$$

This guarantees not only the exact recovery of all possible k-sparse signals, but also ensures a degree of robustness to non-sparse signals that directly depend on how well the signals are approximated by k-sparse vectors. Such guarantees are called *instance optimal* since they relate to the optimal performance for each instance of x [3]. This distinguishes them from guarantees that only hold for some subset of possible signals, such as sparse or compressible signals— the quality of the guarantee adapts to the particular choice of x. These are also commonly referred to as *uniform guarantees* since they hold uniformly for all x. The choice of norms in (3.3) is somewhat arbitrary. The reconstruction error can be measured easily using other p norms. The choice of p, however, will limit what kinds of guarantees are possible and will also potentially lead to alternative formulations of the NSP. Moreover, the form of the right-hand side of (3.3) might seem somewhat unusual in that we measure the approximation error as $\sigma_{k(x)_1}/\sqrt{k}$ rather than simply something like $\sigma_{k(x)_2}$. However, such a guarantee is actually not possible without taking a large number of measurements.

The NSP of order $2k$ is sufficient to establish a guarantee of the form (3.3) for a practical recovery algorithm like the l1 minimization. Moreover, the following adaptation of a theorem demonstrates the imperative need to satisfy the NSP of order $2k$, when any recovery algorithm satisfying (3.3) exists.

Theorem 3.1: *Let $A : R^n \rightarrow R^m$ denote a sensing matrix and $\Delta : R^n \rightarrow R^m$ denote an arbitrary recovery algorithm. If the pair (A, Δ) satisfies (1.6), then A satisfies the NSP of order $2k$.*

Proof: Suppose $h \in N(A)$ and let Λ be the indices corresponding to the $2k$ largest entries of h. Λ is then split into Λ_0 and Λ_1, where $|\Lambda_0| = |\Lambda_1| = k$. Set $x = h_{\Lambda_1} + h_\Lambda$ and $x' = -h_{\Lambda_0}$ so that $h = x - x'$. Since by construction $x \in \Sigma k$, (3.3) can be applied to obtain $x' = \Delta(Ax')$. Moreover, since $h \in N(A)$, the following equation can be drawn:

$$Ah = A(x - x') = 0$$

so that $Ax' = Ax$. Thus, $x' = \Delta(Ax)$. Finally,

$$\|h_\Lambda\|_2 \le \|h\|_2 = \|x - x'\|_2 = \|x - \Delta(Ax)\|_2 \le C\frac{\sigma k(x)_1}{\sqrt{k}} = \sqrt{2}C\frac{\|h_\Lambda c\|_1}{\sqrt{2k}}$$

where the last inequality follows from (3.3).

3.2.1.2 The restricted isometry property

While the NSP is both necessary and sufficient for establishing guarantees of the form (3.3), these guarantees do not account for *noise*. When the measurements are contaminated with noise or have been corrupted by some error such as quantization, it will be useful to consider somewhat stronger conditions. Candès and Tao introduced the following isometry condition on matrices A and established its important role in CS.

Definition 3.2 *A matrix A satisfies the* restricted isometry property *(RIP) if there exists a $\delta k \in (0, 1)$ such that*

$$(1 - \delta_k) \|x\|_2^2 \leq \|Ax\|_2^2 \leq (1 + \delta_k) \|x\|_2^2 \tag{3.4}$$

Hold for all $x \in \sum_k$.

If a matrix A satisfies the RIP of order $2k$, (3.4) can be interpreted as saying that A approximately preserves the distance between any pair of k-sparse vectors. This will clearly have fundamental implications concerning robustness to noise. It is important to note that, in the definition of the RIP, the bounds are assumed to be symmetric about 1. This assumption is considered for convenience in notations. In practice, arbitrary bounds could be considered instead.

$$\alpha \|x\|_2^2 \leq \|Ax\|_2^2 \leq \beta \|x\|_2^2$$

where $0 < \alpha \leq \beta < \infty$. Given any such bounds, A can be scaled to ensure satisfaction of the symmetric bound about 1 in (3.4). In specific, multiplying A by $\sqrt{2/(\beta + \alpha)}$ will result in an A that satisfies (3.4) with constant $\delta_k = (\beta - a)/(\beta + a)$. All the theorems in this chapter are based on the assumption that A satisfies RIP. If A satisfies the RIP of order k with constant δ_k, for any $k' < k$ there exists A satisfies the RIP of order k with constant $\delta_{k'} \leq \delta_k$. Moreover, it is shown that if A satisfies the RIP of order k with a sufficiently small constant, then it will also automatically satisfy the RIP of order γk for certain $\gamma > 1$, though with a somewhat worse constant.

With the null NSP and RIP being satisfied, the perfect sensing matrix can be formed which is efficient in picking up the right samples for reconstruction. Next subsection would show the sensing matrix construction.

3.2.2 Sensing Matrix Constructions

To begin with the construction of sensing matrices, it is straightforward to show that an $m \times n$ Vandermonde matrix V constructed from m distinct scalars has spark $(V) = m + 1$. The spark of a matrix A is the smallest number n to ensure existence of a set of n columns in A which are linearly dependent. It can be mathematically represented as

$$spark(A) = \min_{d \neq 0} \|d\|_0 \text{ such that } Ad = 0$$

But, these matrices are poorly conditioned for large values of n, rendering the recovery problem numerically unstable. Similarly, there are known matrices A of size $m \times m^2$ that achieve the coherence lower bound $u(A) = 1\sqrt{m}$, e.g., for the recovery of the Gabor frame. These constructions restrict the number of measurements needed to recover a k-sparse signal to be $m = O(k^2 - \log n)$. It is also possible to construct matrices of size $m \times n$ that satisfy the RIP of order k, but such constructions also require m to be relatively large. In many real-world settings, these results would lead to an unacceptably large requirement on m. Fortunately, these limitations can be overcome by randomizing the matrix construction. For example, random matrices A of size $m \times n$ whose entries are independent and identically distributed (i.i.d.) with continuous distributions have spark(A) = $m + 1$ with probability one. More significantly, it can also be shown that random matrices will satisfy the RIP with high probability when the entries are chosen according to a Gaussian, Bernoulli, or more generally any sub-Gaussian distribution. If a matrix A is chosen according to a sub-Gaussian distribution with $m = O(k \log(n/k)/\delta_{2k}^k$, A will satisfy the RIP of order $2k$ with probability at least $1 - 2\exp(-c_1\delta_{2k}^k m)$. Furthermore, the coherence can be shown to converge to $\mu(A) = \sqrt{(2\log n)/m}$ when the distribution used has zero mean and finite variance.

The use of random matrices to construct A has a number of additional benefits. First, it is possible to show that the measurements are democratic for random constructions. This implies the possibility to recover a signal using any sufficiently large subset of measurements. Thus, with the use of random A, it is possible to be robust to the loss or corruption of a small fraction of measurements. Secondly, in practice, setting where x is sparse is important with respect to some basis Φ. The satisfaction of the RIP by product of $A\Phi$ is actually required. For example, if A is chosen according to a Gaussian distribution and Φ is an orthonormal basis, one can easily show that $A\Phi$ will also have a Gaussian distribution and will satisfy the RIP with high probability. Although less obvious, similar results hold for sub-Gaussian distributions as well. This property, sometimes referred to as *universality*, constitutes a significant advantage of using random matrices to construct A.

Finally, considering that the fully random matrix approach is sometimes impractical to build in hardware, several hardware architectures have been implemented and/or proposed that enable random measurements to be acquired in practical settings. Examples include the random demodulator, random filtering, the modulated wideband converter, random convolution, and the compressive multiplexer. These architectures typically use a reduced amount of randomness and are modeled by matrices A that have significantly more structure than a fully random matrix. Perhaps somewhat surprisingly, while it is typically not quite as easy as in the fully random case, one can prove that many of these constructions also satisfy the RIP and/or have low coherence. Furthermore, one can analyze the effect of inaccuracies in the matrix A implemented by the system; in the simplest cases, such sensing matrix errors can be addressed through system calibration.

3.3 L1 Minimization

A variety of approaches exist for recovery of a sparse signal x from a small number of linear measurements. Let us consider a natural first approach to the problem of sparse recovery. Given measurements y and the knowledge of the original signal x being sparse or compressible, x can be recovered by solving an optimization problem of the form

$$\hat{x} = \arg\min_z \|z\|_0 \text{ subject to } z \in B(y) \tag{3.5}$$

where $B(y)$ ensures that $\hat{x}$ is consistent with the measurements y. For example, in the case where the measurements are exact and noise-free, it can be set $(y) = \{z : Az = y\}$. When the measurements have been contaminated with a small amount of noise, $B(y) = \{z : \|Az - y\|_2 \le \varepsilon\}$ could be considered instead. In both cases, (3.5) finds the sparsest x that is consistent with the measurements y. The inherent assumption in (3.5) is that x itself is sparse. In the more common setting where $x = \Phi c$, the approach can easily be modified and instead consider

$$\hat{C} = \arg\min_z \|z\|_0 \text{ subject to } z \in B(y) \tag{3.6}$$

where $B(y) = \{z : A\varphi z = y\}$ or $B(y) = \{Z : \|A\varphi z - y\|_2 \le \varepsilon\}$.

The essentially identical nature of (3.5) and (3.6) can be seen by considering $A = A\Phi$. Moreover, in many cases, the introduction of Φ does not significantly complicate the construction of matrices A such that $\tilde{A}$ satisfies the desired properties. Thus, for the rest of this chapter, Φ will be considered as $\Phi = I$. It is important to note, however, that this restriction does impose certain limits in the analysis when Φ is a general dictionary and not an orthonormal basis. For example, in this case, $\|\hat{x} - x\|_2 = \|\phi\hat{c} - \phi c\| \|\hat{x} - x\|_2$ and thus a bound on $\|\hat{c} - c\|_2$, cannot directly be translated into a bound on $\|\hat{x} - x\|_2$ [4].

While it is possible to analyze the performance of (3.5) under appropriate assumptions related to A, it cannot be pursued with the same since the objective function $\|.\|_0$ is non-convex. Hence, (3.5) is potentially very difficult to solve since a non-convex problem might have multiple locally optimal points and identification of the presence or absence of any solution or the global nature of any solution found is a time-consuming process. In fact, one can show that for a general matrix A, even finding a solution that approximates the true minimum is NP-hard. One method for converting this non-convex problem into some solvable form is to replace $\|.\|_0$ with its convex approximation $\|.\|_1$. Specifically, it is considered as follows:

$$\hat{x} = \arg\min_z \|z\|_1 \text{ subject to } z \in B(y) \tag{3.7}$$

Provided that $B(y)$ is convex, (3.7) is computationally feasible. In fact, when $B(y) = \{z : Az = y\}$, the resulting problem can be posed as a linear program [53]. While it is clear that replacing (3.5) with (3.7) transforms a computationally intractable problem into a tractable one, it may not be conditionally true that the solution to (3.7) will be similar to the solution to (3.5). The promotion of sparsity through the use of l_1 can be expected. It is already known that the solutions to the l_1 minimization problem coincided exactly with the solution to the l_2 minimization problem for any $p < 1$ and, notably, was sparse. Moreover, the use of l_1 minimization for exploiting sparsity is the least work of Beurling on Fourier transform extrapolation from partial observations.

Additionally, in a somewhat different context, Logan showed in 1965 that a perfect recovery of a band-limited signal is possible in the presence of *arbitrary* corruptions or errors on a small interval. Again, the recovery method consists of searching for the band-limited signal that is closest to the observed signal in the l_1 norm. Historically, the use of l_1 minimization on large problems finally became practical with the explosion of computing power in the late 1970s and early 1980s. In one of its first applications, it was demonstrated that geophysical signals consisting of spikes could be recovered from only the high-frequency components of these signals by exploiting l_1 minimization. Finally, in the 1990s, signal-processing community developed interest in these approaches for the purpose of finding sparse approximations to signals and images when represented in over complete dictionaries or unions of bases. l_1 minimization received significant attention in the statistics literature as a method for variable selection in regression, known as the Lasso.

Thus, there are a variety of reasons to apprehend that l_1 minimization can provide an accurate method for sparse signal recovery. More importantly, this also constitutes a computationally tractable approach to sparse signal recovery.

In this section, an overview of l_1 minimization is provided from a theoretical perspective. The algorithm for l_1 minimization is discussed in Section 3.3.1.

3.3.1 L1 Minimization Algorithms

The l_1 minimization approach analyzed in Section 3.3 provides an overall framework for recovering sparse signals. Picturesque representation of l_1 minimization is shown in Figure 3.2.

Figure 3.2 shows the weighted l1 minimization; l_1 minimization not only leads to accurate recovery, but, more than that, its formulations are also convex optimization problems for which there exist efficient and accurate numerical solutions. For example, (3.7) with $B(y) = \{z : Az = y\}$ can be posed as a linear program. While these optimization problems could all be solved using a general-purpose convex optimization software, there now also exists a tremendous variety of algorithms designed for finding explicit solution to these problems in the context of CS [5]. This body of literature has primarily focused on the case where $B(y) = \{Z : \|Az - y\|_2 \leq \varepsilon\}$. However, multiple equivalent formulations of this problem do exist. For instance, a majority of l_1 minimization algorithms in the literature have actually considered the unconstrained version of this problem [5]; i.e.,

$$\hat{x} = \arg\min \frac{1}{2} \|Az - y\|_2^2 + \lambda \|z\|_1$$

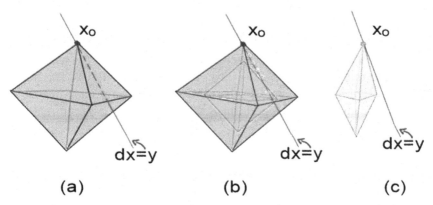

(a)   **(b)**   **(c)**

Figure 3.2 Weighting l_1 minimization for sparse recovery improvement. (A) x_0 is the sparse signal, $\phi x = y$ is the possible set, and l_1 ball of radius $\|x_0\|_{l_1}$. (B) $x \neq x_0$ is present for which $\|x\|_{l_1} < \|x_0\|_{l_1}$. (C) Weighted l_1 ball.

Note that for some choice of the parameter λ, this optimization problem will yield the same result as the constrained version of the problem given by

$$\hat{x} = \arg\min_z \|z\|_1 \text{ subject to } \|Az - y\|_2 \leq \varepsilon$$

However, in general, the value of λ which makes these problems equivalent is unknown a priori. Since it is a more natural parameterization (being determined by the noise or quantization level) in many settings, it is also useful to have algorithms that offer direct solutions to the latter formulation.

3.4 Greedy Algorithms

Compressed sensing (CS) is often the same as that of with l_1-based optimization. However, when choosing an algorithm for a particular application, different properties have to be considered and weighed against one another. Important algorithm properties, such as speed and storage requirements, ease of implementation, flexibility, and recovery performance, have to be weighed. In this chapter, a range of algorithms are provided that help the recovery of the input signal by solving the CS recovery problem [5]. These methods therefore are able to adapt to different conditions prevailing in various applications.

3.4.1 Matching Pursuit (MP)

3.4.1.1 Orthogonal matching pursuit (OMP)

A more sophisticated strategy is implemented in orthogonal matching pursuit (OMP). In OMP, the approximation for x is updated in each iteration by projecting y orthogonally onto the columns of A associated with the current support set $T^{[i]}$. This process is continued till the stopping criterion is reached. OMP therefore minimizes $\|y - A\hat{x}\|_2$ over all $\hat{x}$ with support $T^{[i]}$. The full algorithm is listed in Algorithm 3.1 where $\dagger$ represents the pseudo-inverse operator [6]. The minimization is performed with respect to all of the currently selected coefficients:

$$\hat{x}_{T^{[i]}}^{[i]} = \arg\min \|y - A_{T^{[i]}} \tilde{x}_{T^{[i]}}\|_2^2$$

OMP never re-selects an element, and the residual at any iteration is always orthogonal to all currently selected elements. The computational cost of OMP is dominated by the matrix vector products, when fast transforms are used. Various techniques for solving the least-squares problem have been proposed. These include the SVD method, QR factorization, Cholesky factorization,

Algorithm 3.1 Orthogonal matching pursuit

Input: y, A, k
Initialize: $r^{[0]} = y, \hat{x}^{[0]} = 0, T^{[0]} = \phi$
for i = 1, I := i+1 till stopping criterion is reached, do
$g^{[i]} = A^T r^{[i-1]}$
$j^{[i]} = \arg\max \left| g_j^{[i]} \right| / \left\| A_j \right\|_2$
$T^{[i]} = T^{[i-1]} \bigcup j^{[i]}$

$\hat{x}_{T^{[i]}}^{[i]} = A_{T^{[i]}}^+ y$

$r^{[i]} = y - A\hat{x}^{[i]}$
end for
Output: $r^{[i]}$ and $\hat{x}^{[i]}$

or iterative techniques such as conjugate gradient methods. While OMP is more computationally complex than matching pursuit (MP), it is generally superior in performance, particularly in the context of CS. There are two main problems with applying OMP to large-scale data. First, the computation and storage costs of a single iteration of OMP are quite high for large-scale problems, and second, the selection of one sample at a time means that exactly *k* iterations are needed to approximate *y* with *k* sample of *A*. When *k* is large, this can be impractically slow. Some of the variations discussed below were proposed to specifically address these issues [7].

3.4.1.2 Directional pursuits

Despite both MP and OMP having identical selection strategies and updating their coefficients by minimizing the squared error criterion, $\left| y - Ax^{[i]} \right\|_2^2$, the form of the update is substantially different. OMP minimizes the coefficients for all selected elements at iteration *i*, while in MP, the minimization involves only the coefficient of the most recently selected element [5]. Simple improvements or transformations to these algorithms would affect various other methods that prove efficient in finding the approximation of the input. For example, a relaxed form of MP has been considered in a situation where a damping factor is included.

Consider updating the selected coefficients $\hat{x}_{T^{[i]}}^{[i]}$ along some other, at the *i*-th iteration, yet to be defined, direction $d_{T^{[i]}}^{[i]}$.

$$\hat{x}_{T^{[i]}}^{[i]} = \hat{x}_{T^{[i]}}^{[i-1]} + a^{[i]} d_{T^{[i]}}^{[i]} \tag{3.8}$$

Algorithm 3.2   Directional pursuit

Input: y, A, and k
Initialize: $r^o = y, \hat{x}^{[0]} = 0, T^{[0]} = \phi$
    For i = 1; i:i+1 till stopping criterion is met do
    $g^{[i]} = A^T r^{[i-1]}$
    $j^{[i]} = \arg\max j \left| g_j^{[i]} \right| / \left\| A_j \right\|_2$
    $T^{[i]} = T^{[i-1]} \bigcup j^{[i]}$
    Calculate update direction
    $d_{T[i]}^{[i]}; c^{[i]} = A_{T[i]} d_{T[i]}^{[i]}$ and $a^{[i]} = \dfrac{\langle r^{[i]}, c^{[i]} \rangle}{\left\| c^{[i]} \right\|_2^2}$
    $x_{T[i]}^{[i]} := x_{T[1]}^{[i-1]} + a^{[i]} d_{T[i]}^{[i]}$
    $r^{[i]} = r^{[i-1]} - a^{[i]} c^{[i]}$
    end for
    Output: $r^{[i]}$ and $\hat{x}^{[i]}$

The step size $a^{[i]}$ can be explicitly chosen to minimize the same quadratic cost as before.

$$a^{[i]} = \frac{r^{[i]}, c^{[i]}}{\left\| c^{[i]} \right\|_2^2} \tag{3.9}$$

where $c^{[i]} = A_{T[i]} d_{T[i]}^{[i]}$. When such an update is used along with the standard MP/OMP selection criterion, this directional pursuit is a member of the family of *general matching pursuit* algorithms and shares the same necessary and sufficient conditions for exact recovery as OMP. Note also that both MP and OMP naturally fit in this framework with update directions: $\delta_j[i]$ and $A_{T[i]}^+ g_{T[i]}$, respectively.

    The algorithmic steps involved in directional pursuit are given in Algorithm 3.2. The aim of introducing directional updates is to produce an approximation to the orthogonal projection with a reduced computation cost.

3.4.1.3 Gradient pursuits

Variants of the pursuit algorithms can be formulated for equalizing backslides of the previous ones. A natural choice for the update direction for the previous algorithm is the negative gradient of the cost function.
$\left\| y - A_{T[1]} \tilde{x}_{T[1]} \right\|_2^2$, i.e.

$$d_{T[i]}^{[i]} := g_{T[i]}^{[i]} A_{T[i]}^T \left(y - A_{T[1]} \hat{x}_{T[1]}^{[i-1]} \right) \tag{3.10}$$

Fortunately, (3.10) is already a by-product of the selection process. The vector $g[i]$ (which has already been calculated) is simply restricted to the elements $T[i]$. Using (3.10) as the directional update results in the most basic form of directional pursuit which is called *gradient pursuit* (GP). Directional pursuit sometimes uses local directional updates to exploit the localized structure present in certain dictionaries [5].

The increase in computational complexity over MP is small, and when the sub-matrices of A are well conditioned (i.e., A has a good restricted isometry property), minimizing along the gradient direction can provide a good approximation to solving the full least-squares problem [8]. Assuming A has a small restricted isometry constant δ_k, the condition number κ of the Gram matrix can be restricted to

$$T^{[i]}, \, G^{[i]} = A_{T^{[i]}}^T A_{T^{[i]}}$$

By

$$k(G^{[i]} \leq \left(\frac{1 + \delta_k}{1 - \delta_k} \right) \tag{3.11}$$

for all $T^{[i]}$, $\left| T^{[i]} \right| \leq k$. A worst-case analysis of the gradient line search [9] then shows that for small δ_k, the gradient update achieves most of the minimization:

$$\frac{F(\hat{x}_{T^{[i]}}^{[i]}) - F(\hat{x}_{T^{[i]}}^*)}{F(\hat{x}_{T^{[i]}}^{[i-1]}) - F(\hat{x}_{T^{[i]}}^*)} \leq \left(\frac{k-1}{k+1} \right)^2 \tag{3.12}$$
$$\leq \delta_k^2$$

where $\hat{x}_{T^{[i]}}^*$ denotes the least-squares solution of $F(\hat{x}_{T^{[i]}} = \|y - A\hat{x}_{T^{[i]}}\|_2^2)$. Hence, for small δ_k, the convergence, even of a single gradient iteration, is good.

3.4.1.4 StOMP

StOMP was proposed with the objective of providing good reconstruction performance for CS applications while keeping computational costs low enough for application to large-scale problems [10]. The threshold strategy is as follows:

$$\lambda_{\text{stompl}}^{[i]} - t^{[i]} \left\| r^{[i-1]} \right\|_2 / \sqrt{m}, \tag{3.13}$$

where the authors give the guidance that a good choice of $t^{[i]}$ will usually take a value: $2 \leq t^{[i]} \leq 3$. Theoretical performance guarantees for this method when applied to more general matrices A and more general coefficient values are not available. Furthermore, from a practical point of view, the selection of the parameter t appears critical for good performance. A specific problem that can occur lies in the premature termination of the algorithm when all inner products fall below the threshold. Indeed, mixed results are observed in the range of experiments presented in StOMP. The selection strategy in StOMP is difficult to generalize beyond specific scenarios. Blumensath and Davies therefore proposed an alternative selection strategy that can be more tightly linked to general MP/OMP recovery results based upon a *weak* selection strategy. Weak selection was originally introduced to deal with the issue of infinite dimensional dictionaries where only a finite number of inner products can be evaluated. Weak selection allows the selection of a *single* element $Aj[i]$ whose correlation with the residual is close to maximum:

$$\frac{\left| g_{j[i]}^{[i]} \right|}{\left\| A_{j[i]} \right\|_2} \geq \alpha \max_j \frac{\left| g_j^{[i]} \right|}{\left\| A_j \right\|_2} \tag{3.14}$$

A nice property of weak orthogonal matching pursuit (WOMP) is its inheritance of a weakened version of the recovery properties of MP/OMP. Instead of selecting a *single* element, the *stagewise weak* selection chooses *all* elements whose correlation is close to the maximum. That is, we set the threshold in (3.13) as follows:

$$\lambda_{\text{weak}}^{[i]} = \alpha \max_j \frac{\left| g_j^{[i]} \right|}{\left\| A_j \right\|_2}$$

In practice, variations in selection strategy are complementary to those in directional updates. The combination of CGP and the stagewise weak selection is recommended considering its good theoretical properties as well as good empirical performance. The combination is called *stagewise weak conjugate gradient pursuit* (StWGP).

3.4.1.5 ROMP

Regularized OMP (ROMP) [11, 12] is another multi-element selection method for CS recovery. It groups the inner products g_i into sets J_k such that the elements in each set have a similar magnitude; i.e., they satisfy

$$\frac{[g_i]}{\|A_i\|_2} \leq \frac{1}{r} \frac{[g_j]}{\|A_i\|_2}, \quad \text{for all} \quad i, j \in J_k$$

ROMP then selects the set J_k for which $\sum_{j \in Jk} (|g_j| / \|A_j\|_2)^2$ is the largest. The strategy proposed in [11, 12], for ROMP selection, r was assumed to be 0.5. In this case, the algorithm was shown to have uniform performance guarantees closer to those of ℓ_1-based methods than those that exist for OMP and its derivatives. ROMP has played an important historical role in the research on greedy algorithms, being the first to exhibit uniform recovery guarantees. However, the constants in the theoretical guarantees are significantly larger than those for ℓ_1 minimization and ROMP has been quickly superseded by the thresholding techniques. This combined with the fact that empirically ROMP is not competitive with other pursuit algorithms means that it is generally not considered as a good practical algorithm for CS.

3.4.1.6 CoSaMP

The matching pursuit developed into variety of algorithms among which compressive sampling matching pursuit (CoSaMP) [13] is an important one. CoSaMP algorithm by Needell and Tropp and the subspace pursuit (SP) algorithm by Dai and Milenkovic are very similar and share many of their properties. Hence, discussion on both methods may be considered similar.

General Framework

Both CoSaMP and SP keep track of an active set T of nonzero elements and both add as well as remove elements in each iteration. At the beginning of each iteration, a k-sparse estimate $\hat{x}^{[i]}$ is used for calculating a residual error $y - A\hat{x}^{[i]}$, whose inner products with the column vectors of A are also calculated. The indexes of those columns of A with the k (or $2k$) largest inner products are then selected and added to the support set of $\hat{x}^{[i]}$ to get a larger set $T^{[i+0.5]}$. An intermediate estimate $\hat{x}^{[i+0.5]}$ is then calculated as the least-squares solution Argmin $\tilde{x}_{T^{[i+0.5]}} \|y - A\tilde{x}_{T^{[i+0.5]}}^{[i+0.5]}\|_2$. The largest k elements of this intermediate estimate are now found and used as the new support set $T^{[i+1]}$. In the last step, the CoSaMP algorithm takes as a new estimate the intermediate estimate $\hat{x}^{[i+0.5]}$ restricted to the new smaller support set $T^{[i+1]}$, while SP solves a second least-squares problem restricted to this reduced support. The CoSaMP algorithm, which was introduced and analyzed by Needell and Tropp, is summarized in Algorithm 3.3. The general CoSaMP framework is depicted in Figure 3.3.

Instead of making exact calculations of $A_{T^{[i+0.5]}}^+ y$ in each iteration, which increases computational complexity, a faster approximate implementation of the CoSaMP algorithm is proposed. This fast version replaces the exact least-squares estimate $\hat{x}_{T^{[i+0.5]}}^{[i+0.5]} = A_{T^{[i+0.5]}}^+ y$ with three iterations of a gradient descent or a conjugate gradient solver. Needell and Tropp suggest different strategies for stopping the CoSaMP algorithm. If the RIP holds, then the size of the error $y - A\hat{x}^{[i]}$ can be used to bound the error $x - \hat{x}^{[i]}$, which in turn

Algorithm 3.3   Compressive sampling matching pursuit (CoSaMP)

Input: y, A and k

Initialize: $T^{[0]} = \sup p\left(H_k(A^T y)\right)$, $\hat{x}^{[0]} = 0$

For i=0, i:=i+1, until stopping criterion is met do

$$g^{[i]} = A^T\left(y - A\hat{x}^{[i]}\right)$$

$$T^{[i\,|\,0.5]} = T^{[i]} \bigcup \sup p(g_{2k}^{[i]})$$

$$\hat{x}_{T^{[i+0.5]}}^{[i+0.5]} = A^+_{T^{[i+0.5]}} y,\ \hat{x}_{\overline{T^{[i+0.5]}}}^{[i+0.5]} = 0$$

$$T^{[i+1]} = \sup p(\hat{x}_k^{[i+0.5]})$$

$$\hat{x}_{T^{[i+1]}}^{[i+1]} = \hat{x}_{T^{[i+1]}}^{[i+0.5]},\ \hat{x}_{\overline{T^{[i+1]}}}^{[i+1]} = 0$$

end for

Output: $r^{[i]}$ and $\hat{x}^{[i]}$

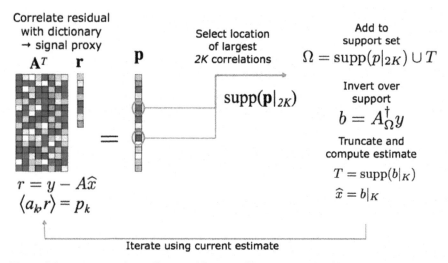

Figure 3.3   Compressed sampling matching pursuit (CoSaMP) algorithm for approximation of input signal.

can be used to stop the algorithm. However, in practice, this relationship is not guaranteed if the RIP holding is unknown. In this case, an alternative would be to stop the iterations as soon as $\left\|\hat{x}^{[i]} - \hat{x}^{[i+1]}\right\|_2$ is small or whenever the approximation error $\left\|y - A\hat{x}^{[i]}\right\|_2 < \left\|y - A\hat{x}^{[i+1]}\right\|_2$ increases. While the first of these methods does not guarantee the convergence of the method, the second approach is guaranteed to prevent instability. However, it is also somewhat too strict in the case in which the RIP holds.

3.4.1.7 Subspace pursuit (SP)

The SP algorithm [14], developed and analyzed by Dai and Milenkovic, is very similar to CoSaMP as shown in Algorithm 3.4.

Here, the same stopping rule based on the difference $y - Ax[i]2 - y - Ax[i + 1]2$ has been proposed by Dai and Milenkovic. This guarantees the method remaining stable, even in a situation in which the RIP condition fails. The main difference between the two approaches is the size of the set added to $T[i]$ in each iteration as well as the additional least-squares solution required in the SP. Furthermore, the possibility of replacing the least-squares solution in CoSaMP with three gradient-based updates implies that it can be implemented much more efficiently than SP [14].

3.5 Iterative Hard Thresholding

IHT [15] is a greedy algorithm that iteratively solves a local approximation to the CS recovery problem

$$\min_{\tilde{x}} \|y - A\tilde{x}\|_2^2 \text{ subject to } \|\tilde{x}\|_0 \leq k \qquad (3.15)$$

Algorithm 3.4   Subspace pursuit (SP)

Input: y, A and k

Initialize: $T^{[0]} = \sup p\left(H_k(A^T y)\right)$, $\hat{x}^{[0]} = A^+_{T^{[0]}} y$

For i=0, i:=i+1, until $\left\|y - A\hat{x}^{[i+1]}\right\|_2 \geq \left\|y - A\hat{x}^{[i]}\right\|_2$ do

$g^{[i]} = A^T(y - A\hat{x}^{[i]})$

$T^{[i+0.5]} = T^{[i]} \bigcup \sup p(g_k^{[i]})$

$\hat{x}^{[i+0.5]}_{T^{[i+0.5]}} = A^+_{T^{[i+0.5]}} y, \ \hat{x}^{[i+0.5]}_{\overline{T^{[i+0.5]}}} = 0$

$T^{[i+1]} = \sup p(\hat{x}_k^{[i+0.5]})$

$\hat{x}^{[i+1]} = A^+_{T+1} y$

end for

Output: $r^{[i]}$ and $\hat{x}^{[i]}$

The local approximation to this non-convex problem can be derived based on the optimization transfer framework of [9]. Instead of optimizing Equation (3.15) directly, a surrogate objective function is introduced as in Equation (3.16).

$$C_k^S(\tilde{x}, z) = \mu \|y - A\tilde{x}\|_2^2 - \mu \|A\tilde{x} - Az\|_2^2 + \|\tilde{x} - z\|_2^2 \qquad (3.16)$$

The advantage of this cost function is that (3.16) can be written as

$$C_k^S(\tilde{x}, z)\alpha \sum_j [\tilde{x}_j^2 - 2\tilde{x}_j(z_j + \mu A_j^T y - A_j^T Az)], \qquad (3.17)$$

Equation (3.17) can be optimized for each $\tilde{x}_j$ independently. If the constraint $\|\tilde{x}\|_0 \leq k$ is to be ignored, (3.17) can be written as

$$x^* = z + \mu A^T(y - Az) \qquad (3.18)$$

At this minimum, the cost function (3.17) has a value proportional to *CS*

$$C_k^S(x^*, z)\alpha \|x^*\|_2^2 - 2\langle x^*, (z + \mu A^T(y - Az))\rangle = -\|x^*\|_2^2. \qquad (3.19)$$

The constraint can therefore be enforced through the choice of the k largest coefficients of x^*, and by setting all other coefficients to zero. The minimum of (3.17) subject to the constraint that $\|\tilde{x}\|o \leq k$ is thus attained at

$$\hat{x} = H_k(z + \mu A^T(y - Az)), \qquad (3.20)$$

where H_k is the nonlinear *projection* that sets all but the largest k elements of its argument to zero. In cases where the k largest coefficients are not uniquely defined, the algorithm is assumed to make a selection from the offending coefficients using a predefined order. This local optimization approach can be turned into an iterative algorithm by setting $z = \hat{x}^{[i]}$ in which case the iterative hard thresholding (IHT) algorithm is obtained as in Algorithm 3.5.

Algorithm 3.5   Iterative hard thresholding (IHT)

Input: y, A, K, and μ
Initialize: $\hat{x}^{[0]} = 0$
For i= 0, i: =i+1, until stopping criterion is met do
$\hat{x}^{[i+1]} = H_k(\hat{x}^{[i]} + \mu A^T(y - A\hat{x}^{[i]}))$
End for
Output: $\hat{x}^{[i]}$

The IHT algorithm is easy to implement and is also computationally efficient. Apart from vector additions, the main computational steps are the multiplication of vectors by A and its transpose, as well as the partial sorting required for the thresholding step. Storage requirements are therefore small and, with the use of structured measurement matrices, multiplication by A and AT can also often be done efficiently. There are many other iterative algorithms coming into existence every day for the recovery of sparse signals measured using compressed sensing procedure.

3.5.1 Empirical Comparisons

A simple problem may be taken for consideration: 10,000 dictionaries of size 128×256 were generated with columns Ai drawn uniformly from the unit sphere. From each dictionary, and at a number of different degrees of sparsity, elements were selected at random and multiplied with unit variance, zero-mean Gaussian coefficients to generate 10,000 different signals per sparsity level [16]. The average performance of various greedy pursuit methods is calculated in terms of the exact recovery of the elements used for generating the signal. The results are shown in Figure 3.4. The results for MP, GP,

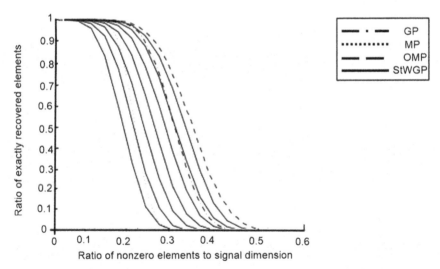

Figure 3.4 Comparison between MP (dotted), OMP (dashed), GP (dash-dotted), and StWGP (solid) in terms of exactly recovering the original coefficients. The solid lines correspond to the following (from left to right): $\alpha = 0.7$, 0.75, 0.8, 0.85, 0.9, 0.95, and 1.0 (CGP) (Figure 3.5).

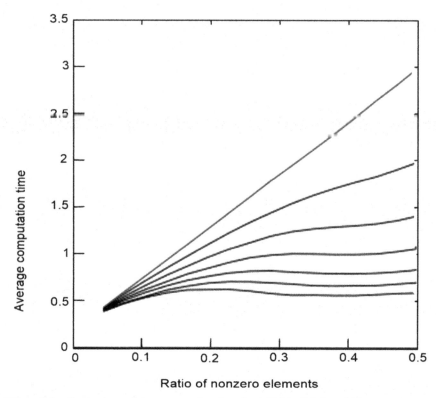

Figure 3.5   Comparison of the computation time for StWGP with the different values of α. From top to bottom, the figure corresponds to $\alpha = 1.0, 0.95, 0.9, 0.85, 0.8, 0.75$, and 0.7.

StWGP, ACGP, and OMP are also shown. All algorithms were stopped after they had selected exactly the number of elements used to generate the signal. It is clear that reducing the number of selected samples reduces the recovery performance. The advantage of this is a reduction in computational cost. This is shown in Figure 3.2. Here, the curves correspond to the following (going from top to bottom): $\alpha = 1.0, 0.95, 0.9, 0.85, 0.8, 0.75$, and 0.7. The top curve indicates that the computational cost for ACGP (StWGP with $\alpha = 1.0$) grows linearly with the number of nonzero coefficients. In contrast, for $\alpha < 1.0$, the computational cost grows much more slowly. It should be noted here that these figures do not fully capture the performance of StWGP since the dictionaries used do not have a fast implementation. However, they do provide a fair relative comparison between different values of α.

3.6 FOCUSS

Choice of a solution based on the smallest 2 norm is not appropriate when a sparse solution is desired. The procedural minimum-l2-norm criterion brings in a solution with many small nonzero entries, which is contrary to the goal of sparsity [17]. Consequently, there is a need to consider the minimization of alternative measures that promote sparsity. In this context of particular interest are diversity measures that measure the lack of sparsity, and algorithms for minimizing these measures meant for obtaining sparse solutions have been proposed. There are many measures of diversity, but a set of diversity measures has been found to produce very good results as applied to the subset selection problem. This is focal underdetermined system solution (FOCUSS).

$$E^{(p)}(x) = sgn(p) \sum_{i=1}^{n} |x[i]|^p, \quad p \leq 1 \tag{3.21}$$

Minimization of this diversity measure leads to the FOCUSS algorithm, which is iterative and produces appropriate intermediate solutions with least error according to

$$x_{k+1} = W_{k+1} (AW_{k+1})^{\dagger} b \tag{3.22}$$

where $W_{k+1} = \text{diag}\left(|x_k[i]|^{1-(p/2)}\right)$, and $\dagger$ is used to denote the Moore–Penrose pseudo-inverse. In each iteration, certain columns get emphasized, while others are de-emphasized. In the end, a few columns survive to represent, which leads to provision of a sparse solution. Interesting insight can be gained when it is viewed as a sequence of weighted minimum-norm problems. FOCUSS iteration is obtained as the minimum norm solution to an underdetermined set of linear constraints [18].

3.7 MUSIC

The multiple signal classification (MUSIC) algorithm is a well-known method in signal processing for estimating the individual frequencies of multiple time-harmonic signals. Mathematically, MUSIC is essentially a method of characterizing the range of the covariance matrix of the signals. MUSIC was originally developed for the estimation of the direction of arrival for source localization. Later, the MUSIC algorithm was extended to imaging of point scatterers. The performance guarantee exhibited by MUSIC is general, but qualitative in nature [19].

Sequential MUSIC is a multiple measurement vector (MMV) problem where a common sparse support matrix is provided for multiple signals. The concept of sequential MMV I is depicted in Figure 3.6.

Let ξ max and ξ min be, respectively, the strengths of the strongest and the weakest (nonzero) scatterers, $\delta_s^{\pm}$ be the (upper/lower) restricted isometry constants (RIC) of order s, and ε *be* the level of noise in the data. If the noise-to-scatterers ratio (NSR) obeys the upper bound

where

$$\frac{\varepsilon}{\xi_{\min}} < \sqrt{(1+\delta_s^+)^2 \frac{\xi_{\max}^2}{\xi_{\min}^2} + (1-\delta_s^-)^2 \tilde{\Delta}} - (1+\delta_s^+)\frac{\xi_{\max}}{\xi_{\min}}$$

$$\tilde{\Delta} = \frac{1}{2} - \frac{1}{2}\frac{1}{\sqrt{\sqrt{2}\gamma_s + 1}}, \qquad \gamma_s = 1 - \frac{\delta_{s+1}^-(1+\delta_s^+)}{2+\delta_s^+-\delta_{s+1}^-} \tag{3.23}$$

then the MUSIC imaging function J^ε with the thresholding rule

$$\{r \in \kappa : J^\varepsilon(r) \geq 2\gamma_s^{-2}\} \tag{3.24}$$

makes an exact recovery of the locations of the s scatterers. Compressed sensing theory comes into play in addressing the dependence of RIC on the frequency, the number and distribution of random sampling directions (or sensors), the number of scatterers, and the inter-scatterer distances.

In the under-resolved case, the δ_s^- tends to 1 and Γ_S tends to zero, rendering the right-hand side of (3.23) approximately

$$\frac{(1-\delta_s^-)^2 \tilde{\Delta}}{2(1+\delta_s^+)\xi_{\max}/\xi_{\min}} \tag{3.25}$$

where $\xi_{\max}/\xi_{\min}$ is the dynamic range of the scatterers. For a NSR smaller than (3.25), the s scatterers can still be perfectly localized by the

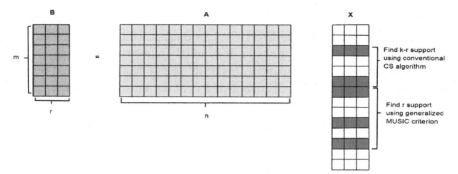

Figure 3.6   Sequential CS-based MUSIC algorithm.

MUSIC algorithm with the thresholding rule (3.24) where the threshold is approximately [19]

$$\frac{2}{(1 - \delta_{s+1}^-)^2} \left(\frac{1 + \delta_s^+}{2 + \delta_s^+}\right)^2 \rangle\rangle 1.$$

Previous observations and numerical results lend support to this super-resolution effect of the MUSIC algorithm. First, let us review the inverse scattering problem and the MUSIC imaging method.

Inverse Scattering

Consider the scattering of the incident plane wave (Figure 3.7).

$$u^i(r) = e^{i\omega r.\hat{d}} \tag{3.26}$$

by the variable refractive index $n(r) = \sqrt{1 + \xi(r)}$ where $\hat{d}$ is the incident direction. The resulting total wave field u satisfies the Helmholtz equation [19]

$$\Delta u(r) + \omega^2(1 + \xi(r))u(r) = 0. \tag{3.27}$$

It is assumed that the wave speed is unity, and hence, the frequency equals the wave number ω. The total field $u = u^i + u^s$ can be written as the sum of the incident field u^i and the scattered field u^s Since $(\nabla^2 + \omega^2)u^i = 0$, the scattered field satisfies

$$-(\nabla^2 + \omega^2)u^s = \omega^2 \xi u, \tag{3.28}$$

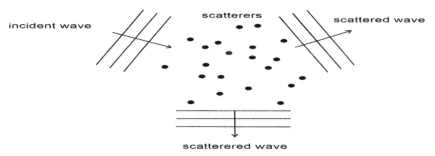

Figure 3.7 Scattering measurement: for each incident wave, the scattering amplitudes in multiple directions are measured.

subject to the radiation condition

$$\lim_{[r]\to\infty}|r|^{(d-1)/2}(\hat{r}.\nabla u^s(r) - 1\omega u^s(r)) = 0, \qquad \hat{r} = r/|r|, \qquad d = 2,3$$

which distinguishes the scattered field from the incident field. Invoking Green's function G, satisfying the radiation condition, of the free-space Helmholtz operator $-(\nabla^2 + \omega^2)$, (3.27) can be converted into the Lippmann–Schwinger integral equation

$$u^s(r) = \omega^2 \int_{R^d} \xi(r')(u^i(r') + u^s(r'))G(r,r')dr'. \tag{3.29}$$

As a result of the radiation condition, Green's function $G(r, r')$ has the following far-field asymptotic:

$$G(r, r') = \frac{e^{i\omega|r|}}{4\pi\,|r|^{(d-1)/2}}(e^{-i\omega r'.\hat{r}} + 0(|r|^{-1})), \quad \hat{r} = r/|r|, \qquad |r| \gg 1,$$

and therefore by (3.29), the scattered field has the asymptotic

$$u^s(r) = \frac{e^{i\omega|r|}}{|r|^{(d-1)/2}}(A(\hat{r}, \hat{d}) + 0(|r|^{-1})), \tag{3.30}$$

where the scattering amplitude A is given by the integral formula

$$A(\hat{r}, \hat{d}) = \frac{\omega^2}{4\pi} \int_{R^d} dr'\xi(r')u(r')e^{-i\omega r'.\hat{r}}. \tag{3.31}$$

The scattering amplitude is the observable data in inverse scattering, and the main objective then is to reconstruct ξ from the knowledge of the scattering amplitude [19]. Note that since u in (3.31) is part of the unknown, the inverse scattering problem is nonlinear. In physical terms, the nonlinearity is the consequence of multiple scattering between different parts of medium in homogeneities. In the Born scattering regime, the total field u on the right-hand side of (3.31) can be replaced by the incident field u^i, linearizing the inverse scattering problem. It can be recalled that the MUSIC algorithm is as applied to localization of point scatterers.

Idealized MUSIC Algorithm

The idealized MUSIC algorithm when the noise and round-off error are absent [19].

Algorithm 3.6   Idealized MUSIC algorithm

Idealized MUSIC algorithm

Input Y, Φ.

Compute the orthogonal projector P onto the null space of Y.

Plot $J(r) = \left| P\phi_r^{-2}, r \in k. \right|$

Identify the singularities of J as the object locations.

Algorithm 3.7   MUSIC algorithm with thresholding

Input $Y^\omega \Phi$ and the sparsity s.

Compute the orthogonal projector P^ω onto the noise subspace of $y^\omega = Y^\omega Y^{\omega *}$

Compute the function $J^\varepsilon(r) = |P^\varepsilon \phi_r|^{-2}$, $r \in$ K.

Output the set of points corresponding to the s highest values of J^ε or equivalently

the set $\{r \in \kappa : J^\varepsilon(r) \geq 2\gamma_s^{-2}\}$ where $\gamma_s = 1 - \frac{\delta_{s+1}^- (1+\delta_s^+)}{2+\delta_s^+ - \delta_{s+1}^-}$ as the object locations

---

3.8 Model-based Algorithms

From a general perception, model-based algorithms work on the basis of a predefined model designed using the requirements and test results in hand. Simple framework of any model based algorithm is given by Figure 3.8.

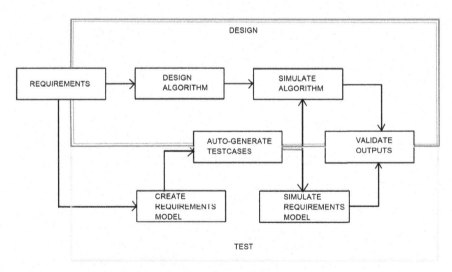

Figure 3.8   General algorithm design procedure on a predefined model.

Not only general applications, but CS recovery algorithms can also be developed as model-based algorithms. By reducing the degrees of freedom of a sparse/compressible signal, signal models provide two important benefits to CS. First, they enable reduction, in some cases significantly, the number of measurements M required to stably recover a signal. Second, during signal recovery, they enable better differentiation of true signal information from recovery artifacts, which leads to a more robust recovery [20].

There exists a model for K sparse signals that enables encoding of structures in a signal model that reduces the number of potential sparse signal supports in α. Then, using the model-based restricted isometry property (RIP), it can be proved that such model sparse signals can be robustly recovered from noisy compressive measurements. Moreover, quantification of the required number of measurements M is done to show that, for some models, M is independent of N. A model-compressible signal is one, whose coefficients α are no longer strictly sparse but have a structured power-law decay. This is meant for establishment of the possibility of robust recovery of model-compressible signals from compressive measurements. CS RIP can be generalized to a new restricted amplification property (RAmP). For some compressible signal models, the required number of measurements M is independent of N. Demonstration of the manner of integrating signal models into two state-of-art CS recovery algorithms CoSaMP and iterative hard thresholding (IHT) can be derived for ensuring practical advantage of the new theory. The main modification is that the nonlinear approximation step in these greedy algorithms is replaced with a model-based approximation. Both new model-based recovery algorithms have provable robustness guarantees for both model-sparse and model-compressible signals.

Conditions for Model-based Recovery
While many natural and manmade signals and images can be described to first order as sparse or compressible, the support of their large coefficients often has an underlying inter-dependency structure. This phenomenon has received only limited attention by the CS community to date. In this section, a model-based theory of CS is introduced that captures such a structure. A model reduces the degrees of freedom of a sparse/compressible signal by permitting only certain configurations of supports for the large coefficient. This allows reduction; in some cases significantly, the number of compressive measurements M is required to stably recover a signal.

Model-Sparse Signals

A K-sparse signal vector x lives in $\sum_K \subset R^N$, which is a union of $\binom{N}{K}$ subspaces of dimension K. Other than its K-sparsity, there are no further constraints on the support or values of its coefficients. A signal model endows the K sparse signal x with additional structure that allows certain K-dimensional subspaces in $\sum_K$ and disallows others.

To state a formal definition of a signal model, let X/Ω represent the entries of x corresponding to the set of indices $\Omega \subseteq \{1, \ldots\ldots, N\}$, and let Ω^C denotes the complement of the set Ω [20].

Definition 3.3: *A signal model M_K is defined as the union of m_k canonical K-dimensional subspaces*

$$M_K = \bigcup_{m=1}^{mk} X_m, \text{ such that } X_m := \{x : x \,|\,\Omega_m \in R^K, x\,|\,\Omega_m^c = 0\},$$

where each subspace Xm contains all signals x with $\sup p(x) \in \Omega_m$. Thus, the model M_K is defined by the set of possible supports $\{\Omega_1, \ldots\ldots, \Omega_{mk}\}$. Signals from $\mathcal{M}_K$ are called K-model sparse. Clearly, $M_k \subseteq \sum_K$ and contains $m_k \leq \binom{N}{K}$ subspaces.

Various types of models are available. Some models account for the fact that the large wavelet coefficients of piecewise smooth signals and images tend to live on a rooted, connected tree structure. Some model accounts for the fact that the large coefficients of sparse signals often cluster together.

Model-based RIP

If it is known that the signal x being acquired is K-model sparse, the RIP constraint on the CS measurement matrix Φ can be relaxed and stable recovery from the compressive measurements $y = \Phi X$ [20] can still be achieved.

Definition 3.4: [5, 6] *An $M \times N$ matrix Φ has the M_k-restricted isometry property (M_k-RIP) with constant δ_{M_k} if, for all $x \in M_k$, the following can be obtained.*

$$(1 - \delta_{M_k}) \|x\|_2^2 \leq \|\Phi x\|_2^2 \leq (1 + \delta_{M_k}) \|x\|_2^2 \tag{3.32}$$

An enlarged union of subspaces must be designed that includes sums of elements in the model for obtaining a performance guarantee for model-based recovery of K-model sparse signals in additive measurement noise.

Definition 3.5: *The B-Minkowski sum for the set M_k, with $B > 1$ an integer, is defined as*

$$M_K^B = \left\{ x = \sum_{r=1}^{B} x^{(r)}, \, with \, x^{(r)} \in M_K \right\}.$$

Define $M_B(x, K)$ as the algorithm that obtains the best approximation of x in the enlarged union of subspaces: M_K^B

$$M_B(x, K) = \arg \min_{\bar{x} \in M_K^B} \|x - \bar{x}\| \, 2.$$

We write $M(x, K) := M_1(x, K)$ when $B = 1$. Note that for many models, it can be assumed $M_K^B \subset M_{BK}$, and so the algorithm $M(x, BK)$ will provide a strictly better approximation than $M_B(x, K)$.

The performance guarantee for model-sparse signal recovery will require that the measurement matrix Φ be a near-isometry for all subspaces in M_K^B for some $B > 1$. This requirement is a direct generalization of the 2K-RIP, 3K-RIP, and higher-order RIPs from the conventional CS theory. Blumensath and Davies have quantified the number of measurements M necessary for a random CS matrix to have the M_K-RIP with a given probability.

Theorem 3.2: Let M_K be the union of m_k subspaces of K-dimensions in R_N. Then, for any t > 0 and any [20]

$$M \geq \frac{2}{c\delta_{MK}^2} \left(In(2m_k) + KIn\frac{12}{\delta_{MK}} + t \right),$$

an $M \times N$ i.i.d. sub-Gaussian random matrix has the M_K-RIP with constant δ_{MK} with probability at least $1 - e^{-t}$. This bound can be used for recovery of the conventional CS result by substituting $m_K = \binom{N}{K} \approx (Ne/K)^K$. The MK-RIP property is sufficient for robust recovery of model-sparse signals.

Model-Compressible Signals
Just as compressible signals are "nearly K-sparse" and thus live close to the union of subspaces Σ_K in R^N, model-compressible signals are "nearly K-model sparse" and live close to the restricted union of subspaces M_K. Compressible signals can be designed in terms of the decay of their K-term approximation error. The ℓ_2 error incurred by approximating $x \in R^N$ by the best model-based approximation in M_K is given by

$$\sigma M_K(x) := \inf_{\bar{x} \in M_K} \|x - \bar{x}\|_2 = \|x - M(x, K)\|_2$$

The decay of this approximation error defines the model compressibility of a signal.

$$m_s = \{x \in R^N : \sigma M_K(x) \le SK^{-1/s}, 1 \le K \le N, S < \infty\}.$$

3.8.1 Model-based CoSaMP

The CoSaMP algorithm can be modified for two reasons. First, it has robust recovery guarantees that are on par with the best convex optimization-based approaches [20].

Second, it has a simple iterative, greedy structure based on a best BK-term approximation (with B a small integer) that is easily modified to incorporate a best BK-term model-based approximation $M_B(K, x)$. Pseudocode for the modified algorithm is given in Algorithm 3.8.

Algorithm 3.8   Model-based CoSaMP

Inputs: CS matrix Φ, measurements y, model M_K
Output: K-Sparse approximation $\hat{x}$ to true signal x

$\hat{x}_0 = 0$, $r = y$, $i = 0$ {initialize}

While halting criterion false do

$\quad i \leftarrow i + 1$

$\quad e \leftarrow \Phi^T r$ {form signal residual estimate}

$\quad \Omega \leftarrow \sup p(M_2(e, K))$ {prune signal residual estimate according to signal model}

$\quad T \leftarrow \Omega \bigcup \sup p(\hat{x}_{i-1})$ {merge supports}

$\quad b/_T \leftarrow \Phi_T^+ y$, $b/_T c \leftarrow 0$ {form signal estimate}

$\quad \hat{m}_i \leftarrow M(b, K)$ {prune signal estimate according to signal model}

$\quad r \leftarrow y - \Phi\hat{x}_i$ {update measurement residual}

end while

$return \ \hat{x} \leftarrow \hat{x}_i$

Algorithm 3.9　Model-based iterative hard thresholding

Inputs: CS matrix Φ, measurements y, model M_K
Outputs: K-Sparse approximation $\hat{x}$

initialize: $\hat{x}_0 = 0$, $r = y$, $i = 0$
While halting criterion false do

$\quad i \leftarrow i + 1$

$\quad b \leftarrow \hat{x}_{i-1} + \Phi^T r$ {form signal estimate}

$\quad \hat{x}_i \leftarrow M(b, K)$ {prune signal estimate according to signal model}

$\quad r \leftarrow y - \Phi\hat{x}_i$ {update measurement residual}

$\quad$ end while

$\quad$ return $\hat{x} \leftarrow \hat{x}_i$

3.8.2 Model-based IHT

The proposed model-based iterative hard thresholding (IHT) [20] is given in Algorithm 3.9. For this algorithm, theorems associated with IHT can be proven with only a few modifications: Φ must have the $M_k^3 - RIP \; with \; \delta_{M_k^3} \leq 0.1$, and the constant factor in the bound changes from 15 to 4. In practice, model-based algorithm converges in fewer steps than IHT and yields highly accurate results in terms of recovery error.

3.9 Non-Iterative Algorithms for Image-Processing Applications

3.9.1 Advantages of Non-Iterative Algorithms

Many real-world signals can be sparsified by invertible transformation, such as wavelets, into a sparse (mostly zero, with K nonzero values at unknown locations) signal. This K-sparse signal can be reconstructed using the K lowest frequency DFT values using Prony's method or MUSIC (2K frequencies are required for complex signals or real-valued images). However, this does not work in practice due to poor conditioning caused by the clustering of the locations of the K nonzero values. The scaling property of the DFT is used for unclustering of these locations and to spread out the frequencies of known DFT values. Shepp-Logan phantom is reconstructed using only 2K DFT values, much fewer than the number required by 1-norm minimization, and using less computation than 1-norm minimization.

A common approach to sparse reconstruction is to compute the minimum 1-norm solution, perhaps by linear programming. If the DFT frequencies ki are randomly chosen, and if enough of them are known, it has been shown that the minimum 1-norm solution is in fact zn. In many practical situations, the luxury of choosing the ki at random is absent. The number of Zki required is $O(K \log N)$ (the exact number is unknown). Other approaches include thresholded Land weber iteration and orthogonal matching pursuit. These are much faster computationally, but require more of the problem in order to compute zn.

3.9.2 Non-Iterative Procedures for Recovery

3.9.2.1 Procedure I

Procedure I [21] is based on the MUSIC algorithm. When the DFT Z_k is known for all $|k| \leq K$, z_n can be reconstructed using any of the well-known array processing techniques such as Prony's method, MUSIC, or ESPRIT. This works well if the locations n_i of nonzero z_n are spaced out in n. The truth is that the sparsified signals tend to have clustered n_i. This makes the problem ill-conditioned. Consider these two extreme cases:

- $n_i = \{0, \ldots K - 1\}$. z_n from $\{Z_k, |k| \leq K\}$ is well known to be very ill-conditioned if $1 << K << N$.
- $n_i = \{0, N K, 2 K N, \ldots (K-1)K N \}$. z_n from $\{Z_k, |k| \leq K\}$ is perfectly conditioned since unknown values of Z_k are just the periodic extension of the given values.

In order to use the deterministic version of MUSIC, n_i must be unclustered to convert a problem like the former to one more like the latter. This may account for the lack of use of this approach in compressed sensing. Algorithm 3.10 provides the step-by-step procedure of recovering signals non-iteratively (Figure 3.9).

3.9.2.2 Procedure II

All the above approaches are iterative, requiring many iterations, and are computationally intensive, or both [22]. The approach of this paper requires only the following:

Note that all these operations are straightforward linear algebra and can be implemented in a non-iterative manner. Of course, iterative methods such as conjugate gradient may be used in some steps, but the overall algorithm is non-iterative, and can even be considered to be a closed-form computation.

Algorithm 3.10 Non-iterative procedure I

- Measure these DFT values X_k of the original x_n:

$$\{X_0, X_{\bar{L}}, X_{2\bar{L}}, X_{3\bar{L}}, \ldots X_{K\bar{L} \bmod (N)}\}$$

- Compute the DFT Z_k of the sparsified signal Z_n:

$Z_k = X_k \Psi k$, where Ψ_k is the DFT of the wavelet ψ_n.

- Form the Hermitian Toeplitz matrix from $\{Z_k\}$.
  These are then $\{Z_0, Z_1, \ldots Z_K\}$ for unclustered Z_n.

- Compute inverse N − point DFT of the null vector
  of the Hermitian Toeplitz matrix. Its elements: S_n.

- The locations n_i of the Zero values of S_n are the
  locations of the nonzero values of the unclustered z_n.

- Reorder the unclustered Z_n to the original z_n.

- Compute $X_k = Z_k / \Psi_k$ from the DFT Z_k of z_n.

- For k such that $\Psi_k = 0$, repeat with other scalings
  of the wavelet ψ_n. Note that X_0 is known.

Algorithm 3.11 Non-iterative procedure II

- The right null-space of H and solution to (1);

- Solution of N $\times$ (K + 1)(N – M + 1) linear system;

- Rank-one factored (K + 1) $\times$ (N – M + 1) matrix;

- An N-point FFT to find nonzero locations of z;

- Solution of an N $\times$ (K + 1) linear system to find z;

- Computation of $x = W z$ (if x is not sparse).

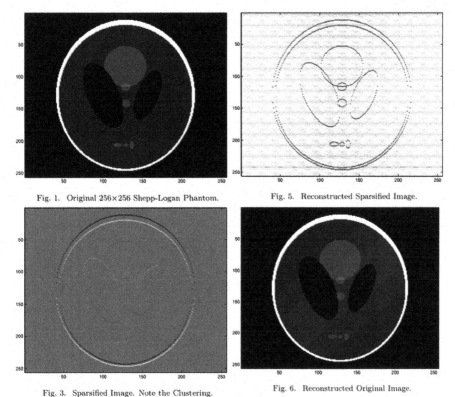

Fig. 1. Original 256×256 Shepp-Logan Phantom.

Fig. 5. Reconstructed Sparsified Image.

Fig. 3. Sparsified Image. Note the Clustering.

Fig. 6. Reconstructed Original Image.

Figure 3.9 Results of applying procedure I to standard phantom test image [21–23].

Derivation of New Algorithm

Let G be an $N \times (N-M)$ matrix of (not necessarily orthogonal) vectors spanning the right null-space of H. Let $\bar{z}$ be the minimum ℓ_2-norm solution to (3.33):

$$HG = 0; \; G = [g^1 |\ldots| g^{N-M}]; \; \bar{Z} = H'(HH')^{-1}y. \tag{3.33}$$

Then, the desired solution z can be written as

$$z = \bar{z} + \sum_{i=1}^{N-M} e_i g^i = \sum_{i=0}^{N-M} e_i g^i \tag{3.34}$$

For some unknown constants $\{e_i\}$, computing the N-point discrete Fourier transform of each column and using the results as first rows of circulant matrices gives

$$\hat{Z} = \sum_{i=0}^{N-M} e_i \hat{G}^i \qquad (3.35)$$

where $\hat{g}_k^i = \sum_{n=0}^{N-1} g_n^i e^{-j2\pi n k/N}, 0 \leq k \leq N-1$ and

$$Z = \begin{bmatrix} \hat{Z}_0 & \cdots & \hat{Z}_{N-1} \\ \ddots & \ddots & \ddots \\ \hat{Z}_{N-1} & \cdots & \hat{Z}_0 \end{bmatrix} \qquad (3.36)$$

$$\hat{g}^i = \begin{bmatrix} \hat{g}_0^i & \cdots & \hat{g}_{N-1}^i \\ \ddots & \ddots & \ddots \\ \hat{g}_{N-1}^i & \cdots & \hat{g}_0^i \end{bmatrix} \qquad (3.37)$$

Using the conjugate symmetry relation, $\hat{g}_{N-k}^i = \hat{g}_k^i$ [22].

The eigenvalues of each circulant matrix $\hat{G}^i$ are the elements $\{g_n^i\}$ of g^i. Because only K elements of z are nonzero, $\hat{z}$ has rank K, and there exists a vector ν of length $K+1$ such that

$$\hat{Z} \begin{bmatrix} V \\ 0 \end{bmatrix} = \sum_{i=0}^{N-M} e_i \hat{G}^i \begin{bmatrix} V \\ 0 \end{bmatrix} = 0 \qquad (3.38)$$

Let $\hat{G}^i$ be the matrix of the first $K+1$ columns of $\hat{G}^i$. Each $\hat{G}^i$ is Toeplitz.

Rewrite this equation as

$$\begin{bmatrix} \hat{G}^0 \ldots \ldots \hat{G}^{N-M} \end{bmatrix} \begin{bmatrix} e_0 v \\ \cdot \\ \cdot \\ \cdot \\ e_{N-M} v \end{bmatrix} = 0 \qquad (3.39)$$

This is a system of N equations in a total of $(N-M+1)$ unknowns $\{e_i\}$ and $(K+1)$ unknowns $\{v_i\}$

- If $N > (N-M+1) + (K+1)$, this is an over-determined system of quadratic equations, which by Bezout's theorem almost surely has only the actual solution assumed to exist. But its solution is difficult.
- If $N > (N-M+1).(K+1)$, this is an over-determined system of linear equations. Solution is easy [22].

The unknowns can be found by solving for $(N - M + 1)(K + 1)$. Arranging these into an $(N - M + 1) \times (K + 1)$ matrix and computing its rank-one decomposition yields $\{e_i\}$ and $\{v_i\}$ to a (irrelevant) scale factor.

Next, make the following definitions:

$$\{z_{i_n}, 1 \leq n \leq K\} = \{z_i : z_i \neq 0, 1 \leq i \leq N\} \tag{3.40}$$

$$[F_{ik}] = \left[e^{-j2x(i-1)(k-1)/N}\right]_{ik} \cdot 1 \leq i, k \leq N \tag{3.41}$$

$$[\bar{F}_{ik}] = \left[e^{-j2x(i-1)(n_k-1)/N}\right]_{ik} \cdot 1 \leq k \leq K \tag{3.42}$$

- $\{z_{i_n}, 1 \leq n \leq K\}$ are the nonzero values of $\{z_i\}$;
- F is the DFT matrix implementing $\bar{Z} = Fz$;
- $\bar{F}$ is a tall $N \times K$ sub-matrix of F.

The $(N \times K)(K \times N)$ factorization is already known as follows:

$$\bar{z} = F \operatorname{diag}[z_i] F^H = \bar{F} \operatorname{diag}[z_{i_n}] \bar{F}^H \tag{3.43}$$

Extending $\bar{F}$ by any other column of F gives

$$0 = \bar{z} \begin{bmatrix} v \\ 0 \end{bmatrix} = \bar{F} \operatorname{diag}[z_{i_n} 0] \bar{F}^H \begin{bmatrix} v \\ 0 \end{bmatrix} \tag{3.44}$$

Since $\bar{F}$ is a tall matrix and $z_{i_n} \neq 0$, we have

$$\bar{F}^H \begin{bmatrix} v \\ 0 \end{bmatrix} = 0 \rightarrow F^H \begin{bmatrix} v \\ 0 \end{bmatrix} = 0 \text{ for } i \in \{i_n\} \tag{3.45}$$

This shows that an inverse N-point DFT of a zero padded v is zero at the locations $\{i_n\}$ of nonzero z_i. The conditioning of the location problem is determined by the condition number of $\bar{F}$. Random distribution of $\{i_n\}$ yields good conditioning: clumping of $\{i_n\}$ and large gaps yields poor conditioning [22]. The DFT of null vector of Toeplitz-structured matrix can be seen as a deterministic version of MUSIC, without computing the autocorrelation first.

3.9.2.3 Procedure III

All the above approaches are iterative, requiring many iterations, and are computationally intensive, or both. The approach of this paper requires only the following:

- Computation of singular value decompositions (SVDs) of $M \times N$ matrices H_1 and $H_2(H = H_1 \otimes H_2)$;
- Computation of left and right null vectors of the $M \times M$ matrix formed from the M^2 elements of y;
- Solution of an $M^2 \times (M-1)^2$ linear system;
- Computation of $x = W'z$(if x is not sparse) [23].

All these operations are straightforward linear algebra and can be implemented in a non-iterative manner. Of course, iterative methods such as conjugate gradient may be used in some steps, but the overall algorithm is non-iterative, and can even be considered to be a closed-form computation.

Derivation of New Algorithm

Review of Kronecker Product

The Kronecker (tensor) product $A \otimes B$ of two $M \times N$ matrices A and B is the $M^2 \times N^2$ matrix

$$[A \otimes B]_{\substack{iM+m \\ jM+n}} = A_{i,j} B_{m,n}, \quad \substack{0 \le i,j \le M-1 \\ 0 \le m,n \le N-1} \tag{3.46}$$

A simple numerical example:

$$\begin{bmatrix} 1 & 2 \\ 3 & 4 \end{bmatrix} \otimes \begin{bmatrix} 5 & 6 \\ 7 & 8 \end{bmatrix} = \begin{bmatrix} 5 & 6 & 10 & 12 \\ 7 & 8 & 14 & 16 \\ 15 & 18 & 20 & 24 \\ 21 & 24 & 28 & 32 \end{bmatrix} \tag{3.47}$$

Relevant properties of the Kronecker product are as follows:

$$(A \otimes B)(C \otimes D) = (AC) \otimes (BD) \tag{3.48}$$

If C and D are inverses of 1D wavelet transforms, and A and B are 1D partial discrete Fourier transform (DFT) matrices, or 1D Toeplitz (convolutional) matrices, $(C \otimes D)$ is the inverse 2D wavelet transform, and $(A \otimes B)$ is the partial 2D DFT [23].

$$vec(A \times B) = (B' \otimes A)vec(X). \tag{3.49}$$

Vec(X) unwraps the matrix X by columns. Applying this to the current problem.

$$vec(Y) = (H_1 \otimes H_2)vec(Z) \leftrightarrow Y = H_2 Z H_1' \tag{3.50}$$

Reformulation of Problem

The original problem is

$$y = H_z = (H_1 \otimes H_2)z \tag{3.51}$$

where z has at most M – 1 nonzero elements. Wrapping vectors y and z into matrices Y and Z:

- M^2 – vector $y \rightarrow (M \times M)$ matrix Y;
- N^2 – vector $z \rightarrow (N \times N)$ matrix Z,

and defining the SVDs of H_1 and H_2 as

$$H_1 = U_1 S_1 V_1; \qquad\qquad H_2 = U_2 S_2 V_2 \tag{3.52}$$

The original problem is equivalent to

$$Y = H_2 Z H_1^{'} = (U_2 S_2 V_2) Z (V_1^{'} S_1 U_1^{'}) \tag{3.53}$$

which in turn becomes

$$R = S_2^{-1} U_2^{'} Y U_1 S_1^{-1} = V_2 Z V_1^{'} \tag{3.54}$$

where the $M \times M$ matrix R is quickly computable from the $M \times M$ matrix Y composed of values of data y.

Since only M – 1 elements of Z are nonzero. Z and hence Y both have rank M – 1. Assuming that no two nonzero elements of Z are in the same row or column, rows and columns containing nonzero elements of Z are known. Let $\bar{n}$ be the right null vector of R:

$$0 = R\bar{n} = (V_2 Z V_1^{'})\bar{n} \rightarrow 0 = Z V_1^{'} \bar{n} \tag{3.55}$$

If the ith row of z is all zero, then the ith element of $Z V_1^{'} \bar{n}$ is zero. If the $(i, j)^{th}$ element of Z is nonzero, the jth row of $V_1^{'}$ is orthogonal to $\bar{n}$. Hence, zeros of $V_1^{'} \bar{n}$ indicate columns of Z with a nonzero element [23].

Repeating this with $R^{'}$ identifies rows of Z with nonzero elements. Hence, rows and columns of Z with nonzero can be identified. Each nonzero element is at the intersection of one of these rows and columns, for a total of $(M - 1)^2$ possible locations of nonzero elements. So M^2 values of y are sufficient to determine which of these elements is nonzero.

3.9.3 NITRA

Non-iterative threshold-based reconstruction algorithm (NITRA) is an efficient and easy way of reconstructing an algorithm with less complexity and in a non-iterative fashion. Upon receipt of y and ϕ, the receiver should apply a robust reconstruction algorithm to these inputs and recover the original input image or video. Many CS reconstruction algorithms iteratively solve LSP where the number of iterations depends upon either sparsity or any comfortable fixed number. The proposed algorithm NITRA is named after its algorithmic procedure which involves no iteration for finding the best match. It uses only transpose function and a thresholding operator β, where β depends upon the measurement vector y. The threshold operator is calculated using Equation (3.56).

$$\beta = \lceil \log_{10} |K \times \max(y)| \rceil \qquad (3.56)$$

where K is the sparsity given as $K = \|x\|_0$. y is considered as an important metric for finding the threshold. This is so because among the three inputs that are to be provided to the receiver, namely ϕ, y and K, y carries the information about the pixel values of the image or video frame in the form of coefficients. Algorithm 3.12 provides the NITRA algorithm which uses these inputs to recover the images and videos [24].

At the transmitter end, the measurement vector y is obtained by finding the inner product of ϕ and x_s as explained in Section 4. In NITRA, ϕ is fixed for every block and hence transmission of ϕ need be done only once to the receiver thus consuming less memory.

According to the NITRA's procedure provided in Algorithm 3.12, the estimation of the original image is made at the receiving end by taking the inner product of ϕ^T and y. Only those values which satisfy the threshold condition

---

Algorithm 3.12   NITRA for reconstruction of image/video at the receiver

NITRA framework for images and videos

At the receiver:
$\beta = \lceil \log_{10} |K \times \max(y)| \rceil$
$x\_new = \phi^T y$
for $i = 1 : N$
    if $x\_new(i) \leq \beta$
$q = 0$;
    else
$q = x\_new(i)$
$recon = \lceil idct(q) \rceil$
Rearrange blocks to frames

β are selected for reconstruction and the others are made zeros, thereby reducing the number of computations. NITRA does not have iterations within every block which is advantageous in the context of reducing computational complexity, execution time, etc., making it suitable for WSNs. After threshold operation, the resultant values are directly transformed back to the real space by taking 2D-IDCT. Since the proposed NITRA uses augmented matrix as sensing matrix which has large number of zeros and unity in leading diagonal, it senses only the necessary information required for perfect reconstruction. NITRA does not need to solve LSP to find the best solution unlike the other conventional CS recovery algorithms which use random matrices which makes solving LSP mandatory. Also, in existing algorithms, the sensing matrix is generated separately for each split block, and hence, the total performance cannot be relied upon in one single execution of the algorithm. The feasibility of reconstruction of images and videos using NITRA, its error bound and accuracy are proven mathematically by deriving the following theorems and lemmas [24].

Error Bound Calculation

***Lemma 1**: For a measurement matrix ϕ which satisfies RIP with sparsity s*

$$\left\|\phi^T y\right\|_2 \leq \sqrt{1 + \delta_s} \left\|y\right\|_2, \tag{3.57}$$

$$(1 - \delta_s) \left\|x_s\right\|_2 \leq \left\|\phi^T \phi x_s\right\|_2 \leq (1 + \delta_s) \left\|x_s\right\|_2 \tag{3.58}$$

And also

$$\left\|(I - \phi^T \phi) x_s\right\|_2 \leq \delta_s \left\|x_s\right\|_2, \tag{3.59}$$

$$\left\|\phi^T \phi x_s\right\|_2 \leq \delta_s \left\|x_s\right\|_2 \tag{3.60}$$

A perfect measurement matrix should satisfy the following lemmas proposed by Needel and Tropp, Proposition 3.5.

***Lemma 2**: If ϕ satisfies the RIP $\left\|\phi x_s\right\|_2 \leq \sqrt{1 + \delta_s} \left\|x_s\right\|_2$, $\forall x : \left\|x_s\right\|_0 \leq s$, then*

$$\left\|\phi x_s\right\|_2 \leq \sqrt{1 + \delta_s} \left\|x_s\right\|_2 + \sqrt{1 + \delta_s} \frac{\left\|x_s\right\|_1}{\sqrt{s}} \tag{3.61}$$

***Lemma 3**: For any "x", let "x_s" be the best approximation to "x". Let $x_r = x - x_s$. Let $y = \phi x + e = \phi x_s + \phi x_r = \phi x_s + \tilde{e}$. If the RIP holds for sparsity "s", then the error $\tilde{e}$ can be bounded by*

$$\left\|\tilde{e}\right\|_2 \leq \sqrt{1 + \delta_s} \left\|x - x_s\right\|_2 + \sqrt{1 + \delta_s} \frac{\left\|x - x_s\right\|_1}{\sqrt{s}} + \left\|e\right\|_2 \tag{3.62}$$

where $\|e\|_2$ is the observation error which is zero. Now that ϕ which satisfies RIP has been obtained, measurement vector y is to be calculated. y is calculated by multiplying $\phi(M \times N)$ and $\hat{x}(N \times 1)$. The resultant vector $y(M \times 1)$ is transmitted to the receiver along with ϕ. Thus, transmission of the entire input image block is reduced to just $M < N$ measurements, resulting in considerable reduction in execution time, complexity, etc. If there are p blocks in a frame/image, then the total number of measurements required for reconstructing the frame/image will be pM.

Since NITRA is a lossy technique of recovery of compressively sensed images and video frames, the perfection in reconstruction must be verified mathematically by checking the error in reconstruction. Using lemmas, the following theorems have been proposed to prove that the error in reconstruction of images and videos by NITRA is within the minimal range. Theorem 3.3 gives the condition to be satisfied for perfect reconstruction. The norm of the difference between the original input vector x and the estimated vector x_r must be less than or equal to the sum of second norm of sparse vector x_K and the estimated error $\tilde{\varepsilon}$. $\tilde{\varepsilon}$ depends upon x, x_K, and e as in Equation (3.64) where e is the assumed error and can be neglected since it tends to zero.

Theorem 3.3: *Considering a noisy observation, $y = \phi x + e$, where x is a vector. Let x_K be the sparse vector with K nonzero elements. NITRA will recover the estimated signal x_r of the input x by satisfying the following condition:*

$$\|x - x_r\|_2 \leq \|x_K\|_2 + \tilde{\varepsilon} \qquad \text{where} \tag{3.63}$$

$$\tilde{\varepsilon} = \|x - x_K\|_2 + \frac{1}{K}\|x - x_K\|_1 + \|\tilde{e}\|_2 \tag{3.64}$$

The accuracy of NITRA for estimating x can be represented by

$$\|x - x_r\|_2 \leq \|x - x_K\|_2 + \frac{1}{\sqrt{K}}\|x - x_K\|_1 + \|\tilde{e}\|_2 \tag{3.65}$$

Equation (3.65) gives the accuracy of NITRA in estimating the original image from the compressed form. The error between the original input vector and the estimated input vector is less than the sum of first and the second norms of the difference between original and the sparse vector. NITRA satisfies this equation exactly, thus providing greater accuracy.

Proof of error Bound in Theorem 1

Initially, NITRA satisfies the condition given by Equation (3.66)

$$\|x - x_r\|_2 \leq \|x_K - x_r\|_2 + \|x - x_K\|_2 \tag{3.66}$$

where x is any input vector. Since RHS of Equations (3.63) and (3.66) are equal, LHS can be equated and $\tilde{\varepsilon}$ is substituted from Equation (3.63). The resultant will be

$$\|x_K - x_r\|_2 \leq \|x_K\|_2 + \frac{1}{\sqrt{K}} \|x - x_K\|_1 + \|\tilde{e}\|_2 \tag{3.67}$$

Substituting Equation (3.67) in Equation (3.66), Equation (3.68) is obtained.

$$\|x - x\|_2 \leq \|x_K\|_2 + \frac{1}{\sqrt{K}} \|x - x_K\|_1 + \|x - x_K\|_2 + \|\tilde{e}\| \tag{3.68}$$

In order to apply *Lemma 3* to represent the error bound $\|\tilde{e}\|_2$, the term $\sqrt{1 + \delta_K}$ is multiplied to equation (3.68) in the following fashion:

$$\sqrt{1 + \delta_K} \|x - x_r\|_2 \leq \sqrt{1 + \delta_K} \|x_K\|_2$$
$$+ \sqrt{1 + \delta_K} \left[\|x - x_K\|_2 + \frac{1}{\sqrt{K}} \|x - x_K\|_1 \right] + \|\tilde{e}\|_2 \tag{3.69}$$

Multiplication of $\sqrt{1 + \delta_K}$ to $\|\tilde{e}\|_2$ is avoided since $\delta_K \ll 1$. Substituting Equation (3.61) in Equation (3.69), the error between the original vector and the reconstructed vector is found to prove Equation (3.63), hence proving that the error by using NITRA for reconstruction is found to be minimum. The steps are as follows:

$$\sqrt{1 + \delta_K} \|x - x_r\| = \sqrt{1 + \delta_K} + \tilde{\varepsilon}$$

$$\|x - x_r\|_2 \leq \|x_K\|_2 + \frac{\tilde{\varepsilon}}{\sqrt{1 + \delta_K}} \tag{3.70}$$

When the difference metric $\delta_K \ll 1$, the denominator of Equation (3.70) tends to unity, and hence, the final equation will be

$$\|x - x_r\|_2 \leq \|x_K\|_2 + \tilde{\varepsilon} \tag{3.71}$$

This proves that the error is far less than the combined errors obtained by adding $\|x_K\|_2$ and the difference terms $\|x - x_K\|_2$ and $\frac{1}{\sqrt{K}} \|x - x_K\|_1$.

Theorem 3.4: *Given a noisy observation $y = \phi x + e$, where x_K is K-sparse vector, if ϕ has RIP, then NITRA will recover an approximation x_r satisfying*

$$\|x_K - x_r\|_2 \leq \|x_K\|_2 + \|e\|_2 \qquad (3.72)$$

The accuracy of the estimation is

$$\|x_K - x_r\|_2 \leq \|e\|_2 \qquad (3.73)$$

Proof of error bound in Theorem 2

The estimated error is dependent on the term $\|x_K - x_{bt}\|_2$. This is the difference between the estimates before and after the thresholding operation. With the help of triangular inequality, the error can be expressed as

$$\|x_K - x_r\|_2 \leq \|x_K - x_{bt}\|_2 + \|x_r - x_{bt}\|_2. \qquad (3.74)$$

After thresholding, x_r becomes the best approximation to x_{bt} than x_K which means

$$\|x_r - x_{bt}\|_2 \leq \|x_K - x_{bt}\|_2 \qquad (3.75)$$

and thus, the error will not exceed twice the value of $\|x_K - x_{bt}\|$. The error is expressed as

$$\|x_K - x_r\|_2 \leq 2\|x_K - x_{bt}\|_2. \qquad (3.76)$$

But $x_{bt} = \phi^T y + e = \phi^T \phi x_K + e$. Using all the above findings, the error is represented as follows:

$$\|x_s - x_r\|_2 \leq 2\|x_s - \phi^T \phi x_K + \phi^T e\|_2$$

$$\leq 2\|x_K - \phi^T \phi x_K\|_2 + 2\|\phi^T e\|_2$$

$$\leq \|(1 - \phi^T \phi) x_K\|_2 + 2\|\phi^T e\|_2 \qquad (3.77)$$

From *Lemma 1*, it can be written that $\|\phi^T e\|_2 = \sqrt{1 + \delta_K}\|e\|_2$ and $\|(I - \phi^T \phi) x_K\|_2 \leq \delta_K \|x_K\|_2$

The term $\|x_K - x_r\|_2$ is nothing but the residue which can be denoted by $\|r_r\|_2$. Therefore,

$$\|x_K - x_r\|_2 \leq 2\delta_K \|x_K\|_2 + 2\sqrt{1 + \delta_K}\|e\|_2 \qquad (3.78)$$

This can also be written as

$$\|x_K - x_r\|_2 \leq a\|x_K\|_2 + b\|e\|_2 \qquad (3.79)$$

where $a = 2\delta_K$ and $b = 2\sqrt{1 + \delta_K}$. The range of δ_K is $0 < \delta_K < 1$. Hence, $\|r_r\|_2$ can be approximated as

$$\|x_K - x_r\|_2 \leq \|x_K\|_2 + \|e\|_2 \tag{3.80}$$

Equation (3.80) proves that the error between the sparse vector and the estimated vector is less than or equal to the sum of norms of sparse vector and the observation error. On the assumption that the observation error is zero, i.e., $\|e\|_2 = 0$, Equation (3.80) can be written as $\|x_K - x_r\|_2 \leq \|x_K\|_2$ or $\|r_r\|_2 \leq \|x_K\|_2$. NITRA was successfully applied to images and videos. Perfect recovery of test images and videos was achieved with higher PSNR and perceptual quality. NITRA exhibited reduced time consumption and less complexity due to perfect selection of measurements through augmented matrix (Figure 3.10).

3.9.4 R3A

Different types of reconstruction algorithms such as OMP, CoSaMP, and StOMP are available for the reconstruction of compressively sensed images by iteratively solving the least-squares problem (LSP). The number of iterations either depends on sparsity or chosen as a fixed number as in many existing algorithms. The large number of iterations involved in many popular reconstruction algorithms makes them complex and time consuming. The proposed recovery algorithm R3A uses no iterations. Hence, the elapsed time, total execution time, and complexity are reduced in the proposed method compared to the other existing methods along with better PSNR [25].

R3A framework, portrayed in Figure 3.11, includes the following steps: At the transmitting side, the input image is converted from RGB to grey values. Then, the image is divided into sub-images of size 4×4 in order to meet the memory requirements. The basis function chosen here to transform the pixel values to other comfortable domains is discrete cosine transform (DCT) as it is simple to implement and gives a performance similar to that of KLT. DCT is applied to every single block for obtaining the sparse vector $\hat{x}$. After sparsification, the measurement vector y must be calculated using ϕ which is obtained using Kronecker product of random Gaussian vector G and the vector I formed using unity followed by zeros.

The Kronecker product is formed with the help of random Gaussian matrix of size $M \times 1$ and a vector of size $1 \times N$ having the first position as unity and the rest of the values as zeros. The combination for Kronecker product was

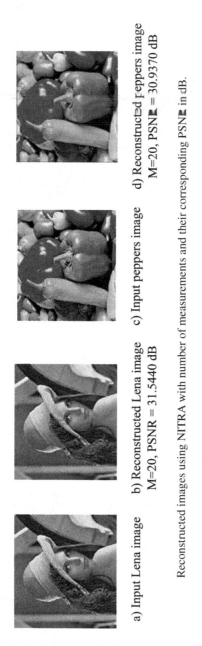

a) Input Lena image   b) Reconstructed Lena image   c) Input peppers image   d) Reconstructed peppers image
M=20, PSNR = 31.5440 dB   M=20, PSNR = 30.9370 dB

Reconstructed images using NITRA with number of measurements and their corresponding PSNR in dB.

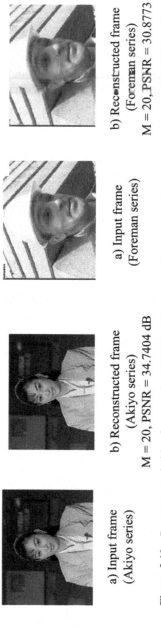

a) Input frame   b) Reconstructed frame   a) Input frame   b) Reconstructed frame
(Akiyo series)   (Akiyo series)   (Foreman series)   (Foreman series)
          M = 20, PSNR = 34.7404 dB   M = 20, PSNR = 30.8773

Figure 3.10 Reconstructed video frames using NITRA with number of measurements and their corresponding PSNR in dB.

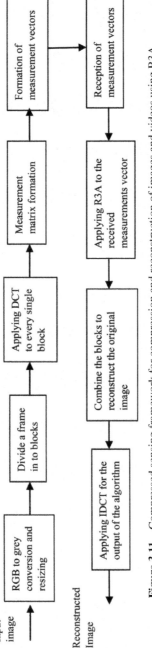

Figure 3.11 Compressed sensing framework for compression and reconstruction of images and videos using R3A.

selected by trial and error, and the combination which gave best results was selected.

$$K_{M \times N} = [G_{M \times 1} \otimes I_{1 \times N}] \tag{3.81}$$

Kronecker product provides two major advantages which are as follows: It is easily implemented in hardware, consumes less time, and exhibits high precision in reconstruction. While using Kronecker matrix, the PSNR of the reconstructed image and the matrix's performance is better than random measurement matrices. The measurement vector y can be found by finding the product of the sparsified signal $\hat{x}$ and ϕ; i.e., $y = \phi \hat{x}$. At the receiver, sparsity k, f, and y are given as input to the R3A algorithm.

R3A Algorithm at the receiver

The following steps describe the procedure for reconstructing the input using R3A [25]:

R3A calculates the transpose of ϕ without solving the least-squares problem. Inner product of ϕ and y provides the reconstructed sparse vector which is given by Equation (3.82).

$$x_K = \phi^T y \tag{3.82}$$

Only very few samples of the recovered sparse vector are selected for further reducing the process time. This is done by using the threshold provided by Equation (3.83):

$$\alpha \leftarrow 10 * \log_{10} |k * \min(y)| \tag{3.83}$$

Algorithm 3.13   Reduced runtime recovery algorithm

Obtain the threshold $\alpha = 10 * \log_{10} |k * \min(y)|$

Calculate the dividing factor
$d = (10 * \log_{10} \sum |y|)/k$

Calculate reconstructed sparse vector
$x_K = \phi^T y$

$\forall i = 1 \ldots N,$

$$q = \begin{cases} x_s(i)/d, & \text{if } x_s(i) \leq \alpha \\ 0, & \text{otherwise} \end{cases}$$

$x_r = IDCT(q)$

The algorithm checks whether the value of x_s is either greater than or less than the threshold. If the value of x_s is greater than threshold, then q is equivalent to x_s/d where $d = (10 * \log_{10} \sum |y|)/k$. Finally, IDCT is applied to q to get back the pixel values of the reconstructed image. Then, the reconstructed values are arranged to 4×4 blocks and these blocks are concatenated to form the original image.

R3A provides better performance guarantees and PSNR similar to the OMP, which is considered to be one of the most prominent CS reconstruction algorithms. R3A also possesses the following properties: It provides near-optimal error guarantees, it is robust to observation noise, it succeeds with a minimal number of observations, and the memory requirement is very low.

3.9.4.1 R3A-based StOMP

The convergence of StOMP is faster than OMP, but it produces less accurate images. Thus, the trade-off can be eliminated with the proposed combinational algorithm. Since R3A-based StOMP algorithm reconstructs the image non-iteratively. It reduces the elapsed time, complexity, and runtime and also produces PSNR, accuracy, and structural similarity better than StOMP [26]. The following steps describe the procedure for reconstructing the input using R3A-based StOMP:

Algorithm 3.14   R3A-based StOMP

Initialize a zero matrix, $Phit = [\,]$

Obtain the threshold
$t = \log_{10} |\max(y)|$

Estimate the intermediate value
$v = |\phi^T \times y| + ((10 \times \log_{10} \sum |y|)/k)$

Save the positions of the values satisfying v > t
$\Gamma \leftarrow \{j : |v_j| > t\}$

Update the empty matrix
$Phit \leftarrow Phit \bigcup \Gamma$

Find the estimated sparse vector
$\hat{x} \leftarrow \phi_{phit} \backslash y$

In R3A-based StOMP, the algorithm initially calculates the inner product of ϕ and y and then adds the inner product to the constant $((10 \times \log_{10} \sum |y|)/k)$. The algorithm then calculates the threshold which is given in Equation (3.84):

$$t = \log_{10} |\max(y)| \qquad (3.84)$$

If the values of the variable v are greater than the threshold, then the positions of these values are stored separately. The corresponding column from the measurement matrix ϕ is taken on the basis of positions and those columns are augmented to the variable empty matrix. Finally, applying pseudo-inverse to the stored measurement matrix columns and measurement vector y gives the reconstructed sparse vector.

R3A-based StOMP eliminates the accuracy and time consumption trade-off in the StOMP and also produces images with better perceptual quality and accuracy than StOMP with reduced time. Also, it succeeds with the minimum number of measurements and provides better performance guarantees and is robust to noise.

3.9.5 SPMT

Split and merge is a commonly used technique for any sorting algorithm since it reduces complexity and brings down correlation errors. Similarly, in CS-based recovery, split and merge process can be applied successfully for the perfect reconstruction of images and videos. Kronecker product used in R3A provides satisfactory results. Yet, perfect extraction of absolute pixel values is barely visible. This leads to the development of augmented matrix ϕ_{aug}, which consists of identity matrix augmented with zero vectors [27]. The reconstruction process is supported by the augmented matrix since the leading diagonals of the identity matrix support a perfect sample selection. Another property of the augmented matrix is that it follows the restricted isometry property (RIP), which is a necessary condition for the projecting matrix for getting perfect reconstruction. The augmented matrix is represented mathematically in (3.85):

$$\phi_{aug} = [I_{M \times M} \mid Z_{(N-M) \times M}] \qquad (3.85)$$

where $I_{M \times M}$ is the $M \times M$ identity matrix and $Z_{(N-M) \times M}$ is a null matrix of size $(N - M) \times M$.

3.9.5.1 SPMT for reconstruction of images and videos

The split, process, and merge technique splits both the projections and the projecting matrix into smaller pieces, processes them separately, and recombines the results to form the estimated image. A set of random values r_M of sparsity size k is generated which represents the support of the measurements that are to be filled in every split y. k requires factorization for finding the number of rows and columns in order to arrange the values of r_M. Even after factorization, the number of rows f_r and number of columns f_c must be such that $f_r < f_c$, since the number of values in each column decides the number of values that should be considered from y for construction every split-y. The formation of support matrix r_M and the SPMT algorithm are shown in Algorithm 3.15.

Once the support matrix is ready, SPMT can be applied to the received measurements y and the measurement matrix ϕ. y is split such that each split-y consists of the values of y corresponding to the positions present in the columns Γ of the support matrix r_K. The split-ϕ_{aug} is found by the same procedure, but instead of selecting only one value, the entire row corresponding to the position in r_K is selected [27]. The intermediate split estimation of the signal is calculated using (3.86), after the completion of splitting of y and φ_{aug}.

$$\hat{x}_{\text{int}}^{(i)} = \varphi_{\text{aug}\Gamma}^{(i)^T} \times y_{\Gamma}^{(i)} \tag{3.86}$$

Algorithm 3.15 Generation of support matrix and SPMT algorithm

Generation of Support Matrix Γ	SPMT Algorithm
1. Generate random values of size k $\quad r = rand(k)$ 2. Factorize k $\quad f = factor\,(k) : f_c > f_r$ $\quad \forall i \in [1 : f_l]$ $\quad\quad$ if $f_i < \max(f)$ $\quad\quad\quad f(i) \quad f_l$ $\quad\quad$ end 3. Form f_r and f_c $\quad f_r - prod\,(f(i))$ $\quad f_c = \max\,(f)$ 4. Form the support r_m $\quad r_K = [r]_{f_r \times f_c}$	1. Split y and ϕ_{aug} $\quad \forall b \in (1 : blocks), i \in (1, f_r), j \in (1, f_c)$ $\quad\quad \Gamma = r_K(i,j)$ $\quad\quad y_{\Gamma}^{(j)} = y\,(\Gamma)$ $\quad\quad \phi_{aug\Gamma}^{(j)} = \phi_{aug}\,(\Gamma)$ 2. Estimate the intermediate values using $y_{\Gamma}^{(j)}$ $\quad$ and $\phi_{aug\Gamma}^{(j)}$ $\quad\quad \forall i \in [1, f_r]$ $\quad\quad\quad \hat{x}_{\text{int}}^{(i)} = \phi_{aug\Gamma}^{(i)^T} \times y_{\Gamma}^{(i)}$ 3. Find the summation of estimated $\quad$ intermediate vectors $\quad\quad \forall j \in [1, f_c]$ $\quad\quad\quad \hat{x}_s = \sum_j \hat{x}_{\text{int}}^{(j)}$ 4. Find the extimated pixel values $\quad\quad \hat{x}_s = IDCT\,([\hat{x}_{\text{int}}])$

Equation (3.86) gives the intermediate results, which must be combined further. The inverse transform function must then be applied for estimating the original pixel values. Since the augmented matrix is used for sensing, the exact information or measurement values necessary for reconstruction are selected, which helps in perfect reconstruction even with least number of measurements.

3.10 Summary

This chapter discussed various recovery procedures which can be categorized into iterative and non-iterative procedures. Various iterative algorithms such as OMP, StOMP, CoSaMP, IHT, ROMP, FOCUSS, and MUSIC were discussed in detail with mathematical backbone. Furthermore, newly introduced non-iterative algorithms such as NITRA, R3A, and SPMT were clearly explained with application to image and video recovery. This chapter has given a simple insight into various recovery procedures available as of now. Many more application-oriented procedures are also evolving every day. The designer can design the recovery algorithms to his purpose and necessity at hand.

References

[1] Holger, R. (2011). "Compressive Sensing and Structured Random Matrices", *Radon Series Comp. Appl. Math*, XX, 1–94, June.

[2] Recht, B., Xu, W., and Hassibi, B. (2011). "Null space conditions and thresholdsfor rank minimization", *Math. Program., Ser. B.* 127, 175–202.

[3] Cohen, A., Dahmen, W., and Devore, R. (2009). "Compressed Sensing and Best K-Term Approximation", *J. Am. Math. Soc.* 22(1), 211–231, January.

[4] Candès, E. J., Eldar, Y. C., Needell, D., and Randall, P. (2011). "Compressed sensing with coherent and redundant dictionaries", *J. Appl. Comput. Harmonic Analy.* 31, 59–73.

[5] Eldar, Y. C., and Kutyniok, G. (2012). "Compressed Sensing: Theory and Applications", (Cambridge: Cambridge University Press).

[6] Tropp, J. A., and Gilbert, A. C. (2007). "Signal Recovery from Random Measurements via Orthogonal Matching Pursuit", *IEEE Trans. Inf. Theor.* 53(12), 4655–4666, December.

[7] Tony Cai, T., and Wang, L. (2011). "Orthogonal Matching Pursuit for Sparse Signal Recovery with Noise", *IEEE Trans. Inf. Theor.* 57(7), 4680–4688, July.

[8] Shewchuk, J. R. (1994). "An Introduction to the Conjugate Gradient Method Without the Agonizing Pain", School of Computer Science Carnegie Mellon University Pittsburgh, August.

[9] Lange, K., Hunter, D. R., and Yang, I. (2000). "Optimization Transfer algorithms using surrogate objective functions", *J. Comput. Graph. Stat.* 9, 1–59.

[10] Donoho, D. L., Tsaig, Y., Drori, I., and Starck, J.-L. (2012). "Sparse Solution of Underdetermined Systems of Linear Equations by Stagewise Orthogonal Matching Pursuit", *IEEE Trans. Inf. Theor.* 58(2), 1094–1120, February.

[11] Needell, D., and Vershynin, R. (2010). "Signal Recovery from Incomplete and Inaccurate Measurements via Regularized Orthogonal Matching Pursuit", *IEEE J. Select. Topic. Signal Process.* 4(2), 310–316, April.

[12] Needell, D., and Vershynin, R. (2008). "Uniform Uncertainty Principle and Signal Recovery via Regularized Orthogonal Matching Pursuit", *Found. Comput. Math.* 9, 317–334, June.

[13] Needell, D., and Tropp, J. A. (2010). "CoSaMP: iterative signal recovery from incomplete and inaccurate samples", *Commun. ACM*, 53(12), 93–100, December.

[14] Dai, W., and Milenkovic, O. (2009). "Subspace Pursuit for Compressive Sensing Signal Reconstruction", *IEEE Trans. Inf. Theor.* 55(5), 2230–2249, May.

[15] Blumensath, T., and Davies, M. E. (2009). "Iterative hard thresholding for compressed sensing", *J. Appl. Comput. Harmonic Anal.* 27, 265–274, May.

[16] Davies, M. E., and Blumensath, T. (2008). "Faster & Greedier: algorithms for sparse reconstruction of large datasets", *ISCCSP 2008, Malta*, pp. 774–779, 12–14 March.

[17] Gorodnitsky, I. F., and Rao, B. D. (1997). "Sparse Signal Reconstruction from Limited Data Using Focuss: A Re-Weighted Minimum Norm Algorithm", *IEEE Tran. Signal Process.* 45(3), 600–616, March.

[18] Rao, B. D. (1996). "Analysis and extensions of the FOCUSS algorithm", *Thirtieth Asilomar Conference on Signals, Systems and Computers*, 2, 1218–1223, 3–6 November.

[19] Fannjiang, A. C. (2011). "The MUSIC Algorithm for Sparse Objects: A Compressed Sensing Analysis", *Inverse Problems*, 27, 1–33.

[20] Baraniuk, R. G., Cevher, V., Duarte, M. F., and Hegde, C. (2010). "Model-Based Compressive Sensing", *IEEE Trans. Inf. Theor.* 56(4), 1982–2001, April.

[21] Yagle, A. E. "Non-Iterative Compressed Sensing Using a Minimal Number of Fourier Transform Values", Department of EECS, The University of Michigan, Ann Arbor, MI 48109-2122, pp: 1–5. http://web.eecs.umich.edu/~aey/sparse/fourier1.pdf

[22] Yagle, A. E. "A Non-Iterative Procedure for Computing Sparse and Sparsifiable Solutions to Slightly Underdetermined Linear Systems of Equations", Department of EECS, The University of Michigan, Ann Arbor, MI 48109-2122, pp: 1–4. http://web.eecs.umich.edu/~aey/sparse/sparse5.pdf

[23] Yagle, A. E. "Non-Iterative Computation of Sparsifiable Solutions to Underdetermined Kronecker Product Linear Systems of Equations", Department of EECS, The University of Michigan, Ann Arbor, MI 48109-2122, pp: 1–4. http://web.eecs.umich.edu/~aey/sparse/sparse6.pdf

[24] Florence Gnana Poovath, J., and Radha, S. (2015). "Non-Iterative Threshold based Recovery Algorithm (NITRA) for Compressively Sensed Images and Videos", *KSII Trans. Internet Inf. Syst.* 9(10), 4160–4176, October.

[25] Florence Gnana Poovathy, J., Deepika, B., and Radha, S. (2015). "Reduced Runtime Recovery Algorithm for Compressively Sensed Images" *Second International Conference on Next Generation Computing and Communication Technologies (ICNGCCT 2015)*, April 22–23, Dubai, UAE.

[26] Deepika, B., Florence Gnana Poovathy, J., Radha, S. (2015). "Reduced Runtime Recovery Algorithm based Stagewise Orthogonal Matching Pursuit", National Conference on Information and Communication Technology (NCICT 2k15), April 10, SSN College of Engineering, Kalavakkam, Chennai.

[27] Florence Gnana Poovathy, J., and Radha, S. (2015). "Non-Iterative Reconstruction Algorithm for Compressively Sensed Images and Videos using Split, Process and Merge Technique (SPMT)", *Maejo Intl. J. Sci. Technol.* (Communicated in December).

4

Compressive Sensing for Audio
and Speech Signals

4.1 Introduction

The principles of compressed sensing (CS) can be applied to sparse decompositions of speech and audio signals. CS offers a significant reduction in the computational complexity. The recorded speech signal is first transformed into a sparse domain using discrete cosine transform (DCT). Once the DCT of the recorded speech signal is found, a window function is to be designed. The function of the window function is to multiply all the components in that window by zero. The signal then becomes sparser, and the resulting signal is ready for CS. This process requires multiplication of random matrix of size K by N. Here, K signifies the level of sparsity and N is the total number of samples in the transformed and windowed function. Random linear projections are made on the sparse signal by using random matrix for taking very few components of the sensed signal. Thus, the signal to be recovered becomes robust for any errors (Figures 4.1 and 4.2).

4.1.1 Issues in Applying CS and Sparse Decompositions to Speech and Audio Signals

A. Dictionary

Complex sinusoids have been accounted for to function admirably for sparse decompositions for a wide range of applications. Their performance to model transient phenomena has been rather ineffective, even when these are modulated sinusoids and also stochastic signal components. The answer for these issues is in generally composite dictionaries (or unions of bases), where modulated sinusoids and even time-domain Kronecker delta functions are incorporated. Besides, voiced speech signals can be modeled not only as sparse in the frequency domain, but also as sparse in the residual domain

121

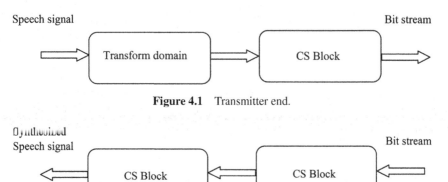

Figure 4.1   Transmitter end.

Figure 4.2   Receiver end.

after the application of linear prediction. This complicates matters somewhat for CS as the measurement matrix should be chosen to ensure its incoherence with the dictionary.

Fortunately, random measurement matrices can generally be expected to have a low coherence with both time-domain spikes and complex sinusoids. Furthermore, random measurement matrices that are universally applicable regardless of the type of dictionary can be constructed [1].

B. Sparsity

Another issue with speech and audio signals is the likelihood of the sparsity of such signals fluctuating substantially over time; a low piano note may contain numerous partials, while a glockenspiel might contain only a few. At one specific time example of a bit of music, a solitary instrument playing only a solitary note might be available, while at different times, numerous instruments playing various notes might be playing in the meantime. This implies it may be either (a) bound the sparsity of the signal by worst-case considerations or (b) somehow estimate the sparsity of the signal over time. The feasibility of the former approach can be illustrated by considering voiced speech. The fundamental frequency of speech signals can be between 60 and 400 Hz. This implies there can be utmost 66 harmonics for a signal sampled at 8,000 kHz. The latter approach would have to be computationally straightforward, or it would some way or defeat the motivation behind the CS. Another aspect which can be taken into consideration is that the sparser, a coefficient vector, can be easily obtained with a larger dictionary. Suppose a signal contains a single sinusoid having a frequency in the middle of the frequencies of two vectors in the dictionary, the contribution of that single sinusoid will then spread to several coefficients. The expected sparsity of the coefficient vector **c** and subsequently the number of samples K required for reconstruction is therefore not only a function of the signal, but also of the dictionary. It is in this way more probable that a single dictionary element will match the signal (or part thereof) if the dictionary is large [1].

C. Noise

When all stochastic signals' contributions cannot be modeled easily using a deterministic function, it must be a noise. Stochastic signal components are inherent and perceptually important parts of both speech and audio signals. Stochastic components occur in speech signals during periods of unvoiced speech or mixed excitation where both periodic and noise-like contributions are present. Similar observations can be made for audio signals. This means that, depending on the application, CS may not be entirely appropriate. On the other hand, if only the tonal parts of the signal are of interest, it may yet be useful. It should also be noted that the characteristics of the noise in speech and audio signals are time-varying, and it can often be observed to vary faster than tonal parts of audio signals. This again suggests that it might be hard to determine the required number of samples a priori and the expected reconstruction quality in LASSO-like reconstructions [1].

Following Figures 4.3 to 4.7 explain the step by step results for obtaining the compressively sensed signal y.

4.2 Multiple Sensors Audio Model

Its all inclusiveness and the way that next to no preparation is done on the sensor side and that an extraordinarily reduced number of estimations are required to make CS an appealing possibility for use in a sensor system where handling force and transmission data transmission are normally constrained. To this end, let us consider a sensing system of L sensors (microphones) around a sound source. Accepting the sound source is omnidirectional and that there arc no reflections, the signal got at the l-th sensor will simply be a deferred and scaled version of the original signal $z(t)$ at the source

$$x_l(t) = \alpha_l z(t - \tau_l) \tag{4.1}$$

where α_l and τ_l are the attenuation and delay at the l-th sensor, respectively. Such a model fits the second joint sparsity model (JSM-2) [2]. Figure 4.8

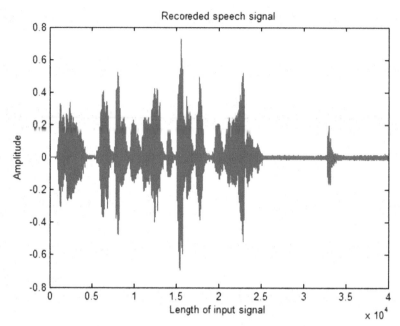

Figure 4.3   Step 1: Original input speech signal *x*.

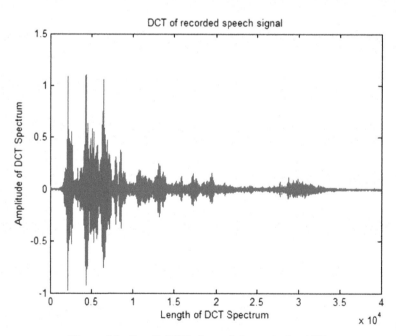

Figure 4.4   Step 2: DCT of recorded speech signal [1].

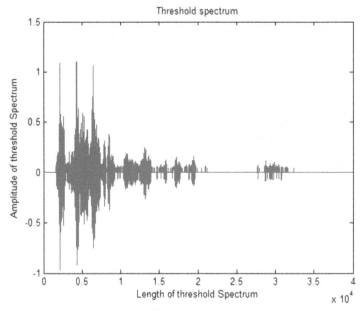

Figure 4.5 Step 3: Threshold spectrum [1].

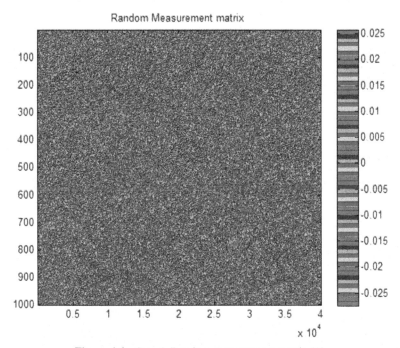

Figure 4.6 Step 4: Random measurement matrix [1].

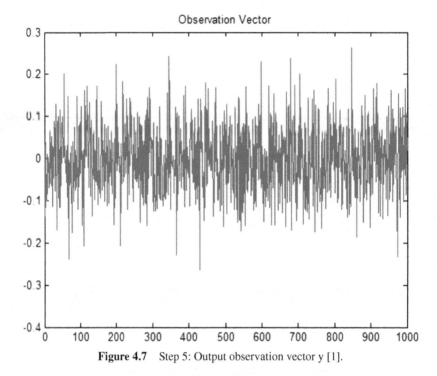

Figure 4.7　Step 5: Output observation vector y [1].

shows an example setup with four microphones. The delay at the l-th sensor is given by $\tau_l = d_l/c$, where d_l is the distance between the source and the l-th sensor, and c is the speed of sound (equal to 344 m/s at 21°C). Assuming a point source model, the attenuation at the l-th sensor is given by $\alpha_l = 1/4\pi d_l^2$ [3].

4.2.1　Reconstruction of Real, Non-Sparse Audio Signals

The performance of multi-sensor CS of audio signals has been discussed using the model mentioned above. The simultaneous orthogonal matching pursuit (SOMP) [4, 5] is used as recovery algorithm which is of course based on orthogonal matching pursuit (OMP) and seeks to recover $z(t)$ using M measurements of each of $x_l(t)$, $l = 1, 2, \ldots, L$. The idea is to exploit the common structure of the signals at each receiver, and much improved performance can be achieved for very sparse signals. There is no similar multi-sensor scheme based on BP, so the investigation is made using a simple scheme for reconstructing the L signals individually using BP and then recombine them. This scheme is called as MS-BP.

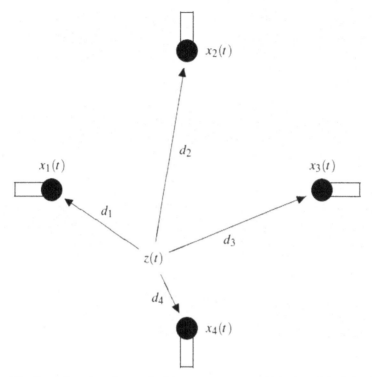

Figure 4.8 Example setup of an audio four-sensor system, $z(t)$ is the original signal at its source, $x_l(t)$ the signal received at the l-th sensor, and d_l the corresponding path length.

4.2.2 Detection and Estimation of Truly Sparse Audio Signals

Consider a sensor network is used for tracking the location of a subject wearing a tracking device that periodically transmits an audio signal. The sensor network would be arranged in some cell-like structure in an outdoor setting, or else, the cells could be rooms of a large building similar to the system considered. Each cell is modeled as in Section 4.2.1. The assumption made here is that each sensor has extremely constrained computational capability, and simply transmits its readings to a data fusion centre (DFC). If CS detection algorithms are to be used in the DFC for minimizing the number of required transmissions, the audio signal must be very sparse, and the system needs to use short pulses of a single frequency. This also ensures that the signals appearing at each sensor are jointly sparse, allowing the use of the incoherent detection and estimation algorithm (IDEA) in the DFC to detect the presence of a signal. This allows detection and performance of estimation significantly with fewer measurements. As IDEA was designed for a single sensor, a multi-sensor version of IDEA is to be developed referred to as MS-IDEA. This involved adapting SOMP rather than OMP. This takes into consideration of intra-signal compression without intra-sensor communication.

Detection in MS-IDEA is performed by testing the highest component after the first iteration of SOMP exceeds a given threshold. Thus, the detection process is very computationally efficient, requiring only one iteration of SOMP. Another efficient procedure called one-component case estimation in MS-IDEA involves the index of the component with the highest value after the first pass of SOMP. Considerably, selection of a large number of measurements is required by the estimation process than the detection process for obtaining a similar level of performance. One can thus envisage a scenario where each cell operates in two different modes. In a detection mode, a minimal number of measurements are transmitted to the DFC. Once a subject has been detected in a particular cell, the DFC can instruct the cell to switch to estimation mode, where more measurements are transmitted to the DFC, enabling estimation of the target.

4.3 Compressive Sensing Framework for Speech Signal Synthesis

If the signal is sparse on one basis, a sparse approximated signal is obtained with samples much less than required by the Nyquist sampling theorem. The CS framework is used for speech synthesis problems by its exciting potential application in signal compression. For an efficient speech compression, the linear prediction coding (LPC) can be used, as the speech is considered to be an AR process. A speech signal is known to be quasi-periodic in its voiced parts; hence, a discrete Fourier transform (DFT) basis will provide a better approximation. Thus, a hybrid dictionary combined with the LPC model and the DFT model as the basis of speech signal is discussed in the subsequent section. The orthogonal matching pursuit (OMP) is employed for computing the sparse representation in the hybrid dictionary domain. The results indicate good performance with the proposed scheme, offering a satisfactory perceptual quality [6].

4.3.1 DFT and LPC Transform Domain

A speech signal consists of two parts: voiced and unvoiced. In the voiced part, the signal demonstrates obvious quasi-period, while in the unvoiced part, the signal is nearly noise. First, the speech signal $x(n) \in R^N$ is given by

$$x(n) = x_v(n) + x_u(n) \qquad (4.2)$$

where $x_v(n)$ and $x_u(n)$ are the voiced part and unvoiced part, respectively. A transform domain is need for demonstrating its period with as few coefficients as possible for the voiced part. So the conventional sinusoid model is satisfactory for describing the voiced speech,

$$x_v(n) = \sum_{l=0}^{K-1} \cos(\omega_l n + \phi_n) \qquad (4.3)$$

When DFT is applied to a sinusoid signal, the frequency domain is very sparse with only two impulses. It proves that a voiced speech has $2K$ prominent coefficients in its frequency domain with relatively small values in other frequencies. Therefore, $x_v(n)$ is $2K$-sparse but not exact. According to the property mentioned above, $x_v(n)$ can be recovered as a $2K$-sparse approximation with an acceptable error. Define $\omega_k \triangleq 2\pi(k-1)/L$ and hence $\Psi_v(n, k) = e^{j\omega_k n}$ is a DFT basis matrix. The CS form representation is showed as

$$x_v = \Psi_v \theta_v \qquad (4.4)$$

The size of L is to be taken care of, which determines the resolution of the frequency domain. The length of the DFT is merely increased for smoothening the spectrum for DFT transform. In CS framework, increasing the size L makes sense, since the $2K$ largest coefficients are acquired instead of all information. As the signal is not periodic, the sinusoid model seems to be of no use for the unvoiced parts. Traditionally, such portions of a speech signal can be well handled using the LPC technique based on the following $AR(p)$ of $x_u(n)$,

$$e(n) = x_u(n) + a1x_u(n-1) + \ldots\ldots\ldots + a_p x_u(n-p) \qquad (4.5)$$

where $a \triangleq [1a_1 \ldots a_p]^T \in R^P$ are the LPC coefficients, while $e(n)$ is the excitation resembling the noise. It can be shown that (4.5) can be rewritten into the following vector matrix form:

$$x_u = \Psi_u e \qquad (4.6)$$

where $\Psi_u \in R^{N \times N}$ is commonly referred to as the synthesis matrix that maps the excitation to the original speech, directly from the unit impulse response $h(n)$ of the system using a transform function

$$H(Z) = \frac{1}{1 + a_1 z^{-1} + \ldots\ldots a_p z^{-p}} = \sum_{n=0}^{+\infty} h(n)z^{-n} \qquad (4.7)$$

In the manner below,

$$
\Psi_u \triangleq
\begin{bmatrix}
h(0) & 0 & \cdots & 0 & 0 \\
h(1) & h(0) & \cdots & 0 & 0 \\
h(2) & h(1) & \cdots & 0 & 0 \\
\cdot & \cdot & \cdot & & \\
\cdot & \cdot & \cdot & & \\
\cdot & \cdot & \cdot & & \\
h(N-1) & h(N-2) & \cdots & h(1) & h(0)
\end{bmatrix}
\tag{4.8}
$$

an important aspect that should be taken into consideration is that Ψ_u is a basis in CS aspect, so every column vector should be normalized, guaranteeing the working recovery algorithm. For the unvoiced parts, a larger error than that in the voiced speech can be predicted. However, this will not have a notable influence on the perceptual quality as they have much lower energy.

4.3.2 Hybrid Dictionary

A DFT basis is employed for approximating voiced parts and constructing an LPC basis for unvoiced parts. This scheme suggests the combination of the two bases into one hybrid dictionary in order to avoid the decision of the voice feature. It follows (4.2), (4.4), and (4.5) that

$$
x = \Psi_v \theta_v + \Psi_u e
\tag{4.9}
$$

Then, let θ_u be the sparse form of the e, it can be derived that

$$
x = \Psi_v \theta_v + \Psi_u \theta_u + r
\tag{4.10}
$$

where r is modeling error. Now, the problem is more like a CS form, with a definition of a coefficient vector for the whole speech $\theta \triangleq [\theta_v \, \theta_u]^T$ a hybrid dictionary $\Psi \triangleq [\Psi_v \, \Psi_u]$ and $x = \Psi\theta + r$. The signal can be recovered from its observations with an estimation error. In fact, the real solution will be better, reflecting adaptation. Specifically, as shown in (4.3), the voiced speech contains noise-like residue, which is thought to be the unvoiced part and will be approximated using the LPC basis, from the perspective of algorithm. Similar situations occur in the unvoiced speech. Generally speaking, both voiced and unvoiced speech signal frames will be regarded as linear combinations of vectors from a DFT as well as an LPC basis in this CS framework. More particularly, in transition between voiced and unvoiced speech, this dictionary will show its dynamic property to offer a better estimation than pure LPC or

DFT basis. Hence, there is no need to judge whether the frame is voiced or unvoiced; the algorithm will select an appropriate combination from the hybrid dictionary adaptively. The OMP algorithm is used, whose mechanism is to choose a column vector in the dictionary most correlated to the signal in each iteration as a new basis, and the iterative number is just the sparsity. Thus, for a fair comparison of the correlation among the entire dictionary, the basis in DFT and LPC should be normalized.

4.3.3 Level of Sparsity

Observed by a measurement matrix whose entries are in i.i.d. Gaussian distribution, a signal of length N with sparse K can be reconstructed with an overwhelming probability if M is in the order of $O(K\log(N))$ within the polynomial time as mentioned before. However, this scheme needs more observations since θ is not exactly sparse. Furthermore, the number of observations is mainly related to the sparsity, which clearly reveals that M can be set adaptively due to its sparsity, for instance, large K for unvoiced part while small for voiced.

4.3.4 Remarks

For both voiced and unvoiced part of the speech signal, the proposed dictionary will provide appropriate basis to approximate the signals. When the signal is recovered from the observations, regardless of where the basis belongs to, the algorithm will adaptively select the best basis. Thus, the residues in speech can be offered a reasonable approximation. The hybrid dictionary can indicate dynamic property for selecting the basis function. In detail, when the processed speech is voiced, more DFT basis will included, while for unvoiced, more LPC basis, bringing an turning away of judgment of voice feature.

4.4 CS Reconstruction of the Speech and Musical Signals

Two different types of audio signals such as musical and speech signals examined in terms of sparsity and performance are mentioned in this section. The comparison observations of the CS reconstruction using different numbers of signal samples are performed in the two domains of sparsity. The successful application of CS to both musical and speech signals is demonstrated, but the speech signals are more demanding in terms of the number of observations. For both type of signals, the DCT domain gives better reconstruction with a small number of observations compared to the Fourier transform domain. Speech signals are more complex in nature, and therefore, they are less sparse in the frequency domain, compared to the pure musical tones. As these signals have short-time stationary property, short-length frames should be considered for CS reconstruction. When observing a small time interval of the speech signal, it is found that it can be considered as sparse. In this way, successful reconstruction is assured using a smaller number of available observations [7].

The spectral content of a certain musical tone consists of several sinusoids located on certain frequencies. The component of the audio signal which is on the lowest frequency is called the basic tone (pitch). The others are called partial tones or harmonics, and they are multiples of basic tone frequency. Since sinusoidal signals satisfy the sparsity property, it can be concluded that the musical tones are convenient for the CS application. These sinusoidal components in the musical signals are called harmonics. The number of harmonics in musical signals certainly affects their sparsity, which consequently affects the number of measurements required for successful reconstruction. The second requirement is called incoherence, which depends on the samples acquisition procedure. The necessary condition for the measurement matrix is it should be incoherent with the transform matrix. The piano tones and vowels are used as a test signals in this section. In the piano tone, the sound is generated by vibration of strings. Pressing a key on the keyboard causes a padded hammer to hit the string and makes them vibrate. The signal is made by vocal chords in the case of speech, which vibrate as the air flows through them. Sounds with different frequencies can be produced by loosening or tensioning. It is not necessary that speech sound should consist of pitch sound and multiples of its frequency. Only it can also contain lots of harmonics with non-correlated frequencies.

Therefore, speech signal is considered to be less sparse in observed domains when compared to the musical tones. Two domains in which signals can be considered as sparse are: DCT and discrete Fourier transform (DFT). By using different types of transform basis, different results can be obtained for the same number of acquired measurements. Comparative analysis regarding these two domains is given in the following. The measurements of the signal are taken from the domain where signal is dense by having suitable sparsity domain. Thus, the measurements of the audio and speech signals are taken randomly from the time domain. The CS measurement matrix Φ is made from the transform domain matrix, by randomly choosing R_ψ rows of the basis matrix Ψ (DFT or DCT matrix).

$$\Psi = \text{DFT } V\Psi = \text{DCT} \tag{4.11}$$

Random selection is performed by random permutation of the vector q, which contains positions from 1 to N. By taking first R_ψ coefficients to form the vector q, it can be defined R_ψ random rows to be selected from the matrix Ψ. Mathematically, this can be described as follows:

$$q = \text{randperm}(N) \tag{4.12}$$
$$B = \text{inv}(Y) \tag{4.13}$$
$$\Phi = B(q(1:R_\psi),:) \tag{4.14}$$

Now, the vector of measurements is given by

$$\Phi = Wx = B(q(1:R_\psi),:)x \tag{4.15}$$

The previous relation can be solved by using $\ell 1$-norm minimization.

4.4.1 Recovery of Audio Signals with Compressed Sensing

CS describes a system in which sampling and compression of a digital signal can be done simultaneously. This means that, when dealing with an analog signal such as audio, the signal must first pass through an ADC and then be sampled. This simplifies that the system has to capture the entire signal before it can be compressed, negating a large portion of the benefit of using compressed sensing. Fortunately, there has been some research done on methods to directly compress the analog signal and create samples which can be recovered using standard CS techniques. One of these systems is described below.

The system described by Kirolos et al. uses a three-step process for generating the samples. The analog signal is first spread through multiplication with a PN sequence which must alternate faster than the Nyquist rate of the signal. The "demodulated" signal is then passed through a low-pass filter and sampled at a rate R based on the sparsity. This signal $y|m|$ can be characterized as a linear transformation of the CS samples (α). For this system, the sampling matrix is composed of two operators: Ψ which maps y to α and Φ which maps the original analog signal x to the discrete signal y. The sampling matrix is described by the following equation:

$$V_{m,n} = \int_{-\infty}^{+\infty} \psi_n(T)p_c(T)h(mM - T)\mathrm{d}T \tag{4.16}$$

where $V_{m,n} \in V$ is the element in the mth row and n^{th} column of the sampling matrix, ψ is the sparsity inducing basis, p_c is the PN sequence, and h is the low-pass filter. Based on this filtering, the original signal x can be recovered by solving the standard ℓ_1 minimization problem using V as the sampling matrix.

4.5 Noise Reduction in Speech and Audio Signals

A compressive sensing method for noise reduction in speech and audio signals [8] is discussed in this section. The noise reduction problem is formulated in the theoretical framework of CS, as a ℓ_1 minimization problem with a linear combination of a constrained term, by adopting a random partial Fourier transform operator. Further, a gradient descend line search (GDLS) algorithm [9] is adopted to solve the optimization problem efficiently.

4.5.1 Data Sparsity of Speech Signals

CS has been successfully applied to medical image processing and also in the domain of speech and audio processing. The CS-based method can be applied to deal with noise reduction for speech and audio signals. Furthermore, one of the most important motivations behind this section is the following argument [8].

Argument 1:
Speech and audio signals are k-sparse, while noise factors are not, and there-fore, these two components are theoretically separable by the method of CS.

The success of the CS theory critically relies on data sparsity of processed signals. For speech and audio signals, data sparsity cannot be attained in the time domain, whereas it is likely to be achieved in some transform domains. Wavelet transform is used for realizing the sparsity of speech signals. For a given number of vanishing moments, the Daubechies wavelets are used in view of their minimal phases and minimal supports. A moderate-scale level of 4 and vanishing moments of 3 are chosen. The ability of wavelet to accurately represent a signal is limited by the number of vanishing moments. In theory, more vanishing moments mean that scaling functions can represent more complex signals accurately. It is also called the accuracy of the wavelet.

4.5.2 Formulation of the Optimization Problem for Speech Noise Reduction

According to the theory of CS, the problem of speech noise reduction in the wavelet transform domain can be formulated as a ℓ_1-norm optimization problem.

Problem 1: Let $\hat{x} = \{\hat{x}(t) : t \in [1, T]\}$ be a noisy speech signal, $\hat{x} = x + e$ where $x = \{x(t)\}$ is the clean speech and $e = \{e(t)\}$ is the noise. Also, let y be the measurement vector; W, a wavelet operator; and F_u, a random Fourier operator. Then, an estimate to the clean speech x can be obtained, i.e., x^*, by solving the following constrained ℓ_1-norm optimization problem,

$$\min \|W.x\|_1 \tag{4.17}$$

Subject to $\|F_u.x - y\|_2 \leq \|F_u.e\|_2 \leq \|F.e\|_2 = \|E\|_2$ where F is the complete Fourier operator and E is the Fourier transform coefficient matrix. The justification of Problem 1 critically relies on Argument 1. Conversely, Argument 1 is proved if the solution to Problem 1 is shown to reduce signal noise effectively.

4.5.3 Solutions to the Optimization Problem

In order to solve Problem 1, it can be converted to a non-constrained form as follows: Problem 2: The estimate of the clean speech x^* can be obtained by solving the following problem:

$$x^* = \arg\min_x \lambda \left[\|F_u.x - y\|_2 + \frac{2\pi}{\gamma} \cdot \|x\|_2 \right] + (1 - \lambda) \|W.x\|_1 \tag{4.18}$$

where λ is a parameter for balancing the contribution of two terms. Problem 2 is a quadratic optimization problem; thus, it can be solved through use of several standard methods, such as GDLS and an iteratively re-weighted least-squares algorithm (IRLS). Here, the GDLS algorithm is adopted to solve Problem 2, which was successfully applied in the case of MRI [9].

4.6 DCT Compressive Sampling of Frequency-Sparse Audio Signals

Using spectral analysis and the properties of DCT, audio signals can be treated as sparse signals in the frequency domain. This is especially true for sounds representing tones. On the other hand, CS has been traditionally used for acquiring and compressing certain sparse images. This section describes the use of DCT and CS for obtaining an efficient representation of audio signals, especially when they are sparse in the frequency domain. In order to obtain a sparse representation in the frequency domain, DCT can be used as signal preprocessor. Then, the subsequent application of CS represents the signals with less information than the well-known sampling theorem [10].

The important properties of DCT are as follows:

1. Decorrelation: The main advantage of signal transformation is the removal of redundancy between neighboring values. This leads to uncorrelated transform coefficients which can be encoded independently.
2. Energy compaction: The efficacy of a transformation scheme can be directly gauged by its ability to pack input data into as few coefficients as possible. This allows the quantizer to discard coefficients with relatively small amplitudes without introducing visual distortion in the reconstructed image. DCT exhibits excellent energy compaction for highly correlated signals.

Compressive sampling needs to deal with speech signals which are only approximately sparse. The issue here is to obtain an accurate reconstruction of such signals from highly under sampled measurements. As a first instance, FFT is applied for obtaining the frequency domain representation. However, this representation has real and complex parts, which result in a difficult reconstruction due to the phase angle changes with the matrix transformations on the compressive sampling program. So the original signal cannot be reconstructed by just applying the inverse FFT. An efficient joint implementation of DCT, a method for obtaining a sparse audio signal representation, and the application of the compressive sampling algorithm to this sparse signal have been tested. The DCT speech signal representation has the ability to pack input data into few coefficients as possible. This allows the quantizer to discard coefficients with relatively small amplitudes without introducing audio distortion in the reconstructed signal. Despite the use of compressive sampling technique primarily for images, reasonable results are achieved due to the preprocessing of the audio signal. This technique can achieve a significant reduction in the number of samples required to represent certain audio signals and therefore a decrease in the required number of bytes for encoding.

4.6.1 Performance of Compressive Sensing for Speech Signal with Combined Basis

In CS framework, reconstruction of a signal relies on the knowledge of measurement matrix and sparse basis used for sensing. Most of the studies so far focus on the application of CS in fields of image processing, speech, astronomy, and radar. This section introduces a new approach called combined basis that is made by separating voiced and unvoiced parts and applying different basis for both parts from given speech and showing detailed comparison of them with LPC basis and orthogonal Gaussian matrix applied on 8 KHz sampled speech signal. Also, it shows improved results of combined DCT and LPC basis compared to LPC and combined DFT and LPC basis. A new idea on speech compression based on compressive sensing by considering a new approach of taking combined basis and its performance is compared with LPC basis by quality assessment parameters such as perceptual evaluation of speech quality (PESQ), signal-to-noise ratio (SNR), and mean-square error (MSE).

4.7 Single-Channel and Multi-Channel Sinusoidal Audio Coding Using CS

CS seeks to represent a signal with a number of linear, non-adaptive measurements. Usually, if the signal is sampled at the Nyquist rate, the number of measurements is much lower than the number of samples needed. CS requires the signal to be *sparse* in some basis in the sense that it is a linear combination of a small number of basis functions in order to correctly reconstruct the original signal. If the sinusoidally modeled part of an audio signal is a sparse signal, it is thus natural to wonder how CS might be used to encode such a signal. The sinusoidal part of an audio signal provides sufficient quality when CS is applied to audio coding, like low-bit rate audio applications. It is shown here for multi-channel audio signals that, except for one primary (reference) audio channel, a simple low-complexity system can be used to encode the sinusoidal model for all remaining channels of the multi-channel recording [11].

4.7.1 Sinusoidal Model

The sinusoidal model was initially used in the analysis/synthesis of speech. A short-time segment of an audio signal $s(n)$ is represented as the sum of a small number of sinusoids with time-varying amplitudes and frequencies. This can be written as

$$s(n) = \sum_{k=1}^{K} \alpha_k \cos(2\pi f_k n + \theta_k) \tag{4.19}$$

where α_k, f_k, and θ_k are the amplitude, frequency, and phase, respectively. To estimate the parameters of the model, one needs to segment the signal into a number of short-time frames and compute a short-time frequency representation for each frame. Consequently, the prominent spectral peaks are identified using a peak detection algorithm (possibly enhanced by perceptual based criteria). Interpolation methods can be used for the enhancement of the accuracy level of the algorithm [12]. Each peak in the lth frame is represented as a triad of the form $\{\alpha_{l,l_0}, f_{l,l_0}, \theta_{l,l_0}\}$ (amplitude, frequency, phase), corresponding to the kth sine wave. A peak continuation algorithm is usually employed for assigning each peak to a frequency trajectory by matching the peaks of the previous frame to the current frame, using linear, amplitude, and cubic phase interpolation. A more accurate representation of audio signals is achieved when a stochastic component is included in the model. This model is usually called the sinusoids plus noise model, or deterministic plus stochastic decomposition. In this model, the sinusoidal part corresponds to the "deterministic" part of the signal due to its structured nature. The remaining signal is the sinusoidal noise component $e(n)$, also referred to here as residual or sinusoidal error signal, which is the "stochastic" part of the audio signal, since it is very difficult to accurately model, but at the same time essential for high-quality audio synthesis. Practically, after the sinusoidal parameters are estimated, the noise component is computed by subtracting the sinusoidal component from the original signal. In this section, encoding the sinusoidal part is discussed.

4.7.2 Single-Channel Sinusoidal Selection

To perform single-channel sinusoidal analysis, the state-of-the-art psychoacoustic analysis based on [13] is discussed. In the ith iteration, the algorithm picks a perceptually optimal sinusoidal component frequency, amplitude, and phase. This choice minimizes the perceptual distortion measure

$$D_i = \int A_i(\omega)\,|R_i(\omega)|^2\,d\omega \qquad (4.20)$$

where $R_i(\omega)$ is the Fourier transform of the residual signal (original frame minus the currently selected sinusoids) after the ith iteration and $A_i(\omega)$ is a frequency weighting function set as the inverse of the current masking threshold energy. One issue with CS encoding is that no further refinement of the sinusoid frequencies can be performed in the encoder, because frequencies which do not correspond to exact frequency bins would result in loss of the sparsity in the frequency domain. This is an important problem, as the imperative need to restrict the sinusoidal frequency estimation to the selection of frequency bins (e.g., following a peak-picking procedure) without the possibility of further refinement of the estimated frequencies in the encoder. This can be reduced by zero-padding the signal frame; in other words, by reducing the bin spacing, the frequency resolution during the parameter estimation can be improved. This can be performed to a limited degree for CS-based encoding as zero-padding increases the number of measurements requiring encoding (and consequently the bit rate). Fortunately, this problem can be partly addressed by employing the "frequency mapping" procedure. Furthermore, the sparsity restriction is not needed to hold after the signal is decoded, and frequency re-estimation can be performed in the decoder, such as interpolation among frames.

4.7.3 Multi-Channel Sinusoidal Selection

In order to perform multi-channel sinusoidal analysis, extension is done in the sinusoidal modeling method which employs a matching pursuit algorithm to determine the model parameters of each frame to include the psychoacoustic analysis of [13]. For the multi-channel case, in each iteration, the algorithm picks a sinusoidal component frequency that is optimal for all channels, as well as channel-specific amplitudes and phases. This choice minimizes the perceptual distortion measure

$$D_i = \sum_c \int A_{i,c}(\omega) \, |R_{i,c}(\omega)|^2 \, d\omega \qquad (4.21)$$

where $R_{i,c}(\omega)$ is the Fourier transform of the residual signal of the cth channel after the ith iteration and is a frequency weighting function set as the inverse of the current masking threshold energy. The contributions of each channel are simply summed to obtain the final measure.

An important question is which masking model will be suitable for multi-channel audio where the different channels have different binaural attributes in the reproduction. In transform coding, a common problem arises due to binaural masking-level difference (BMLD); sometimes, quantization noise that is masked in monaural reproduction is detectable due to binaural release, and using a separate masking analysis for different channels is not suitable for loudspeaker rendering. However, this effect in parametric coding is not well established. Preliminary experiments have been performed using the following: (1) separate masking analysis, i.e., individual $A_{i,c}(\omega)$ based on the masker of channel c for each signal separately, (2) the masker of the sum signal of all channel signals to obtain for all c, and (3) power summation of the other signals' attenuated maskers to the masker of channel c according to

$$A_{i,c}(\omega) = \frac{1}{\left(M_{i,c}(\omega) + \sum_{\substack{k \\ k \neq c}} \omega_k M_{i,k}(\omega)\right)} \tag{4.22}$$

In the above equation, $M(\omega)$ indicates the masker energy, ω_k indicates the estimated attenuation (panning) factor that was varied heuristically, and k iterates through all channel signals excluding c. In this section, the first method is chosen, i.e., separate masking analysis for channels $\omega_k = 0$, for the reason that it cannot find notable differences in BMLD noise unmasking, and that the sound quality seemed to be marginally better with headphone reproduction. The second or third method may be more suitable for loudspeaker reproduction. The use of this psychoacoustic multi-channel sinusoidal model results in sparser modeled signals, increasing the effectiveness of our compressed sensing encoding.

4.8 Compressive Sensing for Speech Signal with Orthogonal Symmetric Toeplitz Matrix

In CS framework, reconstruction of a signal relies on the knowledge of the sparse basis and measurement matrix used for sensing. Most of the studies so far focus on the application of CS in fields of images, radar, astronomy, and speech. A new sensing matrix called orthogonal symmetric Toeplitz matrix (OSTM) generated with binary, ternary, and PN sequence shows a detailed comparison of these with DCT basis applied on 8 KHz sampled speech signal. It shows improved results of OSTM compared to Bernoulli, random, Fourier, and Hadamard matrices [14].

4.8.1 Orthogonal Symmetric Toeplitz Matrices (OSTM)

These are easy to generate as only N numbers need to be stored. Successful implementation of both sampling and reconstruction is a notable feature. This matrix is generated using three sequences binary, ternary, and PN sequence and specially designed for speech signal.

Binary sequence
A binary number is a number expressed in the binary numeral system, or base-2 numeral system, representing numeric values using two digits: 0 and 1.

Ternary sequence

Ternary (sometimes called trinary) is the base-3 numeral system that represents numeric values using the three digits: 0, 1, and 2.

PN sequence

Pseudorandom binary sequences can be generated using "linear feedback shift registers."

The procedure for generating orthogonal symmetric Toeplitz matrix using binary, ternary, and PN sequence:

An $M \times N$ sensing matrix based on OSTM can be constructed like this:

1. Use a given sequence **s** of length N_l, $\sigma = [S1, S2 \ldots \ldots S_N]$.
2. And apply inverse FFT (IFFT) to the sequence to obtain g with length N_l.

$$g = ifft(\sigma) \tag{4.23a}$$

3. Let the elements of g be the first row of OSTM, and follow the circulant property to construct the $N_l \times N_l$ matrix Φ.
4. Choose M rows and normalize it by multiplying $\sqrt{N/M}$ to form the $M \times N$ sensing matrix Φ. After the second step, the orthogonal and Toeplitz nature of $N \times N$ matrix Φ can be proved.

4.9 Sparse Representations for Speech Recognition

Sparse representation (SR) techniques for machine learning applications have become popular in recent years. Since it is not obvious how to represent speech as a sparse signal, sparse representations have received attention only recently from the speech community, where they were proposed originally as a way to enforce exemplar-based representations. As an alternative way of modeling observed data, exemplar-based approaches have also found a place in modern speech recognition. Recent advances in computing power and improvements in machine learning algorithms have made such techniques successful on increasingly complex speech tasks. The goal of exemplar-based modeling is to establish a generalization from the set of observed data accurate inference about the data yet to be observed the "unseen" data. This approach selects a subset of exemplars from the training data to build a local model for every test sample, in contrast to the standard approach, which uses all available training data for building a model before the test sample is seen. Exemplar-based methods, including sparse representation, support vector machines (SVMs), and k-nearest neighbors (kNN) utilize the details of actual training examples when making a classification decision. Since the number of training examples in speech tasks can be very large, such methods commonly use a small number of training examples to characterize a test vector, that is, a *sparse representation*. This approach stands in contrast to such standard regression methods as nearest line techniques, nearest subspace, and ridge regression which utilize information about *all* training examples when characterizing a test vector.

SR classifier can be defined as follows. A dictionary $D = [d_1; d_2 \ldots; d_N]$ is constructed using individual examples of training data, where each $d_i \in Rem$ is a feature vector belonging to a specific class. D is an over-complete dictionary, in that the number of examples n is much greater than the dimension of each d_i (that is, $m \ll N$). To reconstruct a signal y from D, SR requires that equation $y \approx D\beta$, but imposes a sparseness condition on β, meaning that it requires only small number of examples from D to describe y. A classification decision can be made by looking at the values of β coefficients for columns in D belonging to the same class. The objective of this chapter is to explain how sparse representation can be constructed for classification and recognition tasks, and how sparse optimization methods can be exploited in speech and to give an overview of results obtained using sparse representation.

4.9.1 An EBW Compressed Sensing Algorithm

The extended Baum–Welch (EBW) technique was introduced initially for estimating the discrete probability parameters of multinomial distribution functions of HMM speech recognition problems under the maximum mutual information discriminative objective function. Later, EBW was extended to estimating parameters of Gaussian mixture models (GMMs) of HMMs under the MMI discriminative function for speech recognition problems. EBW technique was generalized to the novel line search A-functions (LSAF) optimization technique. A simple geometric proof was provided to show that LSAF recursions result in a growth transformation. The discrete version of EBW invented more than 24 years ago and can be also represented using A-functions. This connection allowed a convergence proof for a discrete EBW to be developed.

4.9.2 Line Search A-Functions

Let $f(x) : U \subset R^n \rightarrow R$ be a real-valued differentiable function in an open subset U. Let $\mathbf{A}_f = \mathbf{A}_f(x, y) : R^n \times R^n \rightarrow R$ be twice differentiable in $x \in U$

for each $y \in U$, where $\mathbf{A}_f$ is an A-function for f if the following properties hold.

1. $\mathbf{A}_f(x, y)$ is a strictly convex or strictly concave function of x for any $y \in U$. (Recall that twice differentiable function is strictly concave or convex over some domain if its Hessian function is positive or negative definite in the domain, respectively.)
2. Hyper planes tangent to manifolds defined by $z = g_y(x) = \mathbf{A}_f(x, y)$ and $z = f(x)$ at any $x = y \in U$ are parallel to each other, that is,

$$\nabla_x A_f(x, y)|_{x=y} = \nabla_x f(x) \qquad (4.23b)$$

A general optimization technique can be constructed on the basis of A-function. A growth transformation is formulated such that the next step in the parameter update that increases $f(x)$ is obtained as a linear combination of the current parameter values and the value $\tilde{x}$ that optimizes the A-function, for which $\nabla_x \mathbf{A}_f(x, y)|_{x=\tilde{x}} = 0$. More precisely, A-function gives a set of iterative update rules with the following "growth" property: let x_0 be some point in U and $U \in \tilde{x}_0 \neq x_0$ be a solution of $\nabla_x A(x, x_0)|_{x=\tilde{x}} = 0$. Defining

$$x_1 = x(\alpha) = \alpha \tilde{x}_0 + (1 - \alpha)x_0 \qquad (4.24)$$

for sufficiently small $|\alpha| \neq 0$ that $f(x(\alpha)) > f(x_0)$, where $\alpha > 0$ if $A(x, x_0)$ concave and $\alpha < 0$ if $A(x, x_0)$ convex. The technique of generating $\tilde{x}$ this way and performing the line search is termed "line search A-function" (LSAF).

4.9.3 An Analysis of Sparseness and Regularization in Exemplar-based Methods for Speech Classification

A. Classification based on exemplars

The goal of classification is to use training data from k different classes for determining the best class to assign to test vector y. First, let us consider taking all training examples n_i from class i and concatenate them into a matrix H_i as columns, in other words $H_i = [x_{i,1}, x_{i,2}, \ldots, x_{i,ni}] \in R^{m \times n}{}_i$, where $x \in R^m$ represents a feature vector from the training set of class i with dimension m. Sufficient number of training examples given from class i shows that a test sample y from the same class can be represented as a linear combination of the entries in H_i weighted by β, that is:

$$y = \beta_{i,1} x_{i,1} + \beta_{i,2} x_{i,2} + \ldots + \beta_i, n_i x_i, n_i \qquad (4.25)$$

However, since the class membership of y is unknown, a matrix H is defined to include training examples from all k classes in the training set. In other words, the columns of H are defined as $H = [H_1, H_2, \ldots, H_k] = [x_{1,1}, x_{1,2}, \ldots, x_{k,n_k}] \in \mathrm{R}^{m \times N}$. N is the total number of all training examples from all classes. The test vector y can be written as a linear combination of all training examples; in other words, $y = H\beta$. This linear system can be solved for β and use information about β for making a classification decision. Specifically, large entries of β should correspond to the entries in H with the same class as y. Thus, one proposed classification decision approach is to compute the l_2 norm for all β entries within a specific class, and choose the class with the largest l_2-norm support.

B. Description of TIMIT

The behavior of various exemplar-based methods on the TIMIT [15] corpus is analyzed. The corpus contains over 6,300 phonetically rich utterances divided into three sets, namely the training, development, and core test set. For testing purposes, the standard practice is to collapse the 48 trained labels into a smaller set of 39 labels. All methods are tuned on the development set, and all experiments are reported on the core test set. The complete experimental setup, as well as the features used for classification, is similar to [16]. First, each frame in the signal is represented by a 40-dimensional discriminatively trained space-boosted maximum mutual information (fBMMI) feature. Each phonetic segment is split into thirds, taking the average of these frame-level features around 3^{rds} and splice them together to form a 120-dimensional vector. This allows us to capture time dynamics into each segment. At each segment, segmental feature vectors to the left and the right are joined together and a linear discriminative analysis (LDA) transform is applied to project 200-dimensional feature vectors down to 40 dimensions. Similar to [16], a neighborhood of closest points to y is found in the training set using a kd-tree. These k neighbors become the entries of H.

4.10 Speaker Identification Using Sparsely Excited Speech Signals and Compressed Sensing

Speaker identification is the task of determining an unknown speaker's identity. In this section, the text-independent speaker identification performed based solely on a speaker's voice is described. Speaker identification is achieved by performing a one-to-many match among the unknown voice signal and the previously available speech database of multiple speakers, assuming that the unknown speaker belongs in this data set. The focus is on the possibility of performing speaker identification by applying the recently proposed compressed sensing theory. The compressed sensing theory for sparsely excited speech signals is applied to the specific problem of speaker identification and is found to provide encouraging results using a number of measurements as low as half of the signal samples. In this manner, compressed sensing theory allows the use of fewer samples for achieving accuracy in identification, which in turn would be beneficial in several sensor network-related applications. Additionally, enforcing sparsity on the excitation signal is shown to provide identification accuracy which is more robust to noise than using the noisy signal samples.

The reasons for examining the applicability of CS theory to the speaker identification problem are twofold. First, the CS theory achieves the reconstruction of a sparse signal using only a fraction of the number of samples dictated by the Nyquist theorem. Therefore, in a sensor network scenario, the measurement operation could be performed locally and the few measurements in each time frame could be transmitted to a base station for further processing. From a different point of view, the second reason is aforementioned sparsity restriction: By forcing the signal to be sparse in some basis, a noisy signal may be more robustly reconstructed. This is similar to signal denoising by low-rank modeling. In this case, the signal sparsity is an important factor, since the CS reconstruction will only be valid for signals which are initially sparse in some domain. Thus, in this second approach, testing of compressed sensing-based speaker identification results in a more robust identification than when directly using the signal's samples to perform the identification. A key question is whether a speech signal can be considered to be sparse in some sense. For audio signals, the sinusoidally modeled component can be considered to be sparse, and CS theory applied to low bit rate audio coding. For speech signals, compressed sensing was recently applied to a sparse representation using the source/filter model for speech coding, and encouraging preliminary results were obtained.

A filter codebook is created with the help of each of the speakers in the database, and the identification process is based on the selection of the speaker in the database corresponding to the codebook which results in the best CS reconstruction. It is shown that the percentage of correct identification using CS theory can reach 80% on average using a number of measurements which are as low as half of the signal's samples. When additive noise is used, the performance of CS-based identification is shown to be quite robust, with reference to a baseline GMM-based approach for this task.

4.10.1 Sparsely Excited Speech

The speech model based on the Nyquist sampled speech sample sequence $x(n)$ being represented by the convolution relation

$$x[n] = h[n] \times r[n] \tag{4.26}$$

where $h[n]$ is the signal domain impulse response of the smooth spectral envelope (which in this discussion is represented using the linear prediction coefficients (LPC)), and $r[n]$ is the residual excitation component. The convolution relation can be expressed in frame by frame matrix form as

$$\mathbf{x} = \mathbf{h}\,\mathbf{E_v}, \tag{4.27}$$

where $\mathbf{h}$ is an $N \times N$ impulse response matrix and $\mathbf{E}_v$ is an $N \times 1$ excitation vector. Linear convolution is considered, and thus, $\mathbf{h}$ is Toeplitz lower triangular. The residual excitation vector is not truly sparse, as for real speech, all of the elements of $\mathbf{E}_v$ will be nonzero. However, the past work showed that $\mathbf{E}_v$ is indeed highly compressible, and thus, (4.27) is a suitable representation of speech for use with compressed sensing. Unfortunately, the basis matrix $\mathbf{h}$ is signal-dependent, and the authors solve this problem by constructing a codebook of L basis matrices from training speech data. Given that $\mathbf{h}$ is formed by the LPC coefficients of the speech signal, L is in fact the codebook size formed using the LPC vectors, represented as line spectral frequencies (LSFs).

4.10.2 GMM Speaker Identification

As a baseline, the implementation of the speaker identification system of [17] is a simple but powerful system that has been shown to successfully perform this task. This is a GMM (Gaussian mixture model)-based system, where for each one of the speakers in the database, a corpus is used to train a GMM of the extracted sequences of (short-time) spectral envelopes. Thus, for a predefined set of speakers, a sufficient volume of training data is assumed to be available, and identification is performed based on segmental-level information only. During the identification stage, the spectral vectors of the examined speech waveform are extracted and classified under one of the speakers in the database, according to a maximum *a posteriori* criterion. More specifically, a group of S speakers in the training data set is represented by S different GMM's $\lambda_1, \lambda_2, \ldots, \lambda_S$, and a sequence (or segment) of n consecutive spectral vectors $X = [\mathbf{x_1 x_2} \cdots \mathbf{x_n}]$ is identified as spoken by speaker $\hat{s}$ based on the following:

$$\hat{s} = \arg \max_{1 \leq q \leq S} p(\lambda_q / X) = \arg \max_{1 \leq q \leq S} \frac{p(X/\lambda_q)p(\lambda_q)}{p(X)} \tag{4.28}$$

For equally likely speakers and since $p(X)$ is the same for all speaker models, the above equation becomes

$$\hat{s} = \arg \max_{1 \leq q \leq S} p(\lambda_q / X) \tag{4.29a}$$

and finally, for independent observations and using logarithms, the identification criterion becomes

$$\hat{s} = \arg \max_{1 \leq q \leq S} \sum_{k=1}^{n} \log p(x_k / \lambda_q) \tag{4.29b}$$

where

$$p(x_k / \lambda_q) = \sum_{i=1}^{M} p_q(\omega_i) N(x_k; \mu_{i,q}, \Sigma_{i,q}) \tag{4.30}$$

where $G_D(x; \mu_i, \Sigma_i)$ denotes a Gaussian density with mean μ_i and covariance Σ_i. This is a text-independent system; *i.e.*, the sentences during the validation stage need not be the same as the ones used for training. The error measure employed is the percentage of segments of the speech recording that identified as spoken by the most likely speaker. A segment in this case is defined as a time interval of pre-specified duration containing n spectral vectors, during which these vectors are collectively classified based on Equation (4.30), to one of the speakers by the identification system. If each segment contains n vectors (n depending on the pre-specified duration of each segment), different segments overlap as shown below, where Segment #1 and Segment #2 are depicted as follows:

$$\overbrace{x_1, x_2, ..., x_n}^{segment\#1}, x_{n+1}, x_{n+2}, ... \qquad \overbrace{x_1, x_2, x_3, ..., x_n}^{segment\#2}, x_{n+1}, x_{n+2}, x_{n+3}, ...$$

The resulting percentages constitute an intuitive measure of the performance of the system. There is a performance decrease when decreasing the segment duration, which is an expected result since the more data available, the better the performance of the system. A large number of segments are also important for obtaining accurate results; it should be noted, though, that an identification decision is made for each different segment, independently of the other segments.

4.10.3 Speaker Identification Using CS

A speaker identification system using CS is proposed by forming a codebook of basis matrices from speech training data for each of the S speakers. This is essentially formed by performing a codebook of the LSF vectors separately for each speaker. In fact, this process is similar to the GMM training for speaker identification and is based on the assumption that LSFs are suitable feature vectors for the classification task.

A simple way to do classification using compressed sensing is to find a basis for each of the C classes of interest and then reconstruct a sparse vector from each of the class bases. The measured signal is then said to come from the class that produced the sparsest recovered vector. This can work well, but requires incoherence of the class bases. In this case, the class bases would be the $\mathbf{h}l$'s for each speaker. Unfortunately, these bases are far from incoherent.

Another method to perform speaker identification:

1. Find a residual excitation vector for each basis matrix from each speaker's codebook using $\hat{r}_{s,l} = \arg\min_{r} \|y - \phi h_{s,l} r\|_2 \quad s.t \; \|r\|_0 = K$.

2. Once these have been found, calculate $d_s = \min_{l} \|y - \phi h_{s,l} \hat{r}_{s,l}\|_2$ which represents the minimum distance between the measurements $\mathbf{y}$ and measurements from the reconstructions from s^{th} speaker's codebook.

3. Let $d_{i,s}$ be the d_s calculated for the i-th frame. The actual speaker s^* in the i-th frame should have the smallest distance, so that

$$d_{i,s^*} < d_{i,s}, \quad \forall_s \neq s^*$$

4. If (3) is true, the correct speaker is chosen, and if not, there is an error.

In practice, the reliability of speaker identification is improved by considering n frames at a time (*i.e.*, a segment as defined in Section 4.10.2). This is based on the fact that the speaker will not change from frame to frame and will rather be constant for a group of frames. Thus, a sliding window is used for determining the most probable speaker as

$$\hat{s} = \arg\min_{s} \sum_{j=i-(n-1)}^{i} d_{i,s} \tag{4.31}$$

Obviously, if $\hat{s} \neq s^*$, then the identification has failed for this particular segment.

4.11 Joint Speech-Encoding Technology Based on Compressed Sensing

At present, audio coders based on sparse decompositions have already been providing excellent compression and used for the MPEG audio coding standards for a long time. However, researchers still hope for a better performance speech-encoding technology.

The CS theory shows that a signal having a sparse or compressible representation in one basis can be recovered from projections onto a small set of measurement vectors that are incoherent with the sparsity basis. The groundbreaking work by Candes *et al.* and Donoho shows that such a signal can be precisely reconstructed from just a small set of random linear measurements (smaller than the Nyquist rate), implying the potential of dramatic reduction in sampling rates, power consumption, and computation complexity in digital data acquisitions.

Despite the achievements in CS theories, there are rather very few studies on the applicability of CS to audio signals, particularly on speech, music, or naturally occurring signals such as animal calls and environmental sounds. All these signals are usually not sparse and have a large number of nonzero components in whatever basis might be used in reconstruction. The CS paradigm was used for audio compression in [18]. Relying on some classical techniques for solving the CS problem, such as basis pursuit and orthogonal matching pursuit, the method derived in [18] constructs a sparse discrete cosine transform representation of the underlying audio signal. At present, there still exists a huge gap between the CS theory and its applications to audio signals. The researchers hope that the speech-encoding method based on compressed sensing can replace the traditional speech-encoding methods. In this section, the joint speech-encoding method based on compressed sensing and traditional pulse-coding modulation (PCM) speech-encoding method are discussed.

CS includes three main steps: sparse representation, measurement, and reconstruction. The signal sparse representation is the fundamental premise of CS implementation, so the optimal sparse bases of the signals have been widely researched. DWT is introduced for solving the problem involved in making a sparse representation of an audio signal; this can approximate smooth functions very efficiently. It can achieve arbitrary high accuracy by selecting appropriate wavelet basis and can concentrate the large wavelet coefficients in the low frequencies. It has also a multi-resolution framework and associated fast transform algorithms. Smaller the number of measurements in compressed sensing, lower is the computational complex. The measurement matrix size is usually decided by the length of the signal, so that the signal can be shortened with some methods when the signal precision meets the demand for some applications. For one-dimensional signal, it is well known that the length of the low-frequency signal can shorten the half of original signal length when the signal finished the one-level wavelet decomposition. Considering the signal approximation reconstruction in speech encoding, the wavelet transform based on lifting scheme was used in this section.

High-frequency and low-frequency coefficients can be acquired through adoption of the wavelet transform. Because the high-frequency coefficients usually are sparse, it can be reconstructed using the CS method. In CS reconstruction, the $\ell1$-norm optimal algorithms were used. For the low-frequency coefficients, which have better approximation with the original speech signal, they were encoded in PCM.

According to above analyses, the new joint speech-encoding scheme is proposed based on compressed sensing theory and PCM technology for reducing the computational complex and improving the compression.

1. Firstly, one-level wavelet decomposition of the speech signal was finished.
2. Secondly, the high-frequency coefficients were reconstructed by compressed sensing method and the low-frequency coefficients were encoded using PCM.
3. Finally, using the inverse adaptive wavelet transform, the speech signal was reconstructed with the encoded low-frequency coefficients and the reconstructed the high-frequency coefficients.

4.11.1 Joint Speech-Encoding Scheme

A major portion of the literature on CS has been concerned with very sparse signals, and very few results have been presented that explore the performance of CS when used with signals that are not truly sparse. There are even fewer studies on the applicability of CS to audio signals, particularly on speech, music, or naturally occurring signals such as animal calls and environmental sounds. All these signals are usually not sparse and have a large number of nonzero components in whatever basis might be used in reconstruction.

A joint speech-encoding scheme based on compressed sensing has been discussed in this section for a study on compressed sensing application in speech codec and for further the codec scheme of improving speech codec performance. The codec scheme of the speech shown in Figure 4.9 includes the compressed sensing encoding and the traditional speech-encoding algorithms.

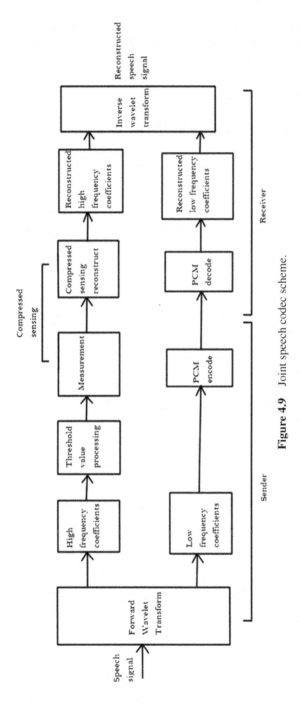

Figure 4.9 Joint speech codec scheme.

In order to reduce the complexity of signal encoding, the original speech finishes the one-dimensional wavelet transform first, and then, the high-frequency and low-frequency coefficients can be acquired. Usually, for the high-frequency coefficients, these values are close to zero. Through setting the hard threshold, the high-frequency coefficients must be sparse, so these coefficients can be reconstructed by CS methods. In this period, the high-frequency coefficients were measured by the random matrix, so the measured values can be obtained at the sender side. According to the wavelet transform performance, for the low-frequency coefficients, it was well known that it has the better approximation of the original speech. The low-frequency coefficients were encoded by PCM method since the length of the low-frequency coefficients was shortened, the PCM complex would be reduced.

In the receiver end, reconstruction of high-frequency coefficients can be done based on the measurement of the measurement of the high-frequency coefficients and by using compressed sensing algorithms. In this section, the CVX which is a Matlab-based modeling system for optimization was used for the reconstructed algorithms in compressed sensing. The reconstructed high-frequency coefficients were prepared for the input coefficients of the inverse wavelet transform. The PCM code of low-frequency coefficients can be decoded by using PCM decode, and then, the reconstructed low frequency can be achieved in the receiver. The speech signal could be reconstructed with the inverse wavelet transform after the reconstruction of the low- and high-frequency components.

4.11.2 Wavelet Transform

Compressed sensing includes three parts: sparse representation, measurement, and reconstruction. If the signal length is bigger, the measurement matrix size is bigger, so the computational complex of the signal reconstruction is higher. The signal length needs shortening for the purpose of reducing the size of measurement matrix in CS. The ability of the wavelet transform to concentrate on the large wavelet coefficients in the low frequencies of the signal is well known. So the high-frequency coefficients are small and close to zero. The high-frequency coefficients are sparse when processed by hard threshold. So the high coefficients can be reconstructed by compressed sensing methods. The signal length in compressed sensing is smaller than that of the original signal.

Wavelet transforms based on lifting schemes have achieved large recognition in recent years. In general, lifting splits a signal into two subsamples, followed by at least two lifting steps: prediction and update; this is shown in Figure 4.10. The implementation process is discussed in the following text.

4.11.3 PCM

PCM is used for speech signal encoding for simplification of the coding process. Pulse code modulation (PCM) is a method of converting an analog message waveform to a digital bit stream of 1's and 0's. For example, PCM is commonly used in telephone exchanges for converting an analog voice signal (300–3,400 Hz) to a 64,000 bit per second data stream. Figure 4.11 shows the PCM generator.

In the PCM generator, the quantizer compares the input with its fixed levels. It assigns any one of the digital levels to the results in minimum distortion or error. The error is called quantization error. This is how the output of the quantizer is called a digital level. If the quantized signal level is binary encoded, the encoder converts the input signal to digits binary word.

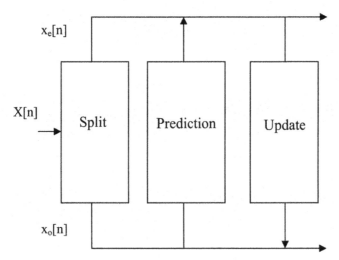

Figure 4.10   Prediction then update scheme.

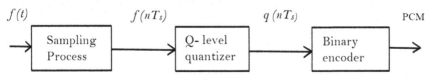

Figure 4.11   PCM modulator.

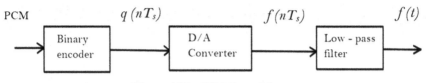

Figure 4.12   PCM Demodulator.

In this section, the input signal of the quantizer is the speech signal, and the sampling process has been omitted. Figure 4.12 shows the block diagram of the PCM receiver. The receiver starts by reshaping the received pulses and then converts the binary bits to analog. Errors arise during quantization at the transmitter due to permanent quantization leading to the inability to reconstruct the original signal perfectly. The quantization error can be reduced by increasing the quantization levels v.

This corresponds to the increase in bits per sample (more information). But increasing bits v increases the signaling rate and requires a large transmission bandwidth. The choice of a parameter for the number of quantization levels must be acceptable with the quantization error. Considering the simulation requirements, as the PCM generator, in PCM demodulator, the binary encoder was considered, and D/A converter and low-pass filter were not adopted in this section [19].

4.12 Applications of Compressed Sensing to Speech Coding Based on Sparse Linear Prediction

A compressed sensing method can be devised for compressing a sparse approximation of speech in the residual domain when sparse linear prediction is involved. The method of computing a sparse prediction residual with the optimal technique based on an exhaustive search of the possible nonzero locations and the well-known multi-pulse excitation, the first encoding technique to introduce the sparsity concept in speech coding is compared. Experimental results demonstrate the potential of compressed sensing in speech-coding techniques, offering high perceptual quality with a very sparse approximated prediction residual. This CS approach leads to a reduction in complexity in obtaining sparse residuals, moving closer to the optimal 0-norm solution while keeping the problem tractable through convex optimization tools and projection onto a random basis. In addition, this section also shows the successful extension of the CS formulation to the case where the basis is not orthogonal, a case which is rarely examined in CS literature. In simulations, the CS-based predictive coding approach provides better speech quality than that of MPE-based methods at roughly the same complexity [20].

4.12.1 Compressed Sensing Formulation for Speech Coding

A. Definition of the transform domain

In speech coding, the transform domain where the representation is required to be sparse is the prediction residual. Considering the simple case in which a linear predictor a of order P that provides a sparse residual is found, the formulation becomes

$$\hat{a} = \arg \min_{a \in R^P} \|x - Xa\|_1 \tag{4.32}$$

where

$$x = \begin{bmatrix} x(N_1) \\ \cdot \\ \cdot \\ x(N_2) \end{bmatrix}, X = \begin{bmatrix} x(N_1 - 1) & \cdots & x(N_1 - P) \\ \cdot & \cdot & \cdot \\ \cdot & \cdot & \cdot \\ \cdot & \cdot & \cdot \\ x(N_2 - 1) & & x(N_2 - P) \end{bmatrix} \tag{4.33}$$

and $\|.\|_1$ is the 1-norm. The start and end points N_1 and N_2 can be chosen in various ways assuming that $x(n) = 0$ for $n < 1$ and $n > N$. An appropriate choice is $N_1 = 1$ and $N_2 = N + P$ (in the case of 2-norm minimization, leading to the autocorrelation and to the Yule-Walker equations). The more tractable 1-norm is used as a linear programming relaxation of the sparsity measure. Given a prediction filter a, the residual vector can be expressed as follows:

$$r = Ax, \tag{4.34}$$

where A is the $N \times N$ matrix that performs the whitening of the signal, constructed from the coefficients of the predictor of order P [21].

Equivalently, it can be written as follows:

$$x = A^{-1}r = Hr, \tag{4.35}$$

where H is the $N \times N$ inverse matrix of A and is commonly referred to as the synthesis matrix [21] that maps the residual representation to the original speech domain. In practice, this inversion is not computed explicitly and **H** is constructed directly from the impulse response h of the all pole filter that corresponds to a. Furthermore, the usual approach is to have $N + P$ columns in H bringing in the effects of P samples of the residual of the previous frame (the filter state/memory). It is important to note that the column vector r will now be composed of $N + P$ rows, but the first P elements belong to the excitation of the previous speech frame and therefore are fixed and do not affect the minimization process. It is now clear that the basis vectors matrix is the synthesis matrix $\psi = H$. It can be written that

$$x = \sum_{i=1}^{K} r_{ni} h_{ni}, \quad \{n_1, n_2, \ldots n_K\} \subset \{1, \ldots N + P\} \tag{4.36}$$

where h_i represents the *i-th* column of the matrix H. The formulation then becomes

$$\hat{r} = \arg \min_{r \in R^N} \|r\|_1 + \gamma \|y - \Phi H r\|_2^2 \tag{4.37}$$

where y = Φx is the speech signal compressed through the projection onto the random basis Φ of dimension $M \times N$. The second term is now the 2-norm of the difference between the original speech signal and the speech signal with the sparse representation, projected onto the random basis. Assuming that

$$\|y - \Phi H r\|_2^2 = \|\Phi(x - Hr)\|_2^2 \approx \|x - Hr\|_2^2 \tag{4.38}$$

the problem in (4.37) can now be seen as a trade-off between the sparsity in the residual vector and the accuracy of the new speech representation $\hat{x} = H\hat{r}$. In order to ensure simplicity in the preceding and following derivations, it is assumed that no perceptual weighting is performed. The results can then be generalized for an arbitrary weighting filter. An important aspect that should be taken into consideration is that if the transformation matrix Φ is not exactly orthogonal, such as in the case of Φ = H, recovery is still possible, as long as the incoherence holds ($\mu(\Phi, H) \approx 1$).

B. Defining the level of sparsity

CS theory states that for a vector x of length N with sparsity level K ($K << N$), $M = O(K \log(N))$, random linear projections of x are sufficient to robustly (i.e., with overwhelming probability) recover x in polynomial time. With a proper random basis, so that Φ and H are incoherent ($\mu(\Phi, H) \approx 1$) [22], as a rule of thumb, four times as many random samples as the number of nonzero sparse samples should be used; therefore, the choice is $M = 4K$. It is now clear that the size of the random matrix Φ depends uniquely on the sparsity level K that is expected as residual vector. Now, the question is how sparse do we expect the residual to be? An interesting case for the choice of K is obtained for voiced speech. In this case, the residual r is a train of impulses. Each impulse is separated by T_p samples, the pitch period of the voiced speech which is inversely proportional to the fundamental frequency f_0. It is now clear that K depends on T_p; for a segment of voiced speech of length N, it can be reasonably assumed to find only N/T_p significant samples in the residual, belonging to the impulse train. A coarse estimation of the integer pitch period T_p can be easily obtained by an open-loop search on the autocorrelation function of the vector x. Then, the number of random projections sufficient for recovering x will be $M = 4(N/T_p)$. In the case of unvoiced speech, the choice of K is not direct; however, a heuristic approach is used where $K = k$ is picked when the improvement in the accuracy of the representation between the choice of $K = k$ and $K = k + 1$ is negligible.

C. Similarities with multi-pulse excitation

In multi-pulse excitation (MPE) coders, the prediction residual consists of K freely located pulses in each segment of length N. This problem is made impractical by its combinatorial nature and a suboptimal algorithm is proposed that the sparse residual is constructed one pulse at a time. Starting with a zero residual, pulses are added iteratively adding one pulse in the position that minimizes the error between the original and reconstructed speech. The pulse amplitude is then found in an analysis-by-synthesis (AbS) scheme. The procedure can be stopped either when the maximum fixed number of amplitudes is found or when adding a new pulse does not improve the quality. MPE provides an approximation to the optimal approach, when all possible combinations of K positions in the approximated residual of length N are analyzed, i.e.,

$$\hat{r} = \arg \min_{r \in R^N} \|x - Hr\|_2^2 \ s.t \ \|r\|_0 = K \tag{4.39}$$

Compressive sensing formulation in (4.37) can then be seen to approximate (4.39), finding a trade-off between the information content of the prediction residual and the quality of the synthesized speech. A new formulation in the context of speech coding based on compressed sensing is introduced. CS formulation based on LASSO has shown to provide an efficient approximation of the 0-norm for the selection of the residual allowing a trade-off between the sparsity imposed on the residual and the waveform approximation error. The convex nature of the problem, and its dimensionality reduction through the projection onto random basis, makes it also computationally efficient. The residual obtained engenders a very compact representation, offering

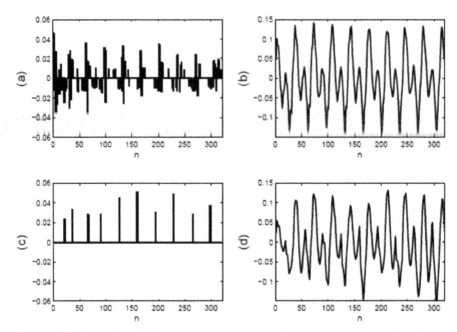

Figure 4.13   Example of CS recovery of the pitch excitation for a segment of stationary voiced speech.

interesting waveform matching properties with very few samples, making it an attractive alternative to common residual encoding procedures. The results obtained also show clearly that CS performs quite well when the basis is not orthogonal, as anticipated in some CS literature. The example of CS recovery of the pitch excitation for a segment of stationary voiced speech is shown in Figure 4.13.

4.13 Summary

The principles of CS can be applied to sparse decompositions of speech and audio signals as it offers a significant reduction in computational complexity. The issue in applying CS and sparse decompositions to speech and audio signals has been discussed. A multi-sensor version of incoherent detection and estimation algorithm (IDEA) called MS-IDEA is developed in the data fusion centre (DFC) for detecting the presence of a signal. This allows intra-signal compression without intra-sensor communication. CS can be successfully applied to both musical and speech signals, but speech signals are more demanding in terms of the number of observations. They are more by complex nature and therefore less sparse in the frequency domain, compared to the pure musical tones. Since sinusoidal signals satisfy sparsity property, it can be concluded that the musical tones are convenient for CS application. It also describes the use of DCT and CS for obtaining an efficient representation of audio signals, especially when they are sparse in the frequency domain. Single-channel and multi-channel sinusoidal audio coding using CS is also discussed. In CS framework, the reconstruction of a signal relies on the knowledge of the sparse basis and measurement matrix used for sensing. A new sensing matrix called orthogonal symmetric Toeplitz matrix (OSTM) generated with binary, ternary, and PN sequence that shows improved results compared to random, Bernoulli, Hadamard, and Fourier Matrices is also discussed in Chapter 4. The text-independent speaker identification that can be performed based solely on a speaker's voice is also discussed.

References

[1] Christensen, M. G., Ostergaard, J., and Holdt Jensen, S. (2009). "On compressed sensing and its application to speech and audio signals." *Signals, Systems and Computers, 2009 Conference Record of the Forty-Third Asilomar Conference on IEEE.*

[2] Candes, E., Romberg, J., and Tao, T. (2006). "Robust uncertainty principles: Exact signal reconstruction from highly incomplete frequency information," *IEEE Trans. Inform. Theory*, 52 (2), 489–509, February.

[3] Griffin, A., and Tsakalides, P. (2008). "Compressed sensing of audio signals using multiple sensors." *Signal Processing Conference, 2008 16th European.* IEEE.

[4] Baron, D., Wakin, M. B., Duarte, M. F., Sarvotham, S., and Baraniuk, R. G. (2005). "Distributed compressed sensing," preprint.

[5] Tropp, J., Gilbert, A., and Strauss, M. (2005). "Simultaneous sparse approximation via greedy pursuit," in *Proceedings of the IEEE International Conference on Acoustics, Speech, and Signal Processing (ICASSP)*, Philadelphia, PA, USA, vol. 5, March, pp. 721–724.

[6] Wang, Y., et al. (2011). "Compressive sensing framework for speech signal synthesis using a hybrid dictionary." *2011 4th International Congress on Image and Signal Processing (CISP)*, Vol. 5. IEEE.

[7] Savic, T., and Albijanic, R. (2015). "CS reconstruction of the speech and musical signals." *arXiv preprint arXiv:1502.01707.*

[8] Wu, D., Wei-Ping, Z., and Swamy, M. N. S. (2011). "A compressive sensing method for noise reduction of speech and audio signals." *2011 IEEE 54th International Midwest Symposium on Circuits and Systems (MWSCAS)*, IEEE.

[9] Lustig, M., Donoho, D., and Pauly, 1. M. (2007). "Sparse MRI: The application of compressed sensing for rapid MR imaging", *Magnet Resonance Med.* 58 (6), 1182–1195.

[10] Moreno-Alvarado, R. G., and Martinez-Garcia, M. (2011) "DCT-compressive Sampling of Frequency-sparse Audio Signals" *Proceedings of the World Congress on Engineering 2011* Vol. II WCE 2011, July 6–8, London, U.K.

[11] Griffin, A., Hirvonen, T., Tzagkarakis, C., Mouchtaris, A., Member, IEEE, and Panagiotis, T., (2011). "Single-Channel and Multi-Channel Sinusoidal Audio Coding Using Compressed Sensing," *IEEE Trans. Audio, Speech, Language Process.* 19 (5), 1382–1395.

[12] Serra, X., and Smith, J. O. (1990). "Spectral modeling synthesis: A sound analysis/synthesis system based on a deterministic plus stochastic decomposition," *Comput. Music J.* 14 (4), 12–24.

[13] van de Par, S., Kohlrausch, A., Heusdens, R., Jensen, J., and Jensen, S. H. (2005). "A perceptual model for sinusoidal audio coding based on spectral integration," *EURASIP J. Appl. Signal Process.* 2005 (1), 1292–1304, Jan.

[14] Siddhi, D., and Naitik N. (2014). "Improved Performance of Compressive Sensing for Speech Signal with Orthogonal Symmetric Toeplitz Matrix" *Intl. J. Signal Process. Image Process. Pattern Recogn.* 7 (4), 371–380 http://dx.doi.org/10.14257/ijsip.2014.7.4.35

[15] Lamel, L., Kassel, R., and Seneff, S. (1986) Speech database development: design and analysis of the acoustic-phonetic corpus. In: *Proceedings of the DARPA speech recognition*, workshop.

[16] Sainath, T. N., Carmi, A., and Kanevsky, D, and Ramabhadran, B. (2010) Bayesian compressive sensing for phonetic classification. In: *Proceedings of the ICASSP.*

[17] Reynolds, D. A., and Rose, R. C. (1995). "Robust text independent speaker identification using Gaussian mixture speaker models," *IEEE Trans. Speech and Audio Process.* 3 (1), 72–83, January.

[18] Griffin, A., and Tsakalides, P. (2008). "Compressed sensing of audio signals using multiple sensors", *Processings of the 16th European signal processing conference (EUSIPCO'08)*, Lausanne, Switzerland.

[19] Gao, G., Shang, L., and Xion, K. (2014). "Study on Joint Speech Encoding Technology based on Compressed Sensing." *Intl. J. Multimedia Ubiquitous Eng.* 9.7, 47–60.

[20] Giacobello, D., et al. (2010). "Retrieving sparse patterns using a compressed sensing framework: applications to speech coding based on sparse linear prediction." *IEEE Signal Processing Letters,* 17.1, 103–106.

[21] Scharf, L. (1991). *Statistical Signal Processing*, (New York: Addison-Wesley).

[22] Candes, E. J., and Romberg, J. (2007). "Sparsity and incoherence in compressive sampling," *Inv. Problem.* 23(3), 969–985.

5

Compressive Sensing for Images

5.1 Introduction

Compressive sensing (CS) requires that the sensed signal is often sparse in some transform domain to enable its recovery from a small number of linear, random, multiplexed measurements. Robust signal recovery is possible from a number of measurements, proportional to the sparsity level of the signal which is much smaller than its dimensionality. Though natural images are not sparse, they are sparse in a specific transform domain. Hence, CS can be used for image-based applications to have energy-efficient and complex representations of smaller magnitude. There are several thrust areas, namely image fusion, image compression, image denoising, and image reconstruction that require data and processing complexity reduction. Compressed sensing can be employed in these areas to enhance the performance with reduced measurements. It can also be employed in specific imaging applications such as magnetic resonance imaging, synthetic aperture radar imaging, passive millimeter wave imaging, and light transport system. This chapter deals with the methodologies and algorithms used for CS-based image applications. Camera architectures uniquely designed for direct acquisition of random projections of the signal without first collecting the pixels/voxels are available for taking compressive measurements. This chapter also provides insight into the compressive imaging architectures such as single pixel camera and lens less imaging. A case study depicting the image transmission using compressed sensing in wireless multimedia sensor networks is also presented.

5.2 Compressive Sensing for Image Fusion

Image fusion is the process of combining two or more images to form one image. The main goal of image fusion is to extract all the important features from all input images and integrate them to form a fused image which is more informative and suitable for human visual perception or computer processing. It has been widely used in earth remote sensing, military reconnaissance, computer vision, medical image recognition, and many other fields. Moreover, the significance of image fusion to military applications is also very obvious, such as the fusion of synthetic aperture radar image and infrared image. The fusion image usually has the following advantages.

1. Improved reliability;
2. Increased information content of image;
3. Highly compatible (different sensors);
4. Strong property of conformity.

Image fusion generally includes three main processes: pretreatment, fusion rules, and fusion image. The flow chart of image fusion is shown in Figure 5.1.

Detailed steps involved in the fusion algorithm employing compressed sensing are as follows:

1. Blocking operation

When dealing with the image of larger size or with multiple images, the volume of data that needs processing is usually very huge. It causes a rather complex computation and a high requirement for hardware implementation. Block processing-based analysis is used for combating this problem. For an image to be fused A, it is divided into many blocks A_i with the size of $n \times n$. By processing these blocks separately, the complexity of computation can be reduced and enables easy hardware realization.

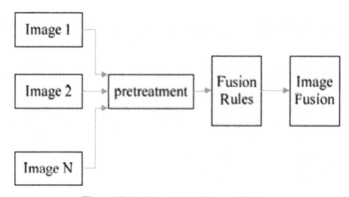

Figure 5.1   Flow chart of image fusion.

2. Translate to vector form

The purpose of translating the blocking image to vector form is to simplify the operation further.

3. Sparse expression of signal

A transform domain that enables the representation the sparseness of signal is identified and is used to describe the signal in sparseness sense and get the sparse coefficients in the transform domain. In general, discrete cosine transform can be used for calculating the sparse coefficient of cosines basics which has a better sparseness degree. According to the above expressions, the vectors to be fused are denoted as xA_i and xB_i. fA_i and fB_i are obtained after the application of discrete cosine transform. The process can be described as follows:

$$\begin{cases} fA_i(k) = \sqrt{\frac{2}{N}} c(k) \sum_{t=0}^{n-1} xA_i(t) \cos \frac{(2t+1)k\pi}{2N} \\ fB_i(k) = \sqrt{\frac{2}{N}} c(k) \sum_{t=0}^{n-1} xB_i(t) \cos \frac{(2t+1)k\pi}{2N} \end{cases} \quad (5.1)$$

In Equation (5.1), $k = 0, 1, \ldots, B - 1$, where n stands for the number of elements in vector and $c(k)$ satisfies the following condition:

$$\begin{cases} c(k) = \frac{1}{\sqrt{2}}, k = 0 \\ c(k) = 1, \quad k \neq 0 \end{cases} \quad (5.2)$$

4. Observation matrix

Irrelevancy is one of the most important factors to be considered in designing the observation matrix. Larger irrelevancy provides better performance. Gaussian random matrix is often adopted, which satisfies the condition of irrelevancy with any given matrix and can provide a satisfactory performance. A Gaussian random matrix with zero mean and variance of 1 is used as the observation matrix, for obtaining the measurements of coefficients.

$$\begin{cases} yA_i = \Phi fA_i \\ yB_i = \Phi fB_i \end{cases} \quad (5.3)$$

5. Fusion rules

Fusion rules decide the fusion performance directly. In this step, the simplest fusion rules based on weighting are used for fusing the observation data. Assuming the fusion data are f_i, then Equation (5.4) is used to calculate it.

$$\begin{cases} f_i(u) = \alpha\, yA_i(u) + \beta yB_i(u), u = 0, 1, \ldots, M - 1 \\ \alpha + \beta = 1 \end{cases} \quad (5.4)$$

This fusion algorithm makes the calculation easy and simpler as it is operated on the measurements rather than the entire image block. It demonstrates the advantages of compressed sensing, which can ensure the performance, that is the same as that of some other complex fusion algorithms.

6. Image reconstruction

Image reconstruction can be achieved by two sub-steps. First, the fusion values have to be recovered for getting the original value of coefficients on cosines basic. BP algorithm, is utilized for the recovery, as it has a better accuracy reconstruction. The second sub-step is calculating the blocking matrix of fusion image by discrete cosine transform. The data that are recovered are on the cosine basic, which need to be translated for fusing data in the original domain. This step is the inverse process of sparse operation.

7. Combination

Arrange these block matrix with the original sequence, fusion matrix can then be obtained by combination. This step is a simple operation of linear combination, which provides a new fusion image C_i [1]. Similarly, image fusion can be performed with the other transform domains too, wherein a suitable fusion algorithm has to be written accordingly.

5.2.1 Multi-Resolution Image Fusion

CS-based image fusion has a number of advantages over conventional image fusion algorithms. It offers computational and storage savings by using a compressive sensing technique. Compressive measurements are progressive in the sense that larger numbers of measurements lead to higher quality reconstructed images. Image fusion can be performed without acquiring the observed signals. Additionally, the recently proposed compressive imaging system [2], which relies on a single photon detector, enables imaging at new wavelengths inaccessible or prohibitively expensive using current focal plane imaging arrays. Multi-resolution decompositions (e.g., pyramid, wavelet, linear) have shown significant advantages in the representation of signals. They capture the signal in a hierarchical manner where each level corresponds to a reduced-resolution approximation. Multi-resolution representations enable fusion of image features separately at different scales. They also produce large coefficients near edges, thus revealing salient information [3]. They also offer computational advantages and appear to be robust. A simple maximum selection (MS) fusion scheme is used for fusing the input images at the pixel level. MS is a widely used fusion rule which considers the maximum absolute values of the wavelet coefficients from the source images as the fused coefficients. The wavelet-based image fusion algorithm consists of two main components. First, the detailed wavelet coefficients are composed using the MS fusion rule:

$$\text{DF} = \text{DM} \text{ with } M = \arg \max_{i=1,\ldots,I} (|Di|) \tag{5.5}$$

where DF is the composite coefficients, DM is the maximum absolute value of the input wavelet coefficients, and I is the total number of the source images. As the approximation and detail images have different physical meaning, they are usually treated differently by the combination algorithm. Then, popular way to construct the fused approximation image AF is as follows:

$$A_F = \frac{1}{I} \sum_{i=1}^{I} A_i \tag{5.6}$$

The unknown signal can be estimated from these compressive measurements to be within a controllable mean-squared error with the algorithm [4] specified in Table 5.1. Image fusion is performed on the compressive measurements rather than on the wavelet coefficients. The reconstruction of the fused image is done by performing a total variation optimization method and an inverse wavelet transform.

5.2.2 Multi-Focus Image Fusion

Due to the limited depth of focus of optical lenses, it is often not possible to acquire an image that contains all relevant objects in focus. Hence, multi-focus image fusion has become an important domain in image fusion. The simplest multi-focus image fusion method is to take pixel-by-pixel average of source images in spatial domain. But, this method causes undesirable side effects such as reduced contrast. Then, multi-resolution techniques are developed for image processing applications. These techniques contain pyramid transform and

Table 5.1 Compressive image fusion algorithm

Algorithm 1 Compressive Image Fusion Algorithm
1. Take the compressive measurements $Y_i, i = 1, \ldots, I$ for the ith input image using the double-star-shaped sampling pattern
2. Calculate the fused measurements using the formula: $Y_F = Y_M$ with $M = \arg \max_{i=1,\ldots,I}(|Y_i|)$.
3. Reconstruct the fused image from the composite measurements Y_F via the total variation optimization method [1].

wavelet transform. A new method for multi-focus image fusion is provided, which combines the source images using compressive sensing theory. Source images are processed and fused by not including all of their pixel information. Such image fusion methodology has the advantage of larger energy saving with reduced computation time and hardware memory. Hence, it can be used in WSN, VSN, and WMSN. The fusion algorithm used is shown in Figure 5.2, and the steps involved are given below [5]:

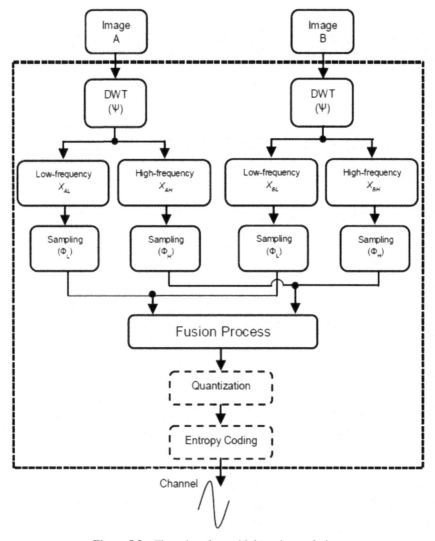

Figure 5.2   Flow chart for multi-focus image fusion.

1. Represent the source images A and B, with their sparse coefficients generated by DWT, Y_A and Y_B, respectively.
2. Choose the random Gaussian matrix $\Phi \in RM \times N$ as the measurement matrix for low-frequency and high-frequency coefficients.
3. Compute the measurements Y_{LA}, Y_{LB}, Y_{HA} and Y_{HB}.
4. Compute the local variance and sum modified Laplacian metrics for low and high band, respectively.
5. Obtain the fused measurements $F_L = [wL_1, wL_2, \ldots, wL_M]^T$ and $F_H = [wH_1, wH_2, \ldots, wH_M]^T$.
6. Before sending to the channels, quantization, and entropy coding of fusion measurement coefficients can be done.

The experimental results of the proposed method are compared with other CS-based approaches [6–8]. The results for clock image are shown in Figure 5.3. Figure 5.3a–b is pairs of multi-focus source images, wherein a camera is first focused on the left and then focused on the right later. Figure 5.3c–e is fused images obtained using maximum selection (MS) [6], the entropy metrics weighted average (EMWV) [7], and standard deviation weight

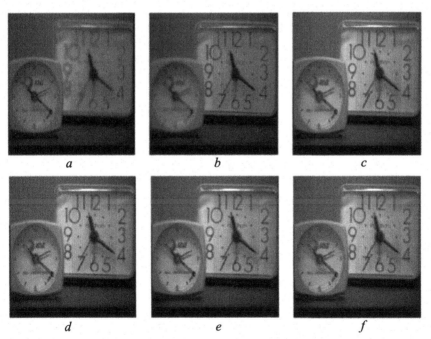

Figure 5.3 Source image "clock." (a) Focus on left. (b) Focus on right. (c) MS result. (d) EMWV result. (e) SDWV result. (f) CS-based fusion algorithm.

average (SDWV) [8]. Figure 5.3f is result of fused image using the CS-based fusion algorithm shown in Figure 5.2.

The proposed method measures and combines low- and high-band coefficients separately, which achieve low memory and better fused image. Thus, the compressive sensing-based fusion of multi-focus images can be employed for reducing energy consumption and bandwidth in transmission. Energy consumption and bandwidth are significant factors in visual sensor networks, which determine the lifetime of sensors.

5.3 Compressive Sensing for Image Compression

In many modern applications, such as camera arrays for surveillance, it is desirable to reduce the complexity and cost of the encoder. More recently, block-based coding methods have been proposed to address the same. The image is divided into non-overlapping blocks of pixels and encoded independently. These methods generally use the same measurement rate for all image blocks, even though the compressibility and sparsity of the individual blocks can be quite different. For effective and simple CS-based image coding, it is desirable to use a block-based strategy that exploits the statistical structure of the image data. The measurement rate of each block can be adapted with respect to the block sparsity. Image blocks are classified into key and non-key blocks and encoded at different rates. However, it is not easy to estimate the sparsity or compressibility of an image block from the CS measurements alone. The encoder makes use of a new adaptive block classification scheme that is based on the mean square error of the CS measurements between blocks. At the decoder, a simple, but effective, side information generation method is used for decoding of the non-key blocks.

A distributed image coding and decoding scheme is utilized which is block-based, makes use of the correlation between neighboring blocks. In block-based CS, an image, X, with dimension $N \times N$ pixels is divided into non-overlapping blocks of $B \times B$ pixels. Let x_i represent the ith block; then, its CS measurements are given by: $y_i = \Phi_B X_i$ where B is a $M_r \times B^2$ measurement matrix. The measurement rate per block is given by $M_b = M_r / B^2$. Since the image data of neighboring blocks are correlated, their CS measurements are also correlated. Hence, image blocks are classified either as key blocks or non-key (WZ) blocks. Key blocks are encoded with a higher measurement rate than non-key blocks, similar to the idea of key frames and non-key frames in video coding [9]. Two block classification strategies, one adaptive and the other non-adaptive, can be adopted.

An image compression algorithm based on DCT-based CS and vector quantization (VQ) can also be used for image compression. This algorithm gives a better PSNR and visual quality when compared with the existing CS-VQ algorithm. The results obtained are even comparable with the JPEG algorithm but only when a small queue size is considered. The basic concept behind the CS states that small collections of non-adaptive linear projections of a sparse signal can efficiently help in the reconstruction of the image through the image data sent to the decoder making use of some optimization procedure.

A novel method for image compression based on compressive sensing using wavelet lifting scheme which is faster, simpler, and also keeping strong edge preservation does exist. It can also provide better compression [10]. The method uses sparse representation based on CDF9/7 wavelet transform. Measurement matrix needs to be one-to-one on all-sparse vectors for ensuring exact recovery of every sparse signal. The Gaussian measurement matrix, Bernoulli measurement matrix, and random orthogonal measurement matrix are used for comparing the best fit of sparse representation of an image by CDF 9/7 wavelet transform. Image reconstruction is done using convex optimization techniques, basis pursuit (BP), and orthogonal matching pursuit (OMP) algorithms.

Discrete wavelet transform (DWT) is applied for sparse representation, while a fast CS measurement taking method is also utilized based on the property of 2D DWT [11]. Unlike the unequally important discrete wavelet coefficients, the resultant CS measurements carry nearly the same volume of information and have minimal effects for bit loss. At the decoder side, one can simply reconstruct the image via $\ell 1$ minimization. Experimental results show that the proposed CS-based image codec without resorting to error protection is more robust compared with existing CS technique and relevant joint source channel coding (JSCC) schemes.

5.4 Compressive Sensing for Image Denoising

Image denoising is an important issue in many real applications. Image denoising can be considered to be recovery of a signal from inaccurately and/or partially measured samples. It is well known that many natural images have a compact representation when expressed in a convenient basis, such as with wavelets. If we can obtain more precise measurements of the original images than the corresponding noisy images, then satisfactory reconstruction of images can be done using CS theory. Thus, a general image denoising framework based on CS is provided as follows:

$$\arg\min_{u} \frac{1}{2} \|\Phi u - \Phi f\|^2 + \lambda \|\Psi u\|_1 \qquad (5.7)$$

where f is the noisy image and Φ is the measurement operator that separates the noise and the latent image. Suitable Φ and Ψ matrices can be chosen for addressing specific images and noising. Several existing methods can be regarded as special cases of following two classes:

Total variation-based algorithms: If Ψ is the gradient operator and Φ is the identical matrix, then Equation (5.7) is the total variation (TV) regularization. If Ψ is the gradient operator and Φ is a wavelet hard thresholding operator, then Equation (5.7) is the method that is proposed by Durand and Froment [12]. The TV regularization and its variants recover an image that has a gradient vector that is as sparse as possible. From the perspective of CS theory, if the image gradient is sparse enough, then the total variation algorithm can recover the image very well. However, most of the images are not sparse in general, unless the images are piecewise constant. Thus, for the choice of Ψ, the translation invariant wavelet transform is more effective than the gradient operator.

Wavelet-based algorithms: If Ψ is the wavelet transform and Φ is an identical matrix, then Equation (5.7) is a sparse analysis approximation denoising method [13] that is an extension of the TV model. If Ψ is a tight frame, then it is equivalent to a wavelet (or translation-invariant wavelet transform) thresholding. If both Ψ and Φ are the same wavelet transform, then Equation (5.7) is wavelets of thresholding. If the regularization term is modified to $\|X\|_0$, then Equation (5.7) is the wavelet (or translation-invariant wavelet) hard thresholding. From the perspective of CS theory, a wavelet hard or soft thresholding is to recover a compressible signal from a noisy signal. CS theory prefers low coherence of the pair (Φ, Ψ), but Φ is equal to Ψ in wavelet thresholding. Adjusting (Φ, Ψ), to be a lower coherence most often lead to a better denoising effect. The "noise" aforementioned refers to Gaussian white noise. Moreover, Poisson noise removal can also be analyzed.

Inexact recovery of a large matrix through matrix completion has provided new insights into the way of recovering missing data among a large set of correlated data. Noisy pixel elimination through matrix completion is analyzed in addition to obtaining noisy sparse representations of a noisy image. In computer vision and image processing, many problems can be formulated as the missing value estimation problem, e.g., image in-painting, video decoding, and video in-painting. The values can be missing due to problems in the acquisition process, or because the user manually identified unwanted outliers. Image denoising has been an active research topic for many years. Since image noise is generally caused by image sensors, amplifiers, ADCs, or maybe even due to quantization. It is imperative that the noise should be handled by an image denoising algorithm. Image denoising problem in general can be modeled as a clean image contaminated by additive white Gaussian noise (AWGN), though modeling in terms of impulse or Poisson noise is also common, exploring K-SVD-based image denoising through low-rank matrix completion is essential. This method incorporates dictionary formation and learning through sparse representation using K-SVD.

Matrix Completion

The K-SVD algorithm is used in exploration of the impact of matrix completion on image denoising. Usually, there exists an underlying structure in the noisy image which can be carried over into a representational space where noisy pixels can be removed for obtaining denoised patches which are very close to the original. The algorithm assumes a partially denoised image obtained from the K-SVD algorithm [14]. Then, the patches of the denoised image are used in the subsequent steps for getting better patches in the reconstructed denoised image. The following steps outline the algorithm:

i. Obtain a partially denoised image using a denoising algorithm, such as K-SVD-based denoising.

ii. Obtain the randomly sampled patches from the partially denoised image across different scales to form different dictionaries.

iii. Train the dictionaries for getting a better compact representation of the randomly sampled dictionaries.

iv. Collect randomly sampled patches from the noisy image and form a randomly sampled dictionary; train it using online dictionary learning algorithm for obtaining a compact trained dictionary. The only difference is that this is done across one scale only.

v. Obtain the sparse representation for a noisy patch and use the sparse coefficients to form a patch from all dictionaries generated from partially denoised image.

vi. Use all the patches from different dictionaries to form a matrix. Remove noisy pixels by comparing the variances of the partially denoised patch and the sparse representation-based patches. In addition to this, thresholds can also be determined using pixel difference between K-SVD denoised patches and noisy patches.

vii. Subject this matrix with missing entries to matrix completion. The recovered matrix represents the completely denoised patch. This process is repeated for all patches of an image.

The next step is sparse representation. Given a noisy patch, a sparse representation of this noisy patch from the noisy dictionary is formed. These coefficients are carried over to form an image patch from all the fifteen dictionaries. These representations individually may represent a recovered image itself. But these are not the best denoised images, since each dictionary can at best represent the original partially denoised patch. Hence, an appropriate method of noise removal is to be undertaken. Based on the variance of the image patches, a different threshold is set to determine pixel values which are far away from a partially denoised image. The noisy image is used for providing an input on the variance of the patch and the variability of individual pixels to aid the pixel removal step. Then, the patches without noisy pixels are arranged to form a larger matrix.

Biological Microscopy Image Denoising
In microscopy, observation of fluorescent molecules is challenged by photo-bleaching and photo-toxicity. These molecules get slowly destroyed by exposure to light. It is, therefore, necessary to stimulate them into fluorescence. However, reducing exposure time drastically deteriorates the signal-to-noise ratio (SNR) and, hence, the image quality. Many denoising methods are available for improving the SNR such as non-local means (NL means), total variation (TV), and nonlinear isotropic and anisotropic diffusion. Denoising is also done by decomposing the data into wavelets, ridgelets, and curvelets bases, and shrinking the obtained transform coefficients. More efficient denoising is possible through sparsity and redundant representations over learned dictionaries, where the image is denoised and a dictionary is trained simultaneously. Hence, a CS-based method to acquire and denoise the data based on statistical properties of the CS optimality, signal modeling, and noise reconstruction is provided.

A discrete signal is considered sparse when it has a large number of zero coefficients on some basis functions. Natural signals measuring discrete events or natural images with smooth and homogeneous objects can be considered as approximately sparse in some basis and be accurately approximated with a small set of coefficients. For biological images, the global total variation is used as the sparsifying transform [15, 16]. This measure is well known in image processing and very popular in variational segmentation problems for its ability to limit high frequencies and to provide regularized segmented regions. This measure is also ideally suited for denoising where the goal is to restore an image with smooth objects and background. Reconstruction is done by minimization of the TV norm [17, 18], which corresponds to a constraint on the number of discontinuities in an image, and the homogeneity of the objects. TV constraint is well suited for biological images, where structures and background provide small gradient values, while a finite set of edges provides high gradient values for drosophila ovocytes.

The signal component $X \in R^N$ has to be recovered in the context of noisy measurements $y = (X + n)$, which is the case for microscopy images corrupted with acquisition noise. If the noise energy is assumed to be bounded by a known constant $\|n\|_{\ell_2} \leq \epsilon$, the transformed signal X is sparse, and $\Phi \in R^{MN}$ is a random matrix sampling x in the Fourier domain, the true signal component x can be recovered nearly exactly using the following convex optimization:

$$\hat{x} = \arg \min_{x \in R^N} \|\Psi x\|_{\ell_1} \ s.t \ \|y - \Phi x\|_{\ell_2} \leq \delta \qquad (5.8)$$

for some small $\delta \geq \varepsilon$. The solution $\hat{x}$ is guaranteed to be within $C\delta$ of the original signal X.

$$\|\hat{x} - x\|_{\ell_2} \leq C\delta \quad \text{with} \quad C > 0. \qquad (5.9)$$

This CS-based estimation framework, with noisy observations and TV spatial constraints, ensures that no false component of $X + n$ with significant energy is created as it minimizes its ℓ_1 norm, which is particularly high for additive random noise components. More specifically, the TV-based spatial sparsity constraint leads to smooth edges and removal of noise components, resulting in an error:

$$\|\hat{x} - x\| \ell_2 \leq \alpha + \beta \qquad (5.10)$$

where α reflects the desired error (responsible for noise removal) from the relaxation of the constrain δ in (5.8) and β reflects the undesired error from smooth edges of signal. If TV represents X efficiently and n inefficiently, the term β vanishes and $\alpha \to C\delta$. For preserving image features, the proposed method decomposes the sparse signal into feature and non-feature regions using statistical analysis and sparse representation. The proposed method computes the measurement signals from the decomposed sparse signals using block-based multiple compressive samplings [19]. The original noise-free image is recovered using orthogonal matching pursuit (OMP) with optimal error tolerance. A three-level DWT is used for decomposing the sparse signal into feature and non-feature regions.

5.5 Compressive Sensing Image Reconstruction

Wavelet-based contourlet transform, block-based random Gaussian image sampling matrix, and projection-driven compressive sensing recovery work in close cooperation in the new process framework to accomplish image reconstruction. Smoothing is achieved via a Wiener filter incorporated into Iterative projected Land weber compressive sensing recovery, yielding fast reconstruction. It works well for normal pictures, infrared images, texture images, and synthetic aperture radar (SAR) images and also has better quality that matches or exceeds that produced other popular ones. Also, smoothing can ensure the goal of improving the quality by eliminating blocking artifacts and quality of reconstruction with smoothing.

In block-based compressive sensing (BCS) [20], the two-dimensional image is divided into $M_n \times M_n$ blocks and sampled with an ordinary random Gaussian matrix. That is, suppose that x_i is a vector representing, in raster-scan fashion, block i of input data X. The corresponding y_i is then $y_i = \Phi_n X_i$, where Φ_n is an $M_n \times n^2$ orthonormal measurement matrix with $M_n = M/N \times B^2$ rounded down to the nearest integer.

$$
\Phi = \begin{bmatrix} \Phi_n & 0 & & 0 \\ 0 & \Phi_n & \cdots & 0 \\ & \vdots & \ddots & \vdots \\ 0 & 0 & \cdots & \Phi_n \end{bmatrix}
\tag{5.11}
$$

BCS has several merits in comparison with the random sampling which is applied to the entire raw data X. First, the measurement operator Φ_n is conveniently stored and employed because of its compact size. Second, the encoder does not need to wait until the entire image is measured, but may send each block after its linear projection. Last, an initial approximation X with minimum mean-squared error can be feasibly calculated due to the smaller size of $\Phi_n(n = 32)$.

Recursive filtering-based procedure can also be used for image reconstruction. At every iteration, the algorithm is excited by injection of random noise in the unobserved portion of the spectrum. Spatially adaptive image denoising filter working in the image domain is exploited for attenuating the noise and reveals new features, details out of the incomplete and degraded observations. The recursive algorithm is a special type of the Robbins-Monro stochastic approximation procedure with regularization enabled by a spatially adaptive filter. Overall, the conventional parametric modeling used in CS is replaced by a nonparametric one. The reconstruction approach is effectively applied to two important inverse problems from computerized tomography: radon inversion from sparse projections and limited-angle tomography [21]. The algorithm allows the achievement of exact reconstruction of synthetic phantom data even from very small number projections. The regularization imposed by the ℓ_0 or ℓ_1 norms is essentially a tool for design of some nonlinear filtering. Spatially adaptive filters sensitive to image features and details can be utilized. Properly designed adaptive filters can yield better results by the formal approach based on formulation of imaging as the variational problem with imposed global constraints. In imaging, the regularizations with global sparsity penalties (such as ℓ_p norms in some domain) often result in inefficient filtering. It is known that a higher quality can be achieved when the regularization criteria are local and adaptive. Hence, it is used in particular in the context of image denoising, where the performance of advanced spatially adaptive (both local and non-local) methods significantly overcomes that of the traditional approaches. The reconstructed phantom images are shown in Figure 5.4.

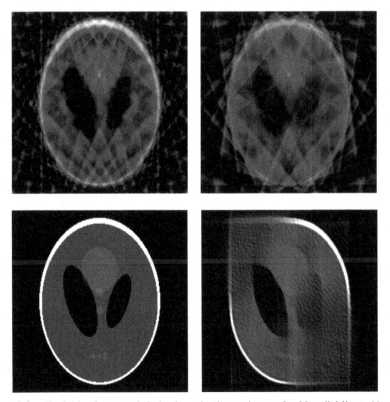

Figure 5.4   Clockwise from top-left: back-projection estimates for 22 radial lines, 11 radial lines, 61 radial lines with limited-angle (90°), and original phantom θ.

Compressive sampling, with deterministic measurement matrices made of chirps, can be applied to images that possess varying degrees of sparsity in their wavelet representations. These matrices come with a very fast reconstruction algorithm whose complexity depends only on the number of measurements n and not on the signal length N. In particular, selecting k columns randomly (i.e., independently from a uniform distribution on all N columns) from the $n \times N$ sensing matrix Φ yields $k \times k$ Gram matrices whose condition numbers are distributed with mean and variance essentially identical to those obtained by using the same procedure on Gaussian matrices. This observation appears to hold over a range of $k < n << N$, compatible with the requirement for random matrices. An efficient reconstruction algorithm that utilizes discrete chirp Fourier transform is used (DCFT) with updated linear least-squares solutions, which is highly suitable for medical images [22]. It has good sparsity properties. The reconstructed images reveal that the algorithm is effective in both reconstruction fidelity and speed.

A chirp signal of length n with chirp rate r and base frequency m has the form

$$v_{r,m}(l) = e^{\frac{2\pi i r l^2}{n} + \frac{2\pi i m l}{n}} \quad r, m, l \; \varepsilon \; Z_n \qquad (5.12)$$

For a fixed n, there are n^2 possible pairs (r, m). The full chirp sensing matrix Φ has size n $\times$ n^2 and its jth column is

$$\Phi_{r,m}(l) = v_{r,m}(l), \quad j = nr + m \; \varepsilon \; Z_{n^2} \qquad (5.13)$$

where $N = n^2$, a k-sparse signal $x \in C^N$ yields a measurement $y = \Phi x \in C^n$, that is, the superposition of k chirp signals. Fast Fourier transform (FFT) is used for recovering x. It detects the nonzero locations, (r_j, m_j) pairs, whose total computational complexity is O($kn \log n$). The magnitudes s_j of the nonzero locations r_j are found by solving the associated least-squares problem.

5.6 Compressive Sensing for Imaging Applications

5.6.1 Compressive Magnetic Resonance Imaging

Magnetic resonance imaging (MRI) is an essential medical imaging tool burdened by an inherently slow data acquisition process. The application of CS to MRI has significant scan time reductions, with benefits for patients and healthcare economics. MRI obeys two key requirements for successful application of CS: (1) Medical imagery is naturally compressible by sparse coding in an appropriate transform domain (e.g., by wavelet transform);

(2) MRI scanners naturally acquire samples of the encoded image in spatial frequency, rather than direct pixel samples. Moreover, MRI-based applications offer significant benefits in imaging speed, leading to improvement in patient care and cost reduction.

Constructing a single MR image commonly involves collection of a series of frames of data called acquisitions. In each acquisition, an RF excitation produces new transverse magnetization, which is then sampled along a particular trajectory in k-space. In principle, a complete MR image can be reconstructed from a single acquisition by using a k-space trajectory that covers the entire region of k-space. This is commonly done in applications such as imaging brain activation. However, for most applications, this results in inadequate image resolution and excessive image artifacts.

Magnetization decays exponentially with time. This limits the useful acquisition time window. Also, the gradient system performance and phys-iological constraints limit the speed at which k-space can be traversed. These two effects combine to limit the total number of samples per acquisition. As a result, most MRI imaging methods use a sequence of acquisitions, with each one sampling a part of k-space. The data from this sequence of acquisitions are then used to reconstruct an image.

Most MR images are sparse in an appropriate transform domain. To begin with, consider angiograms, which are images of blood vessels in the body. These images contain primarily contrast-enhanced blood vessels in void and look sparse to the naked eye. Equivalently, they are already sparse in the pixel domain, so identity transform is used as the sparsifying transform. Some brain images are piecewise smooth, and their gradient field is sparse; the sparsifying transform there is spatial finite-differencing. More complex imagery can be sparsified in more sophisticated domains, such as the discrete cosine transform domain or the wavelet domain—witness the success of JPEG and JPEG 2000, respectively.

Designing a CS scheme for MRI can be viewed as selecting a subset of the frequency domain which can be efficiently sampled, and is incoherent with respect to the sparsifying transform. Sampling trajectories must follow relatively smooth lines and curves. Sampling schemes must also be robust to non-ideal, real-life situations. Non-Cartesian sampling schemes are often sensitive to magnetic field homogeneity, eddy currents, signal decay, hardware delays, and other sources of imperfection. Furthermore, a uniform random distribution of samples in spatial frequency does not take into account the energy distribution of MR images in k-space, which is far from uniform. Most of energy in MR imagery is concentrated close to the center of k-space and rapidly decays toward the periphery of k-space. Therefore, realistic designs for CS in MRI should have variable density sampling with denser sampling near the center of k-space, matching the energy distribution in k-space. Such designs should also create k-space trajectories that are somewhat irregular and partially mimic the incoherence properties of pure random sampling, yet allow rapid collection of data [23].

Represent the reconstructed image by a complex vector m, and let ψ denotes the linear operator that transforms from pixel representation into a sparse representation. Let $F_{S'}$ denote the under sampled Fourier transform, corresponding to one of the k-space undersampling schemes discussed earlier. Reconstructions are obtained by solving the following constrained optimization problem:

$$\begin{aligned} \min \text{imize} \quad & \|\Psi m\|_1 \\ s.t. \quad & \|F_S m - y\|_2 < \in, \end{aligned} \tag{5.14}$$

where y is the measured k-space data from the MRI scanner and ε controls the fidelity of the reconstruction to the measured data. The threshold parameter ε is roughly the expected noise level. Minimizing the ℓ_1 norm of $\|\Psi m\|_1$ promotes sparsity. The constraint $\|F_S m - y\|_2 < \in$ enforces data consistency. Among all solutions which are consistent with the acquired data, Equation (5.11) finds a solution which is compressible by the transform. When finite differences are used for sparsifying transform, the objective in the optimization is effectively the total variation (TV) norm, a widely used objective in image processing. Even if another sparsifying transform is intended, it is often useful to include a TV penalty as well. Such a combined objective seeks image sparsity in both the transform domain and the finite-differences domain, simultaneously. In this case, the optimization is done using Equation (5.15), where λ trades sparsity with finite-differences sparsity.

$$\begin{aligned} \text{minimize} \quad & \|\Psi m\|_1 + \lambda \text{TV}(m) \\ s.t. \quad & \|F_S m - y\|_2 < \in, \end{aligned} \tag{5.15}$$

Application Areas of MRI

Rapid 3D Angiography
Angiography is becoming increasingly popular in the diagnosis of vascular disease. It attempts to image blood vessels in the body and helps detection of aneurysms, vascular occlusions, stenotic disease, and tumor feeder vessels. It also serves to guide surgical procedures and to monitor the treatment of

vascular disease. Often, a contrast agent is injected, significantly increasing the blood signal and enabling rapid data acquisition. In angiography, a significant portion of the diagnostic information comes from imaging the dynamics of the contrast agent bolus. CS is particularly suitable for angiography. Angiograms are inherently sparse images, as they are already sparse in pixel representation and are sparsified even better by spatial finite differences. The need for rapid high temporal and spatial resolution imaging implies the near inevitability of undersampling. CS offers to improve current strategies by significantly reducing the artifacts that result from undersampling.

Whole Heart Coronary Imaging
X-ray coronary angiography is the gold standard for evaluating coronary artery disease. It is, however, invasive. Multi-slice X-ray CT is a non-invasive alternative, but generates high doses of ionizing radiation. MRI is emerging as a non-invasive, non-ionizing alternative. Coronary arteries are constantly subject to heart and respiratory motion, and high-resolution imaging is therefore a challenging task. Heart motion can be handled by synchronizing acquisitions to the cardiac cycle (cardiac gating). Respiratory motion can be mitigated by long scans with navigated breathing compensation, or simply through short breath-held acquisitions. However, breath-held cardiac triggered collection schemes face strict timing constraints and very short imaging windows. The number of acquisitions is limited to the number of cardiac cycles in the breath-hold period. The number of heartbeats per period is itself limited; patients in need of coronary diagnosis cannot be expected to hold their breath for long. Also, each acquisition must be very short for avoiding motion blurring. On top of this, many slices need to be collected to cover the whole volume of the heart. Due to these constraints, traditionally breath-held cardiac triggered acquisitions have limited spatial resolution and only partial coverage of the heart. Compressed sensing can accelerate data acquisition, allowing the entire heart to be imaged in a single breath-hold.

The hardware-efficient spiral k-space trajectory is used for meeting the strict timing requirements. For each cardiac trigger, a single spiral in k-space is acquired for each slice. The heart does move considerably during the imaging period, but as each acquisition is very short, each slice is relatively immune to motion and inter-slice motion is manifested as geometric distortion across the slices rather than blurring. Geometric distortion has little effect on the clinical diagnostic value of the image. Despite the efficiency of spirals, the strict timing limitations make it necessary to undersample k-space twofold. Undersampled variable density spirals can be used for this purpose. Such spirals have an incoherent PSF. When used with linear gridding reconstruction, undersampling artifacts are incoherent and appear simply as added noise. Coronary images are generally piecewise smooth and are sparsified well by finite differences. CS reconstruction can suppress undersampling-induced interference without degrading the image quality.

Brain Imaging

Brain scans are the most common clinical applications of MRI; most such scans are 2D Cartesian multi-slice. For scan time and SNR reasons, the slices are often quite thick, often with large gaps between slices. The ideas of CS promise reduction in collection time while improving the resolution of current imagery. Indeed, by significantly undersampling the existing k-space, some of the saved collection time could be used to collect data from the missing slices and still leave a shorter collection overall. Brain images exhibit transform sparsity in the wavelet domain.

Application to Dynamic Heart Imaging

Dynamic imaging of time-varying objects is challenging in view of the spatial and temporal sampling requirements of the Nyquist criterion. Temporal resolution is often traded off against spatial resolution (or vice versa). Artifacts appear in the traditional linear reconstruction when the Nyquist criterion is violated. Now, consider a special case: dynamic imaging of time-varying objects undergoing quasiperiodic changes, e.g., heart imaging and imaging the hemodynamic response of functional brain activity. Heart motion is quasiperiodic; the time series of intensity in a single voxel is sparse in the temporal frequency domain. At the same time, a single frame of the heart "movie" is sparse in the wavelet domain. A simple transform can exploit both effects: apply a spatial wavelet transform followed by a temporal Fourier transform.

5.6.2 Compressive Synthetic Aperture Radar Imaging

Synthetic aperture radar (SAR) is a radar imaging technology capable of producing high-resolution images of the stationary surface targets and terrain. The main advantages of SAR are its ability to operate at night and in adverse weather conditions, thereby overcoming limitations of both optical and infrared systems. The basic idea of SAR is as follows: As the radar moves along its path, it transmits pulses at microwave frequencies at a uniform pulse repetition interval (PRI) which is defined as 1/PRF, where PRF is the pulse repetition frequency [24]. The reflected energy at any instant can be modeled as a convolution of the pulse waveform with the ground reflectivity function. Each received pulse is preprocessed and passed on to an image formation processor, which produces an image that is a two-dimensional mapping of the illuminated scene. The two-dimensional image formed is interpreted in the dimensions of range and cross-range or azimuth as shown in Figure 5.5.

CS in radar systems has the potential to make the following significant improvements:

- Eliminates the need for the pulse compression matched filter at the receiver
- Reduces the required receiver analog-to-digital conversion bandwidth

Hence, the system needs to operate only at potentially low "information rate" of the radar reflectivity rather than at its potentially high Nyquist rate leading to the design of new, simplified radar systems, shifting the emphasis from

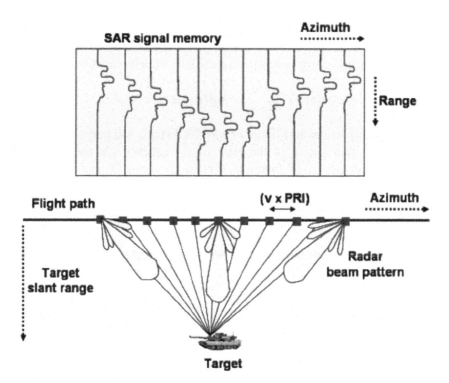

Figure 5.5   Spotlight SAR data collection in 2D.

expensive receiver hardware to smart signal recovery algorithms. In order to illustrate the CS-based radar concept, consider a simplified 1D range imaging model of a target described by $u(v)$ with range variable v. When the transmitted radar pulse $s_T(t)$ interacts with the target by means of a linear convolution, the received radar signal $s_R(t)$ is given by

$$S_R(t) = A \int S_T(t - T) u(T) dT, \qquad (5.16)$$

where the range variable v is converted to time t using $t = 2v/c$, with c the propagation velocity of light and where A represents attenuation arising from propagation and reflection. If the transmitted signal has the property that $s_T(t) * s_T(-t) \approx \delta(t)$ (which is true for PN and chirp signals), then a band-limited measurement of the radar reflectivity $u(t)$ can be obtained by pulse compression, that is, by correlating $s_R(t)$ with $s_T(t)$ in a matched filter. Analog-to-digital conversion occurs either before or after the matched filtering resulting in N Nyquist rate samples. CS-based radar approach is based on two key observations. First, the target reflectivity functions $u(t)$ obtained through the radar process are often *sparse* or *compressible* in some basis. For example, a set of K point targets corresponds to a sparse sum of delta functions as in

$$u(t) = \sum_{-i=1}^{K} a_i \delta(t - k_i) \qquad (5.17)$$

smooth targets are sparse in the Fourier or wavelet domain; and range Doppler reflectivities are often sparse in the joint time–frequency (or ambiguity) domain. Such target reflectivity functions $u(t)$ are good candidates for acquisition via CS techniques. Second, time-translated and frequency-modulated versions of the PN or chirp signals transmitted as radar waveforms $s_T(t)$ form a dictionary (the extension of a basis or frame) that is *incoherent* with the time, frequency, and time–frequency bases that sparsify or compress the above-mentioned classes of target reflectivity functions $u(t)$. This means that PN or chirp signals are good candidates for the rows of a CS acquisition matrix Φ as a "random filter," Matched filter in the radar receiver can be eliminated by combining these observations, and the bandwidth of the receiver A/D converter is lowered using CS principles. Consider the radar system that consists of the following components [25].

The transmitter is the same as in a classical radar; the transmit antenna emits a PN or chirp signal $s_T(t)$. However, the receiver does not consist of a matched filter and high-rate A/D converter but rather only a low-rate A/D converter that operates not at the Nyquist rate but at a rate proportional to the target reflectivity's compressibility. Consider a target reflectivity generated from N Nyquist rate samples $x(n)$ via $u(t) = x\lfloor t/\Delta \rfloor, n = 1, \ldots, N$ on the time interval of interest $0 \leq t < N\Delta$. The radar transmits a PN signal generated from a length N random Bernoulli vector $p(n)$ via $s_T(t) = p\lfloor t/\Delta \rfloor$. The received radar signal $s_R(t)$ is given by Equation (5.15), and it is sampled every $D\Delta$ seconds, where $D = \_N/M\_$ and $M < N$, to obtain the M samples,

$$
\begin{aligned}
y(m) &= s_R(t)_{t=mD\Delta} \\
&= A \int_0^{N\Delta} S_T(mD\Delta - \tau)u(\tau)d\tau \\
&= A \sum_{n=1}^{N} p(mD - n) \int_{(n-1)\Delta}^{n\Delta} u(\tau)d\tau \\
&= A \sum_{n=1}^{N} p(mD - n)x(n),
\end{aligned}
\tag{5.18}
$$

which constitute precisely a scaled version. The PN sequence implements a random filter, and hence, the low-rate samples y contain sufficient information to reconstruct the signal x corresponding to the Nyquist rate samples of the reflectivity $u(t)$ via linear programming or a greedy algorithm. Chirp pulses also yield similar results.

5.6.3 Compressive Passive Millimeter Wave Imaging

Passive millimeter-wave imaging (PMMWI) offers significant advantages over optical visible light and infrared imaging, as millimeter waves are less affected by adverse conditions such as clouds, fog, smoke, and dust. Moreover, PMMWI can be used during both night and day. These advantages make PMMWI an ideal imaging modality for search and rescue, law enforcement, and military applications. Unfortunately, current PMMWI systems suffer from several limitations in terms of the trade-off between signal-to-noise ratio (SNR) and acquisition time. A typical PMMW imager consists of an imaging lens or mirror which focuses the radiation of a distant object onto the antenna of a radiometer. Two types of imaging systems are commonly used: The single-pixel scanning imager uses a movable antenna to mechanically scan the image pixels, while the focal plane imager uses a 2D antenna array to acquire the whole image in a single acquisition. The disadvantage of using an antenna array is that the SNR is poor as each antenna only receives a fraction of the radiation and the spatial resolution is limited due to the minimum aperture size. In addition, a separate radiometer is required for each pixel, which causes such systems to be prohibitively expensive and bulky. On the other hand, the scanning imager receives the full radiation at each location, but the acquisition time for all N pixels of an image can be very large. PMMW imager as shown in Figure 5.6 uses the principle of compressive sensing (CS) for reducing the image acquisition time for a given SNR level. The principle of the imager is similar to the single-pixel CS camera, but instead of a digital micromirror device (DMD), masks with reflective and transmissive elements are used [26]. Incoherent measurements of the unknown PMMW image are obtained by performing subsequent acquisitions with different masks. A system for CS terahertz (THz) imaging is provided which uses a total of 600 different random masks. Handling such a large number acquisition masks poses major difficulties in practice. More recently, a CS THz imaging system using Toeplitz matrix-based masks has been presented.

This mask construction has the advantage of representation of large number of masks by a single acquisition mask. Mask construction is done based on an S-matrix, which is closely related to a Hadamard matrix. This construction allows an efficient representation of N masks using a single mask, from which a subsection of it can be used for each measurement. Nonlinear reconstruction of the unknown PMMW image from incoherent measurements is possible since PMMW images are typically sparse (or compressible) in some transform basis (e.g., wavelets). To reconstruct the PMMW images, a novel Bayesian reconstruction algorithm which exploits the high sparsity inherent in PMMW images has been used. Hierarchical Bayesian

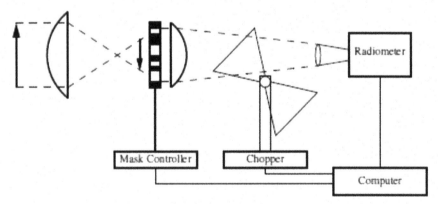

Figure 5.6   Schematic of the proposed PMMW imager.

formulation with sparsity inducing Gaussian priors on the high-pass filter outputs of each pixel is modeled. The algorithm simultaneously estimates the unknown image and the algorithmic parameters from the acquired incoherent measurements.

A two-dimensional mask is placed at the focal plane of the imager for collecting the incoherent measurements of an object, while a collection lens is used for focusing the radiation passing through the mask onto the antenna of a Dicke switched radiometer with a frequency range of 146–154 GHz, corresponding to wavelengths of 1.9–2.1 mm. The mask is used to multiply the radiation at different locations, and the collection lens works effectively as an integrator. Combined, these two elements implement an inner product of the incident radiation with the mask configuration. First, the unknown image x must have a sparse representation in some transform domain. It is reasonable to assume that PMMW images meet this requirement as they are structurally similar to natural images, for which sparse representations exist. Second, the measurement matrix has to be random to ensure coherence of the measurements y. In practice, using general random measurement matrices (such as those drawn from Gaussian or uniform probability distributions) is generally not possible when using a mask-based system, as negative matrix elements cannot be implemented. Even for matrices where all elements are positive, a practical implementation requires masks which let a specified fraction of the radiation pass at each location, which is difficult to achieve. In the proposed system, binary measurement matrices which are often used in practice are utilized, as they can be implemented using masks with transmissive and reflective elements, corresponding to the 1's and 0's in the matrices, respectively. A second problem encountered in practice is that even with a significant decrease in the number of measurements, imaging an object requires M different masks. This causes a major practical problem even for small image sizes, as the number of masks M can be as high as 200 for recovering an image of 1000 pixels. A larger acquisition mask of size $(2p - 1)$ $(2q - 1)$ is utilized, for alleviating this problem, which allows to perform up to N acquisitions by using $p \times q$ sections of the mask for each acquisition. In practice, change of mask is accomplished by simply translating the larger acquisition mask horizontally and vertically, with the beam profile covering a $p \times q$ section of the mask. To construct the larger acquisition mask, a cyclic S-matrix of size $N \times N$ is constructed using the twin prime construction and the elements of the first row of the S-matrix are re-arranged into a $p \times q$ matrix. The larger acquisition mask is then obtained by periodically repeating the $p \times q$ matrix and retaining a matrix of size $(2P - 1)$ $(2q - 1)$.

A new millimeter wave imaging modality with extended depth of field that provides diffraction limited images based on a significant reduction in scan time is also presented. The technique uses a cubic-phase element in the pupil of the system and a nonlinear recovery algorithm to produce images that are insensitive to object distance [27]. This system uses far fewer number of measurements than the conventional systems do and can reduce the scan time significantly. It is a 94-GHz stokes vector radiometer used for mmW measurements. It is a single-beam system that produces images by scanning in horizontal and vertical axis. The radiometer has a thermal sensitivity of 0.3 K with a 30-ms integration time and 1 GHz bandwidth per pixel. A Cassegrain antenna is mounted to the front of the radiometer receiver that has 24"-diameter primary parabolic reflector and a 1.75"-diameter secondary hyperbolic reflector. The position of the hyperbolic secondary is variable.

5.6.4 Compressive Light Transport System

The complexity in the appearance of real-world scenes remains a principal challenge to simulate in computer graphics. Modeling and rendering such scenes under novel lighting with traditional computer graphics is an arduous task which requires talent and experience. As a result, image-based representations have gained popularity, wherein traditional modeling and rendering are replaced by image-based acquisition and relighting. Unfortunately, acquiring this image-based data is time-consuming and storage intensive. For instance, acquiring high-resolution data sets for scenes, such as the bowl of peppers, can require tens of thousands of photographs and gigabytes of storage. Relit images of a scene can be represented using compressive sensing to greatly reduce acquisition time and storage space. Due to the acquisition complexity and storage requirements for such relightable data sets, many different methods have been developed to speed up the acquisition of light transport. Previous image-based techniques rely on sampling methods adaptive methods or techniques using non-trivial measurement patterns or specialized projector camera setups. A central concept in these image-based methods is a reflectance field, an 8D entity that abstracts the light transport through a scene in terms of incident and outgoing illumination on a bounding volume surrounding the scene. Both the incident and outgoing light field are 4D fields; for each position on the bounding volume (2D), all possible directions (2D) are considered. Capturing and handling these 8D fields is difficult. Therefore, most methods consider a reduced approximation. The outgoing light field is reduced to a

2D field by fixing the viewpoint and where the incident light field is also reduced to a 2D field by assuming that incident illumination only varies positionally over the bounding volume. Conventional compressive sensing algorithms typically deal with a single function that needs to be measured and reconstructed. In the case of image-based relighting, many (reflectance) functions are sampled in parallel (i.e., one for each camera pixel), and each of these functions needs to be reconstructed individually using a compressive sensing algorithm. However, this is computationally expensive and ignores spatial relationships between neighboring pixel's reflectance functions. There-fore, a novel hierarchical reconstruction algorithm that is able to extract the compressed signals using less computation and with greater accuracy is required. The standard measurement patterns used in compressive sensing assume an almost perfect acquisition system, where the only source of error is due to the observation of the measurements. In reality, other sources of error do exist. For instance, in image-based relighting, measurements are performed by emitting measurement patterns from a controllable lighting device. This device can also introduce errors such as quantization. Hence, illumination patterns that increase the signal-to-noise ratio of the measurements are designed.

Image-based relighting can be written compactly in matrix notation as $c = T\mathbf{1}$, where T is a $p \times n$ matrix that defines the light transport between n light sources and p camera pixels, c represents these pixels in an observed camera image, stacked in a vector of length p, and l represents the illumination conditions, stacked in a vector of length n [28]. An illumination condition l can consist in any combination of point, directional, and area light sources. Each element in l indicates the emitted radiance of the corresponding light source. The relighting process consists of two stages: First, there is the measurement stage that determines the transport matrix T, by observing the scene under a selection of m different illumination conditions l $j, j \leq m$. Second, there is the relighting stage that computes newly relit virtual observations, given the measured transport matrix from the first stage and a user-defined lighting condition.

For the acquisition stage, multiple illumination conditions $G = [\mathbf{1}0, \ldots, \mathbf{1}m]$ and their corresponding observations $C = [c0, \ldots, cm]$ can also be compactly denoted in matrix notation as $C = TG$. Each row ti of the transport matrix T represents the reflectance function of the ith pixel in the camera image. The observations of the ith pixel are thus governed by an equation similar to $ci, = ti, \cdot G$. This maps directly to a compressive sensing context where L fulfills the role of the measurement ensemble φ and the reflectance function ti corresponds to the discrete signal X. Nevertheless, before compressive sensing can be applied to each reflectance function, reflectance functions need to be either sparse or compressible. Suppose a specific basis B is used and it is orthogonal (i.e., $B^{-1} = B^T$), then C = TG, = $T(BB^T)G$, = $\hat{T}B^T G$, where $\hat{T} = TB$ is the transport matrix expressed in the basis B. By choice of B, it is evident that $\hat{T}$ is compressible and thus suited for measurement by compressive sensing. The illumination pattern is defined as $G = B\varphi$, where φ is one of the theoretical compressive sensing ensembles. This yields $C = \hat{T}(B^T B)\varphi = \hat{T}\varphi$.

The illumination patterns are the measurement (row) vectors φj from the ensemble, projected onto the inverse basis B^T (note that the transpose is due to the pre-multiplication in the definition of $G = B\varphi$. Suppose the measurement ensemble consists of independently and identically distributed Gaussian random variables and the basis is a Haar wavelet basis. Illumination patterns are obtained by applying an inverse Haar wavelet transform on the Gaussian noise vectors. Taking photographs of the scene illuminated by these illumination patterns yields the matrix of observations C. Each row in this matrix (i.e., the observed values for a specific pixel location) yields a vector of measurements that can be used for reconstructing the original reflectance function of that pixel expressed in the basis B.

Brute-force compressive light transport sensing: During acquisition, the illumination patterns as defined earlier are emitted onto the scene and each HDR photograph of the scene is recorded. Next, a reflectance function is inferred for each pixel separately by applying a compressive sensing decoding algorithm to the observations of only that pixel. This brute-force algorithm has the advantage that it is straightforward to implement, and all theoretical properties of compressive sensing are still valid.

Inverse Light Transport Computation
Inverse global illumination was introduced but with the focus on estimating reflectance properties, rather than compensating input lighting patterns. Inverse light transport can be computed by inverting the forward light transport matrix. f-LTM (forward light transport matrix) is first reconstructed from the measurements before deriving the i-LTM from the f-LTM. In the conventional setting, the computation of i-LTM (inverse light transport matrix) mainly involves solving the inverse problem $I_{out} = TI_{in}$, where I_{out} is the observed scene under global illumination and I_{in} is the input light pattern. There are two approaches involved in the solution of the inverse light transport problem. The first approach solves the inverse problem as a system of linear

equations. Effective methods of this approach are the Jacobi method and the Gauss–Seidel method that involve iterative vector–matrix multiplication. The second approach is to invert the matrix T. Matrix inversion ideally corresponds to solving the matrix equation $TT^{-1} = I$. Hence, the second approach has a much higher computational complexity than the first approach. Once T^{-1} is precomputed, given a new I_{in}, I_{out} can be easily obtained with a single-step vector–matrix multiplication. However, inverting a large-size f-LTM is computational and memory intensive, leading to the introduction of various forms of approximation. In contrast to this conventional procedure, CS-based method enables computation of inverse light transport directly from the measurements, without the need for explicitly reconstructing the f-LTM. By skipping the intermediate steps of f-LTM reconstruction and by avoiding the approximate light transport inversion, it is logical that CS-based method will be more computationally efficient and accurate. Pre-computing i-LTM is important when it comes to processing a video or a sequence of images. Without X an explicit i-LTM, the implicit matrix inversion process needs to be repeated for each frame of a video and hence incurs a high computational cost [29].

5.7 Single-Pixel Camera

Single-pixel CS camera architecture is basically an optical computer (comprising a DMD, two lenses, a single photon detector, and an analog-to-digital (A/D) converter) that computes random linear measurements of the scene under view. The image is then recovered or processed from the measurements by a digital computer. The camera design reduces the required size, complexity, and cost of the photon detector array down to a single unit, which enables the use of exotic detectors that would be impossible in a conventional digital camera. The random CS measurements also enable a trade-off between space and time during image acquisition. Figure 5.7 shows the picture of the single-pixel camera [30].

The single-pixel camera is an optical computer that sequentially measures the inner products $y[m] = < X, \varphi_m >$ between an N-pixel sampled version x of the incident light field from the scene under view and a set of two-dimensional (2D) test functions $\{\varphi_m\}$. As shown in Figure 5.8, the light field is focused by biconvex Lens 1 not onto a CCD or CMOS sampling array but rather onto a DMD consisting of an array of N tiny mirrors. Each mirror corresponds to a particular pixel in X and φ_m can be independently oriented either toward Lens 2 (corresponding to a 1 at that pixel in φ_m) or

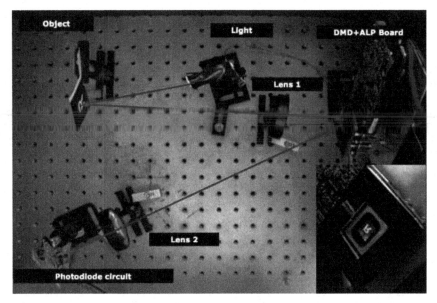

Figure 5.7   Single-pixel CS camera.

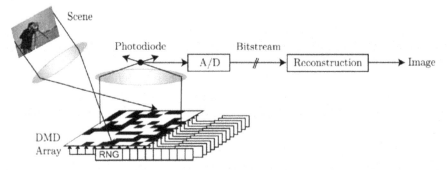

Figure 5.8   Block diagram of single-pixel camera.

away from Lens 2 (corresponding to a 0 at that pixel in φ_m). The reflected light is then collected by biconvex Lens 2 and focused onto a single photon detector (the single pixel) that integrates the product $X[n]\varphi_m[n]$ to compute the measurement $\varphi_m\, y[m] =< X, \varphi m >$ as its output voltage. This voltage is then digitized by an A/D converter. Values of φ_m between 0 and 1 can be obtained by dithering the mirrors back and forth during the photodiode integration time. Estimation and subtraction of the mean light intensity from each measurement is done for obtaining φ_m with both positive and negative values (± 1), which is easily measured by setting all mirrors to the full-on 1 position.

To compute CS randomized measurements $y = \Phi X$, the mirror orientations φ_m are randomly set using a pseudorandom number generator, measure $y[m]$, and repeat the process M times to obtain the measurement vector y. Since the DMD array is programmable, test functions φ_m drawn randomly from a fast transform such as a Walsh, Hadamard, or noiselet transform can be employed.

The single-pixel design reduces the required size, complexity, and cost of the photon detector array down to a single unit, which enables the use of exotic detectors that would be impossible in a conventional digital camera. Examples of detectors include a photomultiplier tube or an avalanche photodiode for low-light (photon-limited) imaging, a sandwich of several photodiodes sensitive to different light wavelengths for multimodal sensing, a spectrometer for hyper spectral imaging, and so on.

In addition to sensing flexibility, the practical advantages of the single-pixel design include the facts that the quantum efficiency of a photodiode is higher than that of the pixel sensors in a typical CCD or CMOS array and that the fill factor of a DMD can reach 90%, whereas that of a CCD/CMOS array is only about 50%. An important advantage to highlight is the fact that each CS measurement receives about N/2 times more photons than an average pixel sensor, which significantly reduces image distortion from dark noise and read-out noise. Theoretical advantages that the design inherits from the CS theory include its universality, robustness, and progressivity.

The single-pixel design falls into the class of multiplex cameras. The baseline standard for multiplexing is classical raster scanning, where the test functions $\{\varphi_m\}$ are a sequence of delta functions $\delta[n - m]$ that turn on each mirror in turn. There are substantial advantages to operating in a CS rather than raster scan mode, including fewer total measurements (M for CS rather than N for raster scan) and significantly reduced dark noise.

Sample image reconstructions from single pixel camera are shown in Figure 5.9. A black-and-white picture of an "R" is used for reconstructing using single-pixel camera. The original dimension of the image is $N = 256 \times 256$. The reconstructed images using total variation minimization from only 2% and 10% measurements are shown in the second and third columns of Figure 5.9, respectively.

One of the main limitations of this architecture is its requirement of the camera to be focused on the object of interest until enough measurements are collected. This may be prohibitive in some applications.

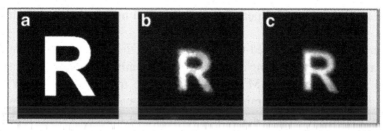

Figure 5.9 Sample image reconstructions. (a) 256×256 original image. (b) Image reconstructed from only 2% of the measurements. (c) Image reconstructed from only 10% of the measurements.

5.8 Lensless Imaging by Compressive Sensing

Bell Laboratories have built a camera that has no lens and uses just a single sensing pixel for taking photographs using compressive sensing, which relies on the assumption that many common measurements have a lot of redundancy [31]. So it is possible to acquire the same data by taking just a few carefully chosen measurements. The skill comes in knowing which measurements to take and how to reassemble them. This technique has the potential to revolutionize traditional optical imaging, which relies on having a lens to create an image and a device to record the light such as an array of pixels or light-sensitive film.

5.8.1 Lensless Imaging Architecture

Bell Laboratories' device is quite simple. It comprises an LCD panel, which acts as an array of apertures that open to allow light to pass through and a single sensor, which can detect three colors of light. Each aperture is addressable so can be opened to allow light to pass through or kept closed. The prototype was built using low cost, commercially available components. There are a number of benefits to this sort of approach to photography. The first is that no lens is used, reducing cost and complexity. Furthermore, no scene will ever be out of focus, and the clarity of images is simply limited by the number of aperture elements. It can also be used for imaging of other light spectra such as infrared or millimeter waves. Some of the apertures on the LCD panel are opened randomly to allow light to pass through. Next, a different array of apertures is opened to allow different parts of the scene to be recorded. This happen a number of times before the correlated snapshots are reassembled into an image.

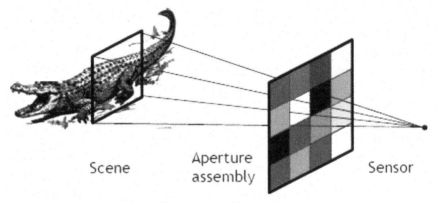

Figure 5.10 Lensless imaging architecture.

The proposed architecture is shown in Figure 5.10. It consists of two components: an aperture assembly and a sensor. The aperture assembly is made up of a two-dimensional array of aperture elements. The transmittance of each aperture element can be individually controlled. The sensor is a single detection element, which is ideally of an infinitesimal size. Each element of the aperture assembly, together with the sensor, defines a cone of a bundle of rays and the cones from all aperture elements are defined as pixels of an image. The integration of the rays within a cone is defined as a pixel value of the image. Therefore, in the proposed architecture, an image is defined by the pixels which correspond to the array of aperture elements in the aperture. An image can be captured by using the sensor to take as many measurements as the number of pixels. For example, each measurement can be made from reading of the sensors when one of the aperture sensor elements is completely open and all others are completely closed, which corresponds to the binary transmittance of 1 (open) or 0 (closed). The measurements are the pixel values of the image when the elements of the aperture assembly are opened one by one in certain scan order. This way of making measurements corresponds to the traditional representation of a digital image pixel by pixel.

5.8.1.1 Compressive measurements
A sensing matrix is first defined for making compressive measurements. Each row of the sensing matrix defines a pattern for the elements of the aperture assembly, while the number of columns in a sensing matrix is equal to the number of total elements in the aperture assembly. Each value in a row of the sensing matrix is used for defining the transmittance of an element of

the aperture assembly. A row of the sensing matrix therefore completely defines a pattern for the aperture assembly and allows the sensor to make one measurement for the given pattern of the aperture assembly. The number of rows of the sensing matrix is the number of measurements, which is usually much smaller than the number of aperture elements in the aperture assembly (the number of pixels). Let the sensing matrix be a random matrix whose entries are random numbers between 0 and 1. The transmittance of each aperture element is controlled to equal the value of the corresponding entry in a row of the sensing matrix for making a measurement. The sensor integrates all rays transmitted through the aperture assembly. The intensity of the rays is modulated by the transmittances before they are integrated. Therefore, each measurement from the sensor is the integration of the intensity of rays through the aperture assembly multiplied by the transmittance of respective aperture element. A measurement from the sensor is hence a projection of the image onto the row of the sensing matrix.

5.8.1.2 Selection of aperture assembly

The architecture of this work is flexible to allow a variety of implementations for the aperture assembly. Liquid crystal sheets [32] may be used for imaging of visible spectrum. Micromirror arrays [33] may be used for both visible spectrum imaging and infrared imaging. When a micromirror array is used, the array is not placed in the direct path between the scene and the sensor, but rather, it is placed at an angle so that the rays from the scene is reflected to the sensor when the micromirrors are turned to a particular angle. Further, when the micromirror array is used, the transmittance is binary, taking the values of 0 and 1. The masks may be used for Terahertz imaging [34, 35] and millimeter wave imaging [36].

5.8.2 Prototype for Lensless Imaging

The imaging device consists of a transport monochrome liquid crystal display (LCD) screen and a photovoltaic sensor enclosed in a light tight box, shown in Figure 5.11. The LCD screen functions as the aperture assembly, while the photovoltaic sensor measures the light intensity. The photovoltaic sensor is a tricolor sensor, which outputs the intensity of red, green, and blue lights. A computer is used for generating the patterns for aperture elements on LCD screen according to each row of the measurement matrix. The light measurements are read from the sensor and recorded for further processing.

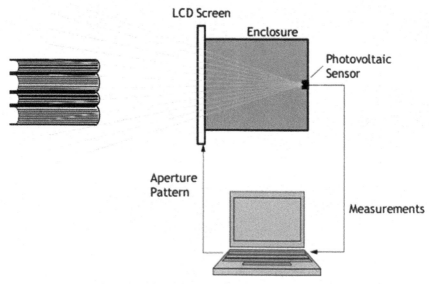

Figure 5.11 Schematic illustration of the lensless compressive image prototype.

The LCD panel is configured for displaying a maximum resolution of $302 \times 217 = 65534$ black or white squares. Since the LCD is transparent and monochrome, a black square means the element is opaque, and a white square means the element is transparent. Therefore, each square represents an aperture element with the transmittance of a 0 (black) or 1 (white). A sensing matrix which is constructed from rows of a Hadamard matrix of order $N = 65536$ is used for capturing compressive measurements. Each row of the Hadamard matrix is permuted according to a predetermined random permutation. The first 65534 elements of a row are then simply mapped to the 65534 aperture elements of the LCD in a scan order from the top to bottom and then from left to right. An "1" in the Hadamard matrix turns an aperture element transparent and a "–1" turns it opaque. The measurements values for red, green, and blue are taken by a sensor at the back of the enclosure box and recorded by the control computer. Only one sensor is used for taking the measurements. A total number of 65534, which corresponds to the total number of pixels of the image, different measurements can be made with the prototype. Fractional of the total possible measurements are only considered for analysis. For example, 25% of measurements means 16384 measurements are taken and used in reconstruction, which is a quarter of the total number

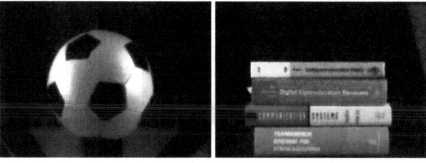

(a) Soccer image, 12.5%   (b) Books image, 25%

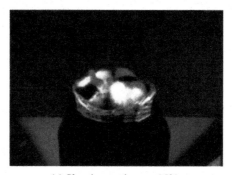

(c) Sleeping cat image, 25%.

Figure 5.12   Images reconstructed from varying measurements.

of pixels (65534). Similarly, 12.5% means 8192 measurements are taken and used in reconstruction. A standard reconstruction method commonly known as L_1 minimization of total variation is used for the image reconstruction from the reduced measurements. The various images reconstructed form varying measurements are shown in Figure 5.12.

5.9 Case Study: Image Transmission in WMSN

Wireless multimedia sensor network (WMSN) is a network of wirelessly interconnected sensor nodes equipped with multimedia devices capable of retrieving video streams, images, audio streams, and scalar sensor data. The availability of inexpensive hardware such as CMOS cameras and microphones has led to the development of WMSNs. WMSNs are able to store, process in real-time, correlate, and fuse multimedia data originated from heterogeneous sources [37]. WMSNs are resource constrained and have high bandwidth demand. It is essential to reduce both computational and communication energy consumptions involved in image transmission to increase the lifetime of these networks. Recently, CS has been widely used, allowing the entire signal to be determined from relatively few linear measurements. It is used for capturing and representing compressible signals at a rate significantly below the Nyquist rate. It simultaneously senses and compresses the data at low complexity. Hence, the CS method applied for reducing the energy consumption in WMSNs is provided.

In the CS-based method, the image is captured, divided into blocks of size 8×8, and sparsified using binary DCT (BinDCT) [38], which is then given to the CS process. The eight-point BinDCT of the C1 configuration is used to obtain the multiplierless transform with dyadic values. BinDCT-C1 needs 23 shifts and 42 additions for obtaining the transformed coefficients [38]. Although there will be a reduction in the reconstruction quality, BinDCT-C1 is adopted in PCS to achieve energy efficiency. CS measurements are obtained by taking the M random projections (Y_2) of the image.

$$Y_2 = \varphi_2 X = \varphi_2 \psi WS = \Theta WS \tag{5.19}$$

The measurements are obtained by multiplying the random $M \times N$ Bernoulli (random ± 1) matrix and DCT coefficient matrix. The DCT coefficient matrix is multiplied by a weighing matrix (W) to extract the requisite sparse elements. W is a binary matrix of size 8×8, having K number of ones along the diagonal in the requisite indices. The measurements are retrieved with only addition operations against floating-point multiplications involved in the conventional measurement process using the random Gaussian matrix. They are encoded (EY_2) using the modified lossless entropy coding algorithm.

In the receiver section, the recovery from the decoded measurements Y_2 is performed using OMP. The reconstruction process is formulated as follows:

$$\hat{S} = \min_S \|S\|_{\ell_1} \text{ subject to, } y_2 = \Theta WS, X = \psi \hat{S} \tag{5.20}$$

IDCT is obtained for the values obtained using OMP, and the image is recovered. The recovered image quality depends on measurements M.

The energy consumption involved in the entire process is the main criteria as the WMSN is operated with batteries on board. The computational energy consumption of the provided method is calculated by using ATMEGA 128 [39] as the target platform. It operates at 8 MHz with an active power consumption of 22 mW. Compilation is performed via WinAVR with the "–O3" optimization setting. The experiments are performed for an 8×8 image block with sparsity level "2", and the results are shown in Table 5.2.

Table 5.2   Computation time and energy for TC (DCT), CCS (DCT), and PCS (BinDCT-C_1) methods for an 8×8 image block on ATMEGA 128 processor

K	Method	DCT Cycles	Measurement Cycles	Encoding Cycles	Total Cycles	Execution Time (ms)	E_{COMP} (μJ)
	TC	10085007	50922	12124	10148053	1268.5	27907
2	CCS	10085007	33007	19625	10137639	1267.2	27879
	PCS	53484	32061	14875	100420	12.5	275

The CS-based method enables an 81.56% reduction in computational energy when compared with the conventional methods. Moreover, it also yields reduced total energy consumption.

5.10 Summary

The use of compressed sensing in image processing can often lead to new insights and solutions. This is evidently true for image-processing applications in wireless sensor networks. A wide variety of imaging applications can also be enhanced through CS. In this chapter, application of CS to various image-processing domains such as fusion, denoising, and reconstruction are provided. In addition, it also provides a concise exposition to several imaging applications.

References

[1] Bai, Q., and Jin, C. (2015). "Image Fusion and Recognition based on Compressed Sensing Theory" *Intl. J. Smart Sensing Intelligent Syst.* 8 (1), 159–180, March.

[2] Wakin, M. B., Laska, J. N., Duarte, M. F., Baron, D., Sarvotham, S., Takhar, D., Kelly, K. F. and R. G. Baraniuk (2006). "An architecture for compressive imaging," in *Proc. of the IEEE Int. Conf. Image Process.* Oct., pp. 1273–1276.

[3] Marr, D. (1982). "Vision: A Computational Investigation into the Human Representation and Processing of Visual Information," W. H. Freeman and Company, New York, pp. 29–61.

[4] Tao Wan Nishan Canagarajah Alin Achim, "Compressive Image Fusion", in the Proceedings of ICIP 2008, pp. 1308–1311.

[5] Kazemi, V., Seyedarabi, H., and Aghagolzadeh, A. (2015). "Compressive multifocus image fusion for visual sensor networks", *Int. J. Image Process. Visual Commun.* 3 (3).

[6] Wan, T., and Qin, Z.C. (2010). "An application of compressive sensing for image fusion," *2010 ACM Int. Conf. Image Video Retrieval*, pp. 3–9.

[7] Luo, X. Y., Zhang, J., Yang, J. Y., and Dai, Q. H. (2009) "Image fusion in compressed sensing," *Proceedings of the 16th IEEE International Conference of Image Processing*, pp. 2205–2208.

[8] Li, X., and Qin, S.Y. (2011). "Efficient fusion for infrared and visible images based on compressive sensing principle," *IET Image Process.* 5 (2), 141–147

[9] Muhammad Yousuf, B., Lai, E.M.-K., and Punchihew, A. (2014). "Compressed sensing-based distributed image compression", *Appl. Sci.* 4, 128–147.

[10] Bhatnagar, D., and Budhiraja, S. (2012). "Image Compression using DCT based Compressive Sensing and Vector Quantization" *Int. J. Comput. Appl.* 50 (20), 34–38, July.

[11] Deng, C., Lin, W., Lee, B.-S. and Lau, C. T. (2010). "Robust Image Compression Based on Compressive Sensing" in *The Proceedings of ICME*, pp. 462–467.

[12] Durand, S., and Froment, J. (2003). "Reconstruction of wavelet coefficients using total variation minimization", *SIAMJ Sci. Comput.* 24 (5), 1754–1767.

[13] Mallat, S. (2008). "A wavelet tour of signal processing. The sparse way", 3rd edn. (New York: Academic Press).

[14] Jin, J., Yang, B., Liang, K., and Wang, X. (2014). "General image denoising framework based on compressive sensing theory", *Comput. Graphics*, 38, 382–391.

[15] Candes, E., and Romberg, J. (2006). "Sparsity and incoherence in Compressive Sampling," *Inverse Problems* 23 (3), 969–985, Nov.

[16] Cohen, A., DeVore, R., Petrushev, P., and Xu, H. (1999). "Nonlinear approximation and the space BV(R2)1," *Am. J. Mathe.* 121, 587–628.

[17] Chan, T. F., Osher, S., and Shen, J. (2001). "The digital TV filter and nonlinear denoising," *IEEE Trans. Image Process.* 10 (2), 231–241, Feb.

[18] Rudin, L. I., Osher, S., and Fatemi, E. (1992). "Nonlinear total variation based noise removal algorithms," *Physica D*, 60, pp. 259–268.

[19] Marim, M. M., Angelini, E. D., Olivo-Marin, J.-C. (2009). A Compressed Sensing Approach for Biological Microscopy Image Denoising. R_emi Gribonval. SPARS'09–Signal Processing with Adaptive Sparse Structured Representations, Apr 2009, Saint Malo, France.

[20] You, H., and Zhu, J. (2015). "Image Reconstruction based on Block-based Compressive Sensing," In: *The Proceedings of the 38th Australasian Computer Science Conference (ACSC 2015)*, Sydney, Australia, pp. 27–30.

[21] Egiazarian, K., Foi, A., and Katkovnik, V. "Compressed Sensing Image Reconstruction via Recursive Spatially Adaptive Filtering" in the *Proceedings of IEEE International Conference on Image Processing, ICIP 2007*, vol. 1, pp. 549–552.

[22] Ni, K., Mahanti, P., Datta, S., Roudenko, S., and Cochran, D. (2009). "Image reconstruction by deterministic compressed sensing with chirp matrices", *Procee. SPIE*, 7497, 2009, doi:10.1117/12.832649

[23] Lustig, M., Donoho, D. L., and Santos, J. M., and Pauly, J. M. (2008). "Compressed Sensing MRI", *IEEE Signal Process. Mag.* March, pp. 72–82.

[24] Patel, V. M., Easley, G. R., Healy Jr., D. M., and Chellappa, R. (2009). "Compressed Sensing for Synthetic Aperture Radar Imaging", In the *Proceedings of 16th IEEE International Conference on Image Processing (ICIP)*, Cairo, pp. 2141–2144.

[25] Baraniuk, R., and Steeghs, P. (2007). "Compressive Radar Imaging", *In the Proceedings of IEEE Radar Conference*, Boston, MA, pp. 128–133.

[26] Babacan, S. D., Luessi, M., Spinoulas, L., Katsaggelos, A. K., Gopalsami, N., Elmer, T., Ahern, R., Liao, S., Raptis, A. (2011). "Compressive Passive Millimeter-Wave Imaging", in the *Proceedings of 18th IEEE International Conference on Image Processing*, pp. 2705–2708.

[27] Patel, V. M., and Mait, J. N. (2012). "Passive Millimeter-Wave Imaging with Extended Depth of Field and Sparse Data", in the *Proceedings of the IEEE International Conference on Acoustics, Speech and Signal Processing (ICASSP)*, pp. 2521–2524.

[28] Peers, P., Mahajan, D. K., Lamond, B., Ghosh, A., Matusik, W., Ramamoorthi, R., and Debevec, P. (2009). Compressive light transport sensing. *ACM Trans. Graph.* 28 (1), Article 3, Jan, 18.

[29] Chu, X., Ng, T.-T., Pahwa, R., Quek, Tony, Q. S., and Huang, T. S. (2011). "Compressive Inverse Light Transport", in the *Proceedings of 22nd British Machine Vision Conference*, pp. 38.1–38.11, MVA Press, September.

[30] Duarte, M. F., Davenport, M. A., Takbar, D., et al. (2008). "Single-pixel imaging via compressive sampling: Building simpler, smaller, and less-expensive digital cameras," *IEEE Signal Process. Mag.* 25 (2), 83–91.

[31] Huang, G., Jiang, H., Matthews, K., and Wilford, P. (2013). "Lensless Imaging by Compressive Sensing", in the *Proceedings of 20th IEEE International Conference on Image Processing (ICIP)*, pp. 2101–2105.

[32] Zomet, A., and Nayar, S. K. (2006). "Lensless Imaging with a Controllable Aperture", *IEEE Conference on Computer Vision and Pattern Recognition (CVPR)*, Jun.

[33] Takhar, D., Laska, J. N., Wakin, M. B., Duarte, M. F., Baron, D., Sarvotham, S., Kelly, K. F., and Baraniuk, R. G. (2006). "A New Compressive Imaging Camera Architecture using Optical-Domain Compression", *Proceedings of IS&T/SPIE Computational Imaging IV*, Jan.

[34] Chan, W. L., Charan, K., Takhar, D., Kelly, K. F., Baraniuk, R. G., and Mittleman, D. M. (). "A single-pixel terahertz imaging system based on compressed sensing," *Appl. Phys. Lett.* 93 (12), 121105–3, Sept. 2008.

[35] Heidari, A., and Saeedkia, D. (2009). "A 2D camera design with a single-pixel detector," in *IRMMW-THz 2009. IEEE*, pp. 1–2.

[36] Babacan, S. D., Luessi, M., Spinoulas, L., Katsaggelos, A. K., Gopalsami, N., Elmer, T., Ahern, R., Liao, S., and Raptis, A. (2011). "Compressive passive millimeter-wave imaging," in *2011 18th IEEE International Conference on Image Processing (ICIP)*, Sept. pp. 2705–2708.

[37] Akyildizian, F., Tommaso, M., Chowdhury Kaushik, R. (2007). A survey on wireless multimedia sensor networks. *Comput. Netw.* 51 (4), 921–960.

[38] Jie, L., Tran Trac, D. (2001). Fast multiplierless approximations of the DCT with the lifting scheme. *IEEE Trans Signal Process* 49 (12), 3032–3044.

[39] Dong-U, L., Hyungjin, K., Steven, T., Mohammad, R., Estrin, D., John, V. (2007). Energy-optimized image communication on resource-constrained sensor platforms. In: *Proceedings of 6th International Conference on Information Processing in Sensor Networks*, USA.

6

Compressive Sensing for Computer Vision

6.1 Introduction

Recent trends indicate that many challenging computer vision and image-processing problems are being solved using compressive sensing and sparse representation algorithms. Image representation, recognition, modeling, enhancement, restoration, analysis, and reconstruction from projections have been a few of the areas which have been viewed from a different angle after the introduction of compressive sensing (CS). CS helps in selecting some important data from a huge volume of data efficiently. The challenging task of computer vision is to develop systems which mimic, represent, and analyze the behavior characterized by human beings.

The systems which aim at understanding and representing such behavior should have highly accurate sensing and acquisition capabilities. This can be achieved by preprocessing for input data formatting and actual methodology of feature formation and analysis, followed by post-processing such as enhancement and restoration. Generally, the systems are application dependent but the basic steps to be followed include image acquisition, preprocessing, image feature extraction, segmentation, and classification. Image acquisition is the first stage involved in any computer vision system. A computational model of a camera, at least for its geometric part, tells how to project a natural 3D scene onto an image and how to project back to 3D from the image. There are different camera models classified according to different criteria such as viewpoint, complexity and imaging type. The two plane model, the fisheye model, and the affine model are some of the commonly used camera models in the computer vision systems. A CCD or a CMOS sensor is invariably used in most of the spatially sampled imaging systems with a predefined set of points defined on the imaging plane which follow the Shannon–Nyquist sampling theorem. Sampling of amplitudes, also known as quantization, and temporal sampling, defined by the frame-rate, are also involved in the acquisition process. Before a computer vision method can be applied to an image for extracting certain features, it is usually necessary to format the data to enable satisfaction of certain criterion required by the method which is called as preprocessing. Feature extraction is an important step for the analysis of image data. It also plays an important role in further post-processing and recognition/classification purposes as well. Feature extraction and selection constitute an active area of research in computer vision, machine learning, data mining, text mining, genomic analysis, image retrieval, etc. Image features have different complexities depending on the input image type. Stable feature selections and optimal redundancy removal exploration of auxiliary data are some of the important challenges associated with feature selection. There are various types of features such as spatial features, transform-based features, edges and boundaries, shape features, and textures.

Various segmentation techniques such as amplitude thresholding, component labeling, boundary-based approaches, region-based clustering, template matching, and texture segmentation are extensively used in image analysis which leads to recognition/classification. Segmentation makes sure that all the irrelevant features are discarded out paving the way for the selection of useful objects of interest. Classification is the final step which quantifies the nature of data and leads to decision-making. The object is classified into different classes. Classification and segmentation are closely intertwined with each one aiding the other in the final outcome. Classification can be supervised or unsupervised. Supervised classification does not depend on a priori probability distribution functions and is based on reasoning and heuristics, whereas, in unsupervised learning, the idea is to identify the clusters or natural groupings in the feature space. A cluster is a set of points in feature space for which their local density is large compared to the density of feature points in the surrounding region. Clustering techniques are useful for image segmentation and also for classification of raw data to establish different classes.

While image acquisition and preprocessing play an important role in acquiring raw input data, image analysis, image restoration, and image enhancement are three important aspects of a computer vision rendering system. Image analysis system which consists of feature extraction, segmentation, and classification/recognition forms the first important step of understanding raw image data. The analyzed data are useful in making decisions in general applications such as video surveillance for event and activity detection, organizing information for content-based data retrieval, for computer human interaction, etc. Analyzing a scene and recognizing all the constituent objects remain the most challenging in a visual task. While computers excel at accurately reconstructing the 3D shape of a scene from images taken from different angles, they cannot name all the objects present in the image. The real world is made of innumerable objects which all occlude one another, have variable poses, and exhibit variability in terms of sizes, shapes, and appearance, making recognition a difficult task. Thus, just performing an exhausting matching against a database of exemplars still remains an extremely difficult problem. The most challenging version of recognition is the general category object recognition. Some techniques rely on the presence of features, while others involve segmenting the image into semantically meaningful regions so as to obtain unique regions for classification. Given such an extremely rich and complex nature of the topic, there is a need to divide the problem into subsequent smaller steps before an effort is made to solve the problems individually.

General object recognition falls into two broad categories, namely the instance recognition and the class recognition. Instance recognition involves recognizing a known 2D or 3D rigid object, potentially being viewed from a novel viewpoint, against a cluttered background and with partial occlusions [1]. Class recognition is a much harder problem of recognizing any instance of a particular object such as animals and any general surrounding object. The harder problems typically are characterized by large data sets. Computational complexity is extremely high if the entire data are to be used for recognition/classification. Compressive sensing would play a significant role in such a scenario. Image data are invariably sparse, leading to representations which can be much less denser than the ones involving large raw inputs. Thus, sparse representation would be able to convert such dense data into sparse data. Sparse signal representation has proven to be an extremely powerful tool for acquiring, representing, and compressing signals. The success is predominantly due to the fact that general audio, image, and video signals have naturally sparse representations in any basis such as DCT and wavelets. This successful technique which has played an extremely important role in classical signal processing for compact representations can also be employed for computer vision applications where contents and semantics of the image are more important than representations and recovery.

6.2 Object Detection Techniques

Moving object detection, defined as extracting the motion part from a video stream, is the basic step for further analysis of video. Every tracking method requires an object detection mechanism either in each frame or when the object first appears in the video. It handles segmentation of moving foreground objects from stationary background objects [2]. It also reduces computation time. Object detection has applications in many areas of computer vision, including image retrieval and video surveillance. Every tracking method requires an object detection mechanism for detecting the object in every frame or when the object first appears in the video. In general, the object detection mechanism makes use of the information in a single frame or uses the temporal information computed from a sequence of frames to reduce the number of false detections. Real-time moving object detection is important for a variety of embedded applications such as security surveillance, traffic monitoring, robotics, video processing, biomedicine, visual tracking, video compression, human–computer interfaces, medical imaging, content-based indexing, and retrieval. The most common techniques are the optical flow method, the segmentation method, the temporal difference method, and the background subtraction method.

6.2.1 Optical Flow

Optical flow methods use the flow vectors of moving objects over time for detection of moving regions in an image [3–5]. Lucas and Kanade [6] have used optical flow for motion detection. Even when the camera is moving, optical flow method is able to detect moving objects in video sequences. It is based on the assumption that intensity I of moving pixel is constant in subsequent frames. It is computed by taking two images at time t and $t + \Delta t$.

$$I(x, y, t) = I(x + \Delta x, y + \Delta y, t + \Delta t) \tag{6.1}$$

Using Taylor series, above equation is expanded to

$$I(x + \delta x, y + \delta y, t + \delta t) = I(x, y, t) + \frac{dI}{dx}\delta x + \frac{dI}{dy}\delta y + \frac{dI}{dt}\delta t + \dots \tag{6.2}$$

Avoiding higher order terms, the equation reduces to

$$\frac{dI}{dx}\delta x + \frac{dI}{dy}\delta y + \frac{dI}{dt}\delta t = 0 \tag{6.3}$$

$$\frac{dI}{dx}\delta x/\delta y + \frac{dI}{dy}\delta y/\delta t + \frac{dI}{dt} = 0 \tag{6.4}$$

$$\frac{dI}{dx}V_x + \frac{dI}{dy}V_x + \frac{dI}{dt} = 0 \tag{6.5}$$

$$I_x.V_x + I_y.V_y = -I_t \tag{6.6}$$

where V_x, V_y represent optical flow vectors and I_x, I_y represent derivatives of the image intensities at coordinate $(x; y; t)$. The values V_x, V_y are used for getting the motion vector for the object detection by applying thresholding technique. The magnitude of motion vector is found as

$$Th = \sqrt{V_x^2 + V_y^2} \tag{6.7}$$

Optical flow method is meant for the calculation of the image optical flow field and doing clustering processing according to the optical flow distribution characteristics of image [7]. This method can get the complete movement information and detect the moving object from the background better; however, a large quantity of calculation, sensitivity to noise, and poor anti-noise performance make it inappropriate for real-time demanding occasions [8].

6.2.2 Temporal Difference

Temporal differencing method uses the pixel-wise difference between two and three consecutive frames in video imagery to extract moving regions. It is a highly adaptive approach to dynamic scene changes; however, it fails to extract all relevant pixels of a foreground object especially when the object has uniform texture or moves slowly [2]. When a foreground object stops moving, temporal differencing method fails in detecting a change between consecutive frames and loses the object. Let $In(x)$ represents the gray-level intensity value at pixel position x and at time instance n of video image sequence I, which is in the range [0, 255]. T is the threshold initially set to a predetermined value. Lipton et al. [2] have developed a two-frame temporal differencing scheme indicating movement of a pixel when it satisfies the following [2]: This method is computationally less complex and adaptive to dynamic changes in the video frames. In the temporal difference technique, extraction of moving pixel is simple and fast. Temporal difference is more sensitive to the threshold value when determining the changes in the difference of consecutive video frames [9]. Temporal difference requires a special supportive algorithm to detect stopped objects.

6.2.3 Background Subtraction

Background subtraction is a methodology broadly used for detecting moving objects in videos streams from static cameras. It is the general process of motion detection that finds the difference of the current image and the background image for detecting the motion region, and it is commonly efficient in delivering data including object information. The keynote of this process lies in the initialization and update of the background image. The efficiency of both can influence the precision of test results. It tries to detect moving regions by subtracting the current image pixel by pixel from a reference background image which is composed by averaging images over time in an initialization period. The pixels where the variation is beyond a threshold value are categorized as foreground. A decent background subtraction algorithm must control the moving objects that first immerse into the background and then become foreground at advanced time. Furthermore, to adapt the real-time requirements of many applications, a background subtraction algorithm must be computationally economical and have little memory requirements, although still being capable to precisely identify moving objects in the video sequence [10].

The background subtraction method uses the difference of the current image and background image for detecting moving objects. After the background image $B(X, Y)$ is obtained, the background image $B(X, Y)$ is subtracted from the current frame $FK(X, Y)$. If the pixel difference is greater than the set threshold, it determines the appearance of the pixels in the moving object, otherwise, as the background pixels. The moving object can be detected after threshold operation. Its expression is as follows:

$$DK(X, Y) = 1 \text{ if } (|FK(X, Y) - B(X, Y)| > \tau)$$
$$= 0 \text{ others} \tag{6.8}$$

where $D(X, Y)$ is a binary image of differential results [10]. But the background subtraction method alone is very sensitive to changes in the external environment. The methods with a background model based on a single scalar value can guarantee adaptation to slow illumination changes, but cannot cope up with multi-valued background distributions. As such, they are prone to errors whenever those situations arise. However, if such errors connect into relatively small blobs, they can be removed from the classified image by a special filter.

6.3 Object-Tracking Techniques

Object tracking is the important issue in human motion analysis. Tracking involves matching detected foreground objects between consecutive frames using different features of object such as motion, velocity, color, and texture. Object tracking is the process to track the object over the time by locating its position in every frame of the video in the surveillance system. In the tracking approach, the objects are represented using the shape or appearance models [11]. The model selected to represent object shape limits the type of motion. For example, only a translational model can be used when an object is represented as a point. In the case where a geometric shape representation like an ellipse is used for the object, parametric motion models such as affine or projective transformations are appropriate [12]. These representations can approximate the motion of rigid objects in the scene. For a non-rigid object, silhouette or contour is the most descriptive representation and both parametric and nonparametric models can be used to specify their motion.

6.3.1 Point Tracking

Point tracking is a robust, reliable, and an accurate tracking method developed by Veenman et al. [13]. This method is generally used for tracking vehicles. This approach requires good accuracy for the detected object and requires deterministic or probabilistic methods [11]. The object is tracked based on a point which is represented in the detected object in consecutive frames, while association of the points is based on the previous object state which can include object position and motion. This approach requires an external mechanism to detect the objects in every frame.

6.3.2 Kernel Tracking

In this approach, the kernel requires shape and appearance of the object [13]. Any feature of object is used for tracking object as kernel like a rectangular template or an elliptic shape with an associated histogram. The object can be tracked after computing the motion of the kernel between consecutive frames. Mean-shift tracking is one of the kernels-tracking method that uses E-kernel. It represents histogram feature based on spatial masking with an isotropic kernel.

6.3.3 Silhouette Tracking

In this approach, silhouette is extracted from a detected object. By shape matching or contour evolution, silhouettes are tracked either by calculating the object region in consecutive frame tracking. Silhouette tracking method makes use of the information stored inside the object region [14] that is appearance density and shape models. Selecting the right features from the object models is important for tracking. The features must be unique so that the objects can be easily distinguished in the feature space. Various features for tracking are as follows: The apparent color of an object is influenced by spectral power distribution of the illuminant and surface reflectance properties of the object [15]. In image processing, the RGB (red, green, blue) color space is usually used to represent color. Object boundaries usually generate strong changes in image intensities [16]. Edge detection is used to identify these changes. An important property of edges is that they are less sensitive to illumination changes compared to color features. The center of mass (centroid) is a vector of 1-by-n dimensions in length that specifies the center point of a region. For each point, it is worth mentioning that the first element of the centroid is the horizontal coordinate (or x-coordinate) of the center of mass, and the second element is the vertical coordinate (or y-coordinate) [14]. Texture is used for classification as well as for tracking purpose. This feature is used for identifying region or object of interest. It is a measurement of the intensity variation of a surface which quantifies properties such as smoothness and regularity [17]. Compared to color, texture requires a processing step for generating the descriptors. Among all features, color and texture features are widely used for tracking the object. Color bands are sensitive to illumination variation.

6.4 Compressive Video Processing

To implement compressed video sensing, each pixel time series is encoded by multiplication with a random matrix having many fewer rows than columns, giving compression combined with encryption. The encoding involves only simple arithmetic, and the compression achieved is within a reasonable multiple of the best which would be possible through computationally intensive means. The encoded data are secure when the encoding matrix is used as a one-time pad. It has also a built-in error-correction feature, in that a small fraction of corrupted samples does not upset the reconstruction process. In hybrid CS schemes, different strategies are used for the coarse-scale and fine-scale signal content. We apply these notions to video by storing a conventionally sampled low-resolution temporal stream and in addition to a CS representation of the full-temporal resolution. The reconstruction of the stream can then be rather inexpensive when viewed at low resolution; for example, in routine use, yet high-resolution segments can be made available on demand [18].

An inexpensive way to compress and encode the video data is carried out by multiplying with a non-square random matrix. The encryption matrix can be viewed as a one-time pad that is completely secure, and the compression effect comes within a reasonable factor of the best possible compression available using much more sophisticated processing. Such an encryption algorithm requires a huge amount of pad, but the pad (i.e., random matrix) does not have to be transmitted; the sender and receiver can know this in advance. So there is no significant burden caused by this. The CS stream is highly compressed. At the same time, there is an automatic error-correction effect: The encrypted compressed stream is immune to occasional erasures, corruption, and packet loss.

The scheme has the general character of being inexpensive at the encoder side and expensive at the decoder side. While the capture and encoding of the data are simple, the decoder side involves convex optimization. Hence, video compressive sensing process could be useful for sensor networks in which low power devices need to cheaply capture and send data at a low rate, while being immune to spying and to error bursts in transmission. On the other hand, it is useful for very variable data rate cameras, creating video streams which typically are watched at regular speed but contain an embedded stream allowing reconstruction of a much finer time-resolution sequence on demand [18].

In the CS process, measurements are captured instead of individual samples, which are random in nature. Each measurement combines data from samples widely distributed across the data stream. Moreover, the number of measurements is smaller than the number of samples and is comparable to but somewhat larger than the minimal number of measurements needed to characterize the signal. The driving idea is that the signal should be compressible when represented in a fixed basis such as the wavelet basis, with a relatively few large coefficients.

Consider a video sequence that consists of frames represented as $z_j \in R^N$ formed from the pixels of frame j of the video sequence, for $j = 1, 2, \ldots, J$, where J is the total number of frames and N is the total number of pixels in a frame. Let $Z = [z_1, z_2, \ldots, z_j] \in R^{N_1 \times J}$ be a video formed from video frames and the total number of pixels in a video be $N_1 J$. Each frame is made sparse by using an orthonormal basis of $N \times 1$ vectors $\{\Psi_i\}_{i=1}^{N}$. Using a basis matrix of dimension $N \times N$ with $\{\Psi_i\}$ as columns, the signal can be expressed as

$$z_j = \Psi_i r_j \tag{6.9}$$

where r_j is a $N \times 1$ sparse vector with K nonzero coefficients. The signal Z_j is said to be sparse if it has very few nonzero coefficients. This signal is multiplied with the measurement matrix of dimension $\Phi_j \in R^{M \times N}$ to obtain the measurement vector y_j using (6.10).

$$y_j = \Phi_j z_j \qquad (6.10)$$

The measurement vector y_j is of length M, which is much smaller than the total number of pixels in the frame. The measurement matrix of dimension $M \times N$ is generated by calculating the minimum number of measurements required to reconstruct the signal using (6.11).

$$M \geq K \log(N/K) \qquad (6.11)$$

where K denotes the sparsity level and N is the total number of samples in a frame [19]. Figure 6.1 shows the detailed block diagram of the VCS framework. The input is the video sequence from which the frames of size $N \times N$ are separated, with each frame divided into blocks of size $n \times n$. Sparsity is an underlying concept in most of today's filtering and compression applications.

A signal is said to be sparse or compressible when the transform coefficient vector has a small number of large-amplitude coefficients and a large number of small-amplitude coefficients. In this paper, the DWT–DCT hybrid approach is used for a sparse representation that combines the advantages of both DCT and DWT.

6.4.1 CS Based on the DCT Approach

Each block is sparsified using DCT, as described in (8), and converted into a single vector $n^2 \times 1$ called a sparse vector as it has only a few nonzero coefficients. The number of nonzero elements in the sparse vector represents the sparsity level K_1. Given an image block $f(i_1, i_2)$ of size $n \times n$, the forward and inverse DCT are defined as [20].

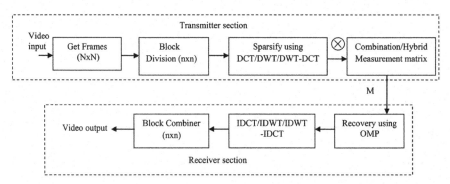

Figure 6.1 Video compressed sensing framework.

$$F(u,v) = \frac{2}{n} c_u c_v \sum_{i_1=0}^{n-1} \sum_{i_2=0}^{n-1} f(i_1, i_2) \cos\left[\frac{\pi(2i_1+1)u}{2n}\right] \cos\left[\frac{\pi(2i_2+1)v}{2n}\right]$$

(6.12)

$$f(i_1,i_2) = \frac{2}{n} c_u c_v \sum_{u=0}^{n-1} \sum_{v=0}^{n-1} F(u,v) \cos\left[\frac{\pi(2i_1+1)u}{2n}\right] \cos\left[\frac{\pi(2i_2+1)v}{2n}\right]$$

(6.13)

where u, v $= 0, 1, \ldots, n-1$ and $c_u, c_v = \begin{cases} \dfrac{1}{\sqrt{2}}, u, v = 0 \\ 1, u, v \neq 0 \end{cases}$

When DCT is applied to the block, most of the significant information is concentrated in a few nonzero coefficients only, thereby achieving a higher compression rate.

6.4.2 CS Based on the DWT Approach

DWT is applied to each block as described in (6.14) and (6.16), which results in four bands (LL, LH, HL, and HH). Given an image block $f(i_1, i_2)$ of size $n \times n$, the forward and inverse DWT are defined as [21].

$$W_\varphi(j_0, k_1, k_2) = \frac{1}{\sqrt{n^2}} \sum_{i_1=0}^{n-1} \sum_{i_2=0}^{n-1} f(i_1, i_2) \varphi_{j_0, k_1, k_2}(i_1, i_2)$$

(6.14)

$$W_\psi^q(j_1, k_1, k_2) = \frac{1}{\sqrt{n^2}} \sum_{i_1=0}^{n-1} \sum_{i_2=0}^{n-1} f(i_1, i_2) \psi^{q_{j_1, k_1, k_2}}(i_1, i_2)$$

$$f(i_1,i_2) = \left\{ \frac{1}{\sqrt{n^2}} \sum_{i_1} \sum_{i_2} W_\varphi(j_0, k_1, k_2) \varphi_{j_0, k_1, k_2}(i_1, i_2) \right.$$
$$\left. + \frac{1}{\sqrt{n^2}} \sum_{q=H_1,V_1,D_1} \sum_{j_1=0}^{\infty} \sum_{k_1} \sum_{k_2} W_\psi^q(j_1, k_1, k_2) \psi^q_{j_1, k_1, k_2}(i_1, i_2) \right\}$$

(6.15)

where $W_\varphi(j_0, k_1, k_2)$, $W_\psi^q(j_1, k_1, k_2)$ denote the approximation coefficients and detail coefficients, respectively, $\varphi_{j_0, k_1, k_2}(i_1, i_2)$, $\psi^{q_{j_1, k_1, k_2}}(i_1, i_2)$ denote the scaling and wavelet functions, respectively, $j_0 = 0$, $j_1 \geq j_0$ represents the scale, and $q = \{H_1, V_1, D_1\}$ indicates the index of LH, HL, and HH bands. The LL band contains the approximation coefficients, and the other three bands contain the detail coefficients. The LH, HL, and HH bands are sparse, that is, converted into a sparse vector. DWT has multi-resolution capability, thus achieving better quality.

6.4.3 CS Based on the Hybrid DWT–DCT Approach

This approach combines the advantages of both DCT and DWT, achieving a better quality with a higher compression rate. DWT is applied to each block using (6.14) and (6.16), which results in four bands, and DCT is then applied to the LH, HL, and HH bands using (6.12), achieving more sparsity. These bands with a few nonzero coefficients are converted into a single vector called the sparse vector to which the measurement matrix is applied for obtaining the measurements.

6.5 Compressive Sensing for Background Subtraction

CS-based background subtraction directly recovers background subtracted images using CS for communication-constrained multi-camera computer vision problems. The CS theory is used for recovering object silhouettes (binary background subtracted images) when the objects of interest occupy a small portion of the camera view, i.e., when they are sparse in the spatial domain. The background subtraction is considered as a sparse approximation problem and provides different solutions based on convex optimization and total variation. This method is used for learning and adapting a low-dimensional compressed representation of the background image, which is sufficient to determine spatial innovations; object silhouettes are then estimated directly using the compressive samples without any auxiliary image reconstruction. Recovery of the simultaneous appearance of the objects using compressive measurements may require reconstruction of an auxiliary image.

Background subtraction is fundamental in automatically detecting and tracking moving objects with applications in surveillance, teleconferencing [22, 23], and even 3D modeling [24]. Usually, the foreground or the innovation of interest occupies a sparse spatial support, as compared to the background and may be caused by the motion and appearance change of objects within the scene. A background subtraction algorithm can be performed by obtaining object silhouettes on a single-image plane or multiple-image plane. In all applications that require background subtraction, the background and the test images are typically fully sampled using a conventional camera. After the foreground estimation, the remaining background images are either discarded or embedded back into the background model as part of a learning scheme [23]. This sampling process is inexpensive for imaging at the visible wavelengths as the conventional devices are built from silicon, which is sensitive to these wavelengths. However, when sampling at other optical wavelengths is desired, it becomes quite expensive to obtain estimates at the same pixel resolution as new imaging materials are needed. For example, a camera with an array of infrared sensors can provide night vision capability but can also be highly expensive more than the same resolution CCD or CMOS cameras. Recently, a prototype of single-pixel camera (SPC) was proposed based on the new mathematical theory of CS [25]. It is known that CS can only be applied on sparse signals.

For computer vision applications, it is known that natural images can be sparsely represented in the wavelet domain [26]. Then, according to the CS theory, it is possible to recover the scene by solving a convex optimization problem, by taking random projections of a scene onto a set of test functions that are incoherent with the wavelet basis vectors. Moreover, the resulting compressive measurements are robust against packet drops over communication channels with graceful degradation in reconstruction accuracy, as the image information is fully distributed.

Compared to conventional camera architectures, the SPC hardware is specifically designed for exploiting the CS framework for imaging. An SPC fundamentally differs from a conventional camera by (i) reconstructing an image using only a single optical photodiode (infrared, hyper spectral, etc.) along with a digital micromirror device (DMD) and (ii) combining sampling and compression into a single non-adaptive linear measurement process. An SPC can directly scale from the visual spectra to hyper spectral imaging with only a change of the single optical sensor. Moreover, enabled by the CS theory, an SPC can robustly reconstruct the scene from much fewer measurements than the number of reconstructed pixels which define the resolution, given that the image of the scene is compressible by an algorithm such as the wavelet-based JPEG 2000.

In the case of conventional cameras, the raw images are sent to the central processing location where the CS process is carried out. This exacerbates the communication bandwidth requirements. In more sophisticated approaches, the cameras transmit the information within the background subtracted image, which requires an even smaller communication bandwidth than the compressive samples. However, the embedded systems needed to perform reliable background subtraction are power hungry and expensive. In contrast, the compressive measurement process requires only cheaper previously determined set of test functions. In this way, the compressive measurements require comparable bandwidth to transform coding of the raw data. They trade off expensive embedded intelligence for more computational power at the central location, which reconstructs the images and is assumed to have unlimited resources. The communication bandwidth and camera hardware limitations make reconstruction highly desirable for the sparse foreground innovations within a scene without any intermediate image reconstruction. The background subtracted images are represented sparsely in the spatial image domain, and hence, the CS reconstruction theory should be applicable for direct recovery of the foreground. For natural images, wavelets are used as the transform domain. Pseudorandom matrices provide an incoherent set of test functions for recovering the foreground image. The original signal is then reconstructed using a perfect recovery algorithm.

6.6 Compressive Sensing for Object Detection

With background subtraction, the location, shape, and appearance of the objects given a test image over a known background must be recovered. The background, current, and difference images are denotes as x_b, x_c, and x_d, respectively. The difference image is obtained by pixel-wise subtraction of the background image from the current image.

6.6.1 Sparsity of Background Subtracted Images

Assuming that x_b and x_c are typical real-world images in the sense that, when wavelets are used as the sparsity basis for x_b, x_c, and x_d, these images can be well approximated with the largest K coefficients with hard thresholding [27], where K is the sparsity level. The images x_b and x_t differ only on the support of the foreground, which has a cardinality of $P = |Sd|$ pixels with $P << N$. It is assumed that the images have uniform complexity in space. The sparsity of the real-world images are modeled as a function of their size: $K_{scene} = K_b = K_c = (\lambda_0 \log N + \lambda_1)N$, where $(\lambda_0, \lambda_1) \in R^2$. It is also assumed that the difference image is also a real-world image on a restricted support and its sparsity as $K_d = (\lambda_0 \log P + \lambda_1)P$. The number of compressive samples "M" necessary to reconstruct x_b, x_c, and x_d in N dimensions are then given by $M_{scene} = M_b = M_t \approx K_{scene} \log (N/K_{scene})$ and $M_d \approx K_d \log (N/K_d)$. When $M_d < M_{scene}$, a smaller number of samples is needed to reconstruct the difference image than the background or foreground images.

6.6.2 The Background Constraint

The multiple compressive measurements y_{bi} ($M \times 1$, $i = 1, \ldots, B$) of training background images x_{bi}, where x_b is their mean, are considered. Each compressive measurement is a random projection of the entire image, whose distribution is an i.i.d. Gaussian distribution with a constant variance $y_{bi} \sim N(y_b, \sigma^2 I)$ where the mean value is $y_b = \Phi x_b$. When the scene changes to include an object which was not part of the background model, the compressive measurements are taken and a test vector $y_t = \Phi x_c$ is obtained where $x_d = x_c - x_b$ is sparse in the spatial domain. Generally, the sizes of the foreground objects are relatively small compared to the size of the background image; the distribution of the literally background subtracted vector is modeled as $yd = yt - yb \sim N(\mu d, \sigma^2 I)$ ($M \times 1$) where μd is the mean. The appearance of the objects constructed from the samples yd would correspond to the literal subtraction of the test frame and the background; however, their silhouette is preserved. The number of samples "M" in y_b is greater than Md but is not necessarily greater than or equal to Mb or Mt; hence, it may not be sufficient to reconstruct the background. However, the background image xb still satisfies the constraint $y_b = \Phi xb$. To be robust against small variations in the background and noise, the distribution of the ℓ_2 distances of the background frames around their mean yb is considered:

$$\|y_{bi} - y_b\|_2^2 = \sigma^2 \sum_{n=1}^{M} \left(\frac{y_{bi}(n) - yb(n)^2}{\sigma} \right)^2 \tag{6.16}$$

When M is greater than 30, this sum can be well approximated by a Gaussian distribution due to the central limit theorem. Then, it is straightforward to show that when there is $\|y_{bi} - y_b\|_2^2 \sim N(M\sigma^2, 2M\sigma^4)$ a test frame with a foreground object, the same distribution becomes $\|y_t - y_b\|_2^2 \sim N\left(M\sigma^2 + \|\mu_d\|_2^2, 2M\sigma^4 + 4\sigma^2 \|\mu_d\|_2^2 \right)$ since σ^2 scales the whole distribution and $1/M << 1$. The logarithm of the ℓ_2 distances can be approximated quite accurately with a Gaussian distribution. That is, since $u \ll 1$ implies $1 + u \approx e^u$, $N\left(M\sigma^2, 2M\sigma^4 \right) = M\sigma^2 N\left(1, \frac{2}{M} \right) = M\sigma^2 \left(1 + \sqrt{\frac{2}{M}} N(0,1) \right) \approx M\sigma^2 \exp\left\{ \sqrt{\frac{2}{M}} N(0,1) \right\}$. This derivation can also motivated by the fact that the square root of the chi-squared distribution can be well approximated by a Gaussian [28]. Hence, this can be used to approximate

$$\log \|y_{bi} - y_b\|_2^2 \sim N\left(\mu_{bg}, \sigma_{bg}^2 \right) \tag{6.17}$$

where μ_{bg} is the mean and σ^2 bg is the variance term, which does not depend on the additive noise in pixel measurements. It is observed that the background image needs to satisfy (6.17) in order to cope with the small variations of the background and the measurement noise. However, the samples $y_d = y_t - y_b$ can be used for recovering the foreground objects.

6.6.3 Object Detector Based on CS

Before reconstruction, it is necessary to ascertain the presence of any differences in the test image from the background. The $\ell 2$ distance of yt from yb can be subsequently approximated by

$$\log \|y_t - y_b\|_2^2 \sim N(\mu_t, \sigma_t^2) \tag{6.18}$$

When the object is small, σ_t^2 should be on the same order size of σ_{bg}^2, while μ_t is different from μ_{bg} as shown in (6.18). The optimal detector would be a simple threshold test for testing the hypothesis for the presence of any new object. When σ_t^2 is significantly different from σ_{bg}^2, the optimal test can be a two-sided threshold test [29]. A constant times the standard deviation of the background as a threshold is used and declare that there is a new object if

$$\left| \log \|y_t - y_b\|_2^2 - \mu_{bg} \geq c\sigma_{bg} \right| \tag{6.19}$$

6.6.4 Foreground Reconstruction

During the reconstruction, the actual appearance of the objects is lost as the obtained measurements also contain information about the background. Although it is known that the subtracted image is a sum of two components that exclusively appear in x_b and x_t, it is difficult, if not impossible, to unmix them without taking enough measurements to recover x_b or x_t. Hence, if the appearances of the objects are needed, a straightforward way to obtain them would be to either reconstruct the test image by taking enough compressive samples and then use the binary foreground image as a mask, or reconstruct and mask the background image and then add the result to the foreground estimate.

6.6.5 Adaptation of the Background Constraint

Two types of changes in a background are drifts and shifts. A background drift consists of gradual changes that occur in the background such as illumination changes in the scene. They may result in immediate unwanted foreground estimates. A background shift is a major and sudden change in the definition of the background, such as a new vehicle parked within the scene. Adapting to background shifts at the sensing level is quite difficult considering the requirement of high-level logical operations because high-level logical operations are required, such as detecting the new object and deciding that it is uninteresting. However, adapting to background drifts is essential for a robust background subtraction system as it has immediate impacts on the foreground recovery [30].

6.7 Compressive Sensing for Object Recognition

Object recognition is a process for identifying a specific object in a digital image or video. Object recognition algorithms rely on matching, learning, or pattern recognition algorithms using appearance-based or feature-based techniques. Common techniques include edges, gradients, histogram of oriented gradients (HOG), Haar wavelets, and linear binary patterns. Object recognition is useful in applications such as video stabilization, automated vehicle parking systems, and cell counting in bio-imaging. Object recognition problem can be defined as a labeling problem based on models of known objects. Formally, given an image containing one or more objects of interest (and background) and a set of labels corresponding to a set of models known to the system, the system should assign correct labels to regions, or a set of regions, in the image. The object recognition problem is closely related to the segmentation problem. Segmentation is not possible without at least a partial recognition of objects and, at the same time, object recognition is not possible without segmentation [31].

An object recognition system must have the following components to perform the task:

- Model database
- Feature detector
- Hypothesizer
- Hypothesis verifier

The model database contains all the models known to the system. The information in the model database depends on the approach used for the recognition. It can vary from a qualitative or functional description to precise geometric surface information. In many cases, the models of objects are abstract feature vectors. A feature is some attribute of the object that is considered important in describing and recognizing the object in relation to other objects. Size, color, and shape are some commonly used features [31].

The feature detector applies operators to images and identifies locations of features that help in forming object hypotheses. The features used by a system depend on the types of objects to be recognized and the organization of the model database. Using the detected features in the image, the hypothesizer assigns likelihoods to objects present in the scene. This step is used for reducing the search space for the recognizer using certain features. The model base is organized using some type of indexing scheme to facilitate the elimination of unlikely object candidates from possible consideration. The verifier then uses object models for verifying the hypotheses and refines the likelihood of objects. The system then selects the object with the highest likelihood, based on all the evidence, as the correct object.

All object recognition systems use models either explicitly or implicitly and employ feature detectors based on these object models. Components used in the hypothesis formation and verification components vary in their importance in different approaches to object recognition. Some systems use only hypothesis formation and then select the object with the highest likelihood as the correct object. Pattern classification approaches constitute a good example of this approach. Many artificial intelligence systems, on the other hand, rely little on the hypothesis formation and do more work in the verification phases. In fact, one of the classical approaches, template matching, bypasses the hypothesis formation stage entirely.

An object recognition system must select appropriate tools and techniques. Many factors should be considered in the selection of appropriate methods for a particular application. The central issues that should be considered in designing an object recognition system [31] are as follows:

- Object or model representation: For some objects, geometric descriptions may be available and may also be efficient, while, for another class, one may have to rely on generic or functional features. The representation of an object should capture all relevant information without any redundancy and should organize this information in a form that allows easy access by different components of the object recognition system.

- Feature extraction: It is important to decide the features required to be detected and the manner of their reliable detection. Most features can be computed in two-dimensional images but they are related to three-dimensional characteristics of objects. Due to the nature of the image formation process, computation of reliable features is easy with some, but difficult with others.

- Feature–model matching: There are many features and numerous objects in most object recognition tasks. An exhaustive matching approach will solve the recognition problem but may be too slow to be useful. Effectiveness of features and efficiency of a matching technique must be considered in developing a matching approach.

- Hypotheses formation: The hypothesis formation step is basically a heuristic one for reducing the size of the search space. This step uses knowledge of the application domain to assign some kind of probability or confidence measure to different objects in the domain. This measure reflects the likelihood of the presence of objects based on the detected features.

- Object verification: The presence of each object with likelihood of happening can be verified by using its models. One should examine each plausible hypothesis for verifying the presence of the object or for ignoring it. If the models are geometric, it is easy to precisely verify objects using camera location and other scene parameters. In other cases, it may not be possible to verify a hypothesis.

In object recognition, given a set of labeled training samples, the task is to identify the class to which a test sample belongs. Suppose there are L distinct classes and a set of n training images per class. One can extract an N-dimensional vector of features from each of these images [32].

Let $B_k = [X_{k1}, \ldots, X_{kj}, \ldots, X_{kn}]$ be an $N \times n$ matrix of features from the kth class, where x^{kj} denote the feature from the jth training image of the kth class. A new matrix or dictionary B, as the concatenation of training samples from all the classes, is defined as

$$B = [B_1, \ldots, B_L] \in R^{N \times (n.L)}$$
$$= [X_{11}, \ldots, X_{1n} \,| X_{21}, \ldots, X_{2n} |\, |\ldots| X_{L1}, \ldots, X_{Ln}] \quad (6.20)$$

Consider an observation vector $y \in R_N$ of unknown class as a linear combination of the training vectors as

$$y = \sum_{i=1}^{L} \sum_{j=1}^{n} r_{ij} X_{ij} \quad (6.21)$$

with coefficients $r_{ij} \in R$. /////The above equation can be written more compactly as

$$y = Br \quad (6.22)$$

where

$$= [r_{11}, ..., r_{1n} | r_{21}, ..., r_{2n} | .....| | r_{L1}, ..., r_{Ln} |]^T$$

and T denotes the transposition operation. It is assumed that, given sufficient training samples of the kth class, B_k, any new test image $y \in RN$ that belongs to the same class will lie approximately in the linear span of the training samples from the class k. This implies that most of the coefficients not associated with class k will be close to zero and r is a sparse vector [32]. In order to represent an observed vector $y \in RN$ as a sparse vector r, one needs to solve the system of linear equations. If r is sparse enough and B satisfies certain properties, then the sparsest α can be recovered by solving the following optimization problem.

$$\hat{r} = \arg\min_{r'} ||r'||_1 \text{ subject to } y = Br' \qquad (6.23)$$

When noisy observations are given, basis pursuit denoising (BPDN) can be used for approximating r

$$\hat{r} = \arg\min_{r'} ||r'||_1 \text{ subject to } ||y = Br'||_2 \leq \varepsilon, \qquad (6.24)$$

where the observations are of the following form

$$y = Br + \eta \qquad (6.25)$$

with $\eta||_2 \leq \varepsilon$.

Given an observation vector y from one of the L classes in the training set, one can compute its coefficients $\hat{r}$. One can perform classification based on the fact that high values of the coefficients $\hat{r}$ will be associated with the columns of B from a single class. This can be done comparing how well the different parts of the estimated coefficients, $\hat{r}$, represent y. The minimum of the representation error or the residual error can then be used for identifying the correct class. The residual error of class k is calculated by keeping the coefficients associated with that class and setting the coefficients not associated with class k to zero. This can be done by introducing a characteristic function, $\Pi_k : R_n \rightarrow R_n$, that selects the coefficients associated with the kth class as follows [32]:

$$r_k(y) = ||y\, B\Pi_k(\hat{r})||_2 \qquad (6.26)$$

Here, the vector Π_k has value one at locations corresponding to the class k and zero for other entries. The class, d, which is associated with an observed vector, is then declared as the one that produces the smallest approximation error [32].

$$d = \arg \min_{k} r_k(y) \qquad (6.27)$$

For sparse representation-based classification, it is important to have the capability to detect and then reject the test samples of poor quality. To decide whether a given test sample has good quality, one can use the notion of sparsity concentration index (SCI) proposed in [33]. The SCI of a coefficient vector $r \in R^{(L.n)}$ is defined as

$$\mathrm{SCI}(r) = \frac{\frac{L.\max\|\Pi_i(r)\|_1}{\|r\|_1} 1}{L1} \qquad (6.28)$$

SCI takes values between 0 and 1. SCI values close to 1 correspond to the case where the test image can be represented approximately by using only images from a single class. The test vector has enough discriminating features of its class and so has high quality. If SCI = 0, then the coefficients are spread evenly across all classes. So the test vector is not similar to any of the classes and has of poor quality. A threshold can be chosen to reject the images with poor quality. For instance, a test image can be rejected if *SCI* (r̂) $< \lambda$ and otherwise accepted as valid, where λ is threshold between 0 and 1.

6.8 Compressive Sensing Target Tracking

It is a challenging task to develop effective and efficient appearance models for robust object tracking due to factors such as pose variation, illumination change, occlusion, and motion blur. Despite numerous algorithms having been proposed in the literature, object tracking remains a challenging problem due to appearance change caused by pose, illumination, occlusion, and motion, among others. An effective appearance model is of prime importance for the success of a tracking algorithm that has been attracting much attention in recent years [34–36]. Tracking algorithms can be generally categorized as either generative [34, 35] or discriminative [36] based on their appearance models. Generative tracking algorithms typically learn a model to represent the target object and then use it for searching for the image region with minimal reconstruction error. Black et al. [34] learn an off-line subspace model to represent the object of interest for tracking. The IVT method [37] utilizes an incremental subspace model to adapt appearance changes. Recently, sparse representation has been used in the ℓ_1 tracker where an object is modeled by a sparse linear combination of target and trivial templates [38]. However, the computational complexity of this tracker is rather high, thereby limiting its applications in real-time scenarios.

Li et al. [39] further extend the ℓ_1 tracker by using the orthogonal matching pursuit algorithm for solving the optimization problems efficiently. Despite much demonstrated success of these online generative tracking algorithms, several problems remain to be solved. First, numerous training samples that cropped from consecutive frames are required for learning an appearance model online. Since there are only a few samples at the outset, most tracking algorithms often assume that the target appearance does not change much during this period. However, when there is a significant change in the appearance of the target at the beginning, the drift problem is likely to occur. Second, when multiple samples are drawn at the current target location, there is a likelihood of drift as the appearance model needs to adapt to these potentially misaligned examples [40]. Third, these generative algorithms do not use the background information which is likely to improve tracking stability and accuracy. Discriminative algorithms pose the tracking problem as a binary classification task in order to find the decision boundary for separating the target object from the background. Avidan [36] extends the optical flow approach with a support vector machine classifier for object tracking. Collins et al. [41] demonstrate the ability to learn the most discriminative features online to separate the target object from the background. Grabner et al. [42] propose an online boosting algorithm to select features for tracking. However, these trackers [36, 42] only use one positive sample (i.e., the current tracker location) and a few negative samples when updating the classifier. The updation of the appearance model with noisy and potentially misaligned examples often leads to the tracking drift problem. Grabner et al. [43] propose an online semi-supervised boosting method to alleviate the drift problem in which only the samples in the first frame are labeled and all the other samples are unlabeled. Babenko et al. [40] introduce multiple instance learning into online tracking where samples are considered within positive and negative bags or sets. Recently, a semi-supervised learning approach [44] is developed in which positive and negative samples are selected via an online classifier with structural constraints.

6.8.1 Kalman Filtered Compressive Sensing

The signal being tracked, $\{X_t\}_{t=0}^{\infty}$, is assumed to be both sparse and have a slowly changing sparsity pattern. Given these assumptions, if the support set of x_t, T_t, is known, the relationship between x_t and y_t can be written as follows:

$$y_t = \Phi_{Tt}(X)Tt + W_t \tag{6.29}$$

Above, Φ is the CS measurement matrix, and Φ_{Tt} retains only those columns of Φ whose indices lie in T_t. Likewise, $(X_t)Tt$ contains only those components corresponding to T_t. Finally, w_t is assumed to be zero-mean Gaussian noise. If x_t is assumed to also follow the state model $x_t = x_{t-1} + v_t$ with v_t zero-mean Gaussian noise, then the MMSE estimate of x_t from y_t can be computed using a Kalman filter instead of a CS decoder [45].

This is handled by using the Kalman filter output to detect changes in T_t and re-estimate it if necessary. $\tilde{y}t, f = y_t - \Phi\hat{X}$, the filter error, is used to detect changes in the signal support via a likelihood ratio test given by

$$\tilde{y}_t', f \sum \tilde{y}_t, f\tau \tag{6.30}$$

where τ is a threshold and Σ is the filtering error covariance. If the term on the left-hand side exceeds the threshold, then changes to the support set are found by applying a procedure based on the Dantzig selector. Once T_t has been re-estimated, $\hat{x}$ is re-evaluated using this new support set. Kalman filtered compressive sensing is useful in surveillance scenarios when objects under observation are stationary or slowly moving. Under such assumptions, this method is able to perform signal tracking with a low data rate and low computational complexity [45].

Assume, for now, that the support set at $t = 1$, T_1, is known. Consider the situation where the first change in the support occurs at a $t = t_a$, i.e., for $t < t_a$, $T_t = T_1$, and that the change is an addition to the support. This means that for $t < t_a$, a regular KF is used, which assumes the following reduced order measurement and system models: $y_t = A_T (x_t)_T + w_t$, $(x_t)_T = (x_{t-1})_T + (v_t)_T$, with $T = T_1$. The KF prediction and update steps for this model are as follows:

$$\hat{x}_0 = 0, \; P_0 = 0,$$

$$\hat{x}_{t|t-1} = \hat{x}_{t-1}$$
$$(P_{t|t-1})_{T,T} = (P_{t-1})_{T,T} + \sigma^{2sys}I$$
$$K_{t,T} \triangleq (P_{t|t-1})_{T,T}A_T' \sum, \sum A_T(P_{t|t-1})_{T,T}A_T' + \sigma_{obs}^2 I$$
$$(\hat{x}_t)_T = (\hat{x}_{t|t-1})T + K_{t,T}[y_t - A\hat{x}_{t|t-1}]$$
$$(\hat{x}_t)_{TC} = (\hat{x}_{t|t-1})T^C = (\hat{x}_{t-1})T^C$$
$$(P_t)_{T,T} = [I - K_{t,T}A_T](P_{t|t-1})_{T,T}$$

Detecting if addition to support set occurred. The Kalman innovation error is $\tilde{y}_t$, $y_t - A\,\hat{x}_t|_{t-1}$. For $t < t_a$, $\tilde{y}_t = [A(x_t - \hat{x}_t|_{t-1}) + w_t] \sim N(0, \sum_{ie}, t)$. At $t = t_a$, a new set, Δ, gets added to the support of x_t, i.e., $y_t = A_T (x_t)_T + A_\Delta (x_t)_\Delta + w_t$, where the set Δ is unknown. Since the old model is used for the KF prediction, at $t = t_a$, $\tilde{y}_t$ will have nonzero mean, $A_\Delta (x_t)_\Delta$, i.e.,

$$\tilde{y}_t = A_\Delta(x_t)_\Delta + \tilde{w}_t = A_{T^c}(x_t)_{T^\cup} + \tilde{w}_t, \text{ where} \qquad (6.31)$$
$$\tilde{w}_t \triangleq [A_T(x_t - \hat{x}_t|_{t-1})_T + w_t] \sim N(0, \sum')$$

where $\Delta \subseteq T^c$ is the currently set undetected nonzero. Thus, the problem of detecting the addition of a new set or otherwise gets transformed into the problem of detecting if the Gaussian distributed $\tilde{y}_t$ has nonzero or zero mean. Note that $A_\Delta (x_t)_\Delta = A_{T^c} (x_t)_{T^c}$ and, thus, the generalized likelihood ratio test (G-LRT) for this problem simplifies to detecting if the weighted innovation error norm, IEN, $\tilde{y}'_t - 1$, i.e., $t^\frown y_t$ threshold. Alternatively, one can apply G-LRT to the filtering error that can be written as follows [45]:

$$\tilde{y}_{t,f} = A_\Delta(x_t)_\Delta + A_T(x_t - \hat{x}_t)_T + w_t$$
$$= [I - A_T K_{t,T}]A_\Delta(x_t)_\Delta + \tilde{w}_{t,f}, \tilde{w}_{t,f} \triangleq [I - A_T K]\tilde{w}_t \qquad (6.32)$$
$$= \tilde{w}_{t,f} \sim N(0, \textstyle\sum), \sum \triangleq [I - A_T K_{t,T}] \sum [I - A_T K_{t,T}]$$

6.8.2 Joint Compressive Video Coding and Analysis

Cossalter et al. [46] consider a collection of methods via which systems utilizing compressive imaging devices can perform visual tracking. Of particular note is a method referred to as joint compressive video coding and analysis, in which the tracker output is used for improving the overall effectiveness of the system. Instrumental to this method is work from theoretical CS literature which proposes a weighted decoding procedure that iteratively determines the locations and values of the (nonzero) sparse vector coefficients. Modifying this decoder, the joint coding and analysis method utilizes the tracker estimate to directly influence the weights. The result is a foreground estimate of higher quality compared to one obtained via standard CS decoding techniques. The weighted CS decoding procedure calculates the foreground estimate via

$$\hat{f} = \min_\theta ||W\theta||_1 \text{s.t.} ||y^f - \phi\theta \leq \sigma, \qquad (6.33)$$

where $y^f = y - y^b$, W is a diagonal matrix with weights $[w(1)\dots w(N)]$ and σ captures the expected measurement and quantization noise in $\mathbf{y}^f$. Ideally, the weights are selected according to

$$w(i) = \frac{1}{|f(i) + \varepsilon|}, \qquad (6.34)$$

where $f(i)$ is the value of the ith coefficient in the true foreground image. Of course, these values are not known in advance, but the closer the weights are to their actual value, the more accurate $\hat{f}$ becomes. The joint coding and analysis approach utilizes the tracker output in selecting appropriate values for these weights. The actual task of tracking is accomplished using a particle filter. The state vector for an object at time t is denoted by $z_t = [c_t \; s_t \; u_t]$ where s_t represents the size of the bounding box defined by the object appearance; c_t, the centroid of this box; and $\mathbf{u}_t$, the object velocity in the image plane. A suitable kinematic motion model is utilized to describe the expected behavior of these quantities with respect to time, and foreground reconstructions are used to generate observations.

Assuming the foreground reconstruction $\hat{f}_t$ obtained via decoding the compressive observations from time t is accurate, a reliable tracker estimate can be computed. This estimate, $\hat{z}_t$, can then be used to select values for the weights $[w(1) \ldots w(N)]$ at time $t +1$. If the values of their weights are ideal, the value of $\hat{f}_{t-1}$ obtained from the weighted decoding procedure will be of higher quality than that obtained from a more generic CS decoder.

6.8.3 Compressive Sensing for Multi-View Tracking

Visual camera networks are becoming popular with the increasing presence of cameras for surveillance, medical, and smart room applications. It is important to design distributed algorithms that scale with the number of cameras and demand low communication overheads. Design helps video cameras in capturing large volume of data that is rich in structure and highly redundant. Further, in the context of smart cameras that have local processing capability, it is possible to preprocess the acquired imagery before transmitting it over the communication channel. This allows us to extract the specific information from the videos over certain interesting epochs. The inherent structure in the acquired imagery can be used to transmit only small number of measurements for addressing communication requirements. The linear dependence of the desired parameters (location, visual hull) on the compressed measurements allow for direct reconstruction using ℓ_1-minimization-based recovery algorithms [47, 48]. In prior work on multi-camera localization, silhouettes (foreground likelihood values) have been used for tracking occluding objects from multiple cameras [49]. Similarly, silhouettes have also been used for 3D voxel reconstruction [50], but the complexity of these methods increases linearly in the number of cameras. Three-dimensional voxels have been used to recognize an activity in a scene [51], to estimate pose and register the body parts [52].

In this section, the method for reconstructing 3D voxels using silhouette images from multiple cameras is discussed. Due to the finite resolution of the observation region over which the object parameters are estimated, it is observed that the method is scalable with the number of cameras. Silhouette images are used in computer vision for various applications such as tracking, activity recognition, and building 3D models using voxels. Silhouette images can be considered as sparse matrices where few pixels are in the foreground and most in the background. The sparsity of the silhouette images corresponds to the sparsity of object parameters. The sparsity of silhouette images is utilized, using their compressed samples, for direct recovery of the object parameters in a multi-view setting. Here, the multi-view tracking problem is formulated first and then the formulation of 3D voxel reconstruction task is considered [53].

A common approach involves all the cameras sending the data to a central location where it is used for detecting and tracking objects and looking out for abnormal activities. For the purpose of tracking, it is sufficient to send the random projections of the silhouette image vectors obtained from background subtraction, locally computed at the cameras. Suppose it is assumed that the observation region O (which can be either 2D or 3D space) is being observed by synchronized cameras $c = 1, \ldots, C$. Most part of the region is visible to all the cameras. At any frame $f \in \{1, 2, \ldots, F\}$, the background subtracts silhouette images I_{fc} (of size N row $\times$ N col) at each of the cameras. The foreground in the image is the moving object which is of interest in tracking and on which it is focused for performing the analysis. The foreground is sparse in the image plane. This implies that in the corresponding observation region O, the objects or people corresponding to the image foreground are sparse; i.e., the area (or volume) occupied by the moving objects or people is very small compared to the area (or volume) of the observation region. The silhouette image and the corresponding objects in the observation region are related by a linear transformation. For simplicity, consider a 2D observation region O where camera c is provided with homography H_c between the world plane and the image plane [53].

Assume that, for some frame f, the cameras are observing the 2D observation space O. The region is divided into non-overlapping, tightly packed subregions $n = 1, \ldots, N$ where (x_n, y_n) are the coordinates of the representative point of the subregion n, known at the cameras. In tracking applications, it is necessary to localize the objects to one of the regions. Assume that camera c observes the background subtracted silhouette image I_{fc} (foreground has value 1 and background 0) with image coordinates (u, v). The vectors x and y'_c associated with the object location on the 2D plane and the silhouette image are defined respectively as x is a $N \times 1$ vector with $x(n) = 1$ if object is present in the subregion n and 0 otherwise. x is the indicator function of the objects at points on the ground plane and y' c is the indicator function of foreground at corresponding points on the silhouette image at camera c. Typically, in tracking scenarios, x is sparse since the objects of interest occupy a small area in the observation region. For every frame f, the position of the objects, in other words our desired variable x, must be known. But, what the cameras instead observe is the background subtracted silhouette image I_{fc} from which they construct the $N \times 1$ vector y'_c

$$y'_c(i) = I^f_c(u_i, v_i), \tag{6.35}$$

where the image coordinates (u_n, v_n) of camera c are related to the coordinate (x_n, y_n) of the representative point n by (it should be noted that since (u_n, v_n) take integer values the right hand side can be rounded).

$$\begin{bmatrix} u_n \\ v_n \\ 1 \end{bmatrix} \sim H_c \begin{bmatrix} x_n \\ y_n \\ 1 \end{bmatrix}, \tag{6.36}$$

The equation relating x and y'_c is then given by

$$y'_c = A_c x, \tag{6.37}$$

where A_c is an identity matrix in this setting. Given O, each camera can compute y'_c. A simple projection of the silhouette from a single camera on the ground plane gives an estimate of the object location, but this is not accurate since the parts of the object in parallax do not register under the homography projections. For accurate estimation of the position of the objects on the ground plane from y'_c, information from multiple cameras is used. For this, cameras need to transmit the information to a centralized location where the computation can be performed. Noting that x is sparse, the amount of data to be transmitted to the central processing center can be significantly reduced by projecting the signals into lower dimensions and recovering it using the principle of CS [53].

Assume the vector x to be K-sparse. This means that one can randomly project the vector y'_c into lower dimensions using the $M \times N$ projection matrix with entries from Gaussian distribution $N(0, 1/N)$. Errors are introduced in the signal y'_c due to errors in silhouettes, rounding off errors, and when not all subregions in O are visible to camera c. The errors are modeled $\phi_c y_c'$ as additive white Gaussian (AWG) noise. The resulting equation is

$$y_c = \Phi_c y_c' + e_c \qquad (6.38)$$

The cameras transmit yc to the central location where they are stacked to form vector y

$$\begin{bmatrix} y_1 \\ y_2 \\ \vdots \\ y_c \end{bmatrix} = \begin{bmatrix} \Phi_1 \\ \Phi_2 \\ \vdots \\ \Phi_c \end{bmatrix} x + \begin{bmatrix} e_1 \\ e_2 \\ \vdots \\ e_c \end{bmatrix} \qquad (6.39)$$

Resulting in

$$y = \Phi x + e \qquad (6.40)$$

The vector x is recovered by solving (6.40). For any given frame, after compressing, it is essential to send compressed measurement values compared to samples without compression. Also, the recovery of x is a function of N allowing the algorithm to scale in the number of cameras. For simplicity, rectangular subregions $n = 1, \ldots, N$, are considered with the representative points forming a $N_1 \times N_2$ rectangular grid. The problem described above easily extends to 3D voxel reconstruction. Assume a 3D O being observed by C cameras. At frame f, the cameras observe silhouette images I_{fc}. It is also assumed that the projection matrix Pc is known at camera c. Unlike multi-view ground plane tracking, in 3D voxel reconstruction, it is needed to recover all the three coordinates of the object for reconstructing the 3D shape. Divide the 3D observation region into N sufficiently dense subregions which are non-overlapping and tightly packed. Here, the representative point of the subregion n has coordinates (x_n, y_n, z_n). Again, for simplicity, assume that the subregions are cuboidal volumes called voxels and the representative points form a $N1 \times N2 \times N3$ grid. The subregions are denser compared to tracking and the object occupies a lot more subregions than it did in tracking scenario. Similarly, define $N \times 1$ vectors x and y_c'. $x(n) = 1$ if object occupies subregion n and 0 otherwise. Obviously, if $x(i) = 1$, it would have $y_c'(i) = I_{fc}(u_i, v_i) = 1$ where

$$\begin{bmatrix} u_i \\ v_i \\ 1 \end{bmatrix} = P_c \begin{bmatrix} x_i \\ y_i \\ z_i \\ 1 \end{bmatrix}, \qquad (6.41)$$

and (x_i, y_i, z_i) are the coordinates of the grid point in voxel i. All the voxels whose projection onto the image plane of camera c intersects with the silhouette of image I_{fc} are assigned to be occupied—i.e., $y_c'(i) = 1$. Thus, for any camera, the number of voxels assigned as occupied is greater than the number of truly occupied voxels. Hence, for finding the true voxel occupation x, the silhouettes from multiple cameras are used. In the region O, the volume occupied by the object is assumed to be sparse implying a sparse x. To recover x from y_c', $c = 1, \ldots, C$ follow exactly the recovery procedure adopted in multi-view tracking [53].

6.8.4 Compressive Particle Filtering

The compressive particle filtering algorithm for tracking application was developed by [54]. It is assumed that the system uses a sensor that is able to collect compressive measurements. The goal is to obtain tracks without having to perform CS decoding. That is, the method solves the sequential estimation problem using the compressive measurements directly. Specifically, the algorithm is a modification to the particle filter. First, the system is formulated in state space, where the state vector at time t is given by

$$S_t = [S_t^x \; S_t^y \; S_t^x \; S_t^y \psi_t]^T. \tag{6.42}$$

$[S_t^x, \; S_t^y]$ and $(S_t^x, \; S_t^y)$ represent the object position and velocity in the image plane, respectively, and ψ_t is a parameter specifying the width of an appearance kernel. The appearance kernel is taken to be a Gaussian function defined over the image plane and centered at $(S_t^x, \; S_t^y)$ with i.i.d. component variance proportional to ψ_t. That is, given s_t, the jth component of the vectorized image, $\mathbf{z}_t$, is defined as

$$z_t^j(s_t) = C_t \exp\left\{ -\psi_t \left(\begin{bmatrix} s_k^x \\ s_k^y \end{bmatrix} - r^j \right) \right\}, \tag{6.43}$$

where $\mathbf{r}_j$ specifies the two-dimensional coordinate vector belonging to the jth component of $\mathbf{z}_t$ The state equation is given by

$$s_t + 1 = ft(s_t, v_t) = Ds_t + v_t, \tag{6.44}$$

where

$$D = \begin{bmatrix} 1 & 0 & 1 & 0 & 0 \\ 0 & 1 & 0 & 1 & 0 \\ 0 & 0 & 1 & 0 & 0 \\ 0 & 0 & 0 & 1 & 0 \\ 0 & 0 & 0 & 0 & 1 \end{bmatrix}$$

And $v_t \sim N(0, \text{diag}(\alpha))$ for a preselected noise variance vector α. The observation equation specifies the mapping from the state to the observed compressive measurements y_t. If Φ is the CS measurement matrix used to sense z_t, this is given by

$$y_t = \Phi z_t(s_t) + w_t, \tag{6.45}$$

where w_t is zero mean Gaussian measurement noise with covariance Σ. With the above specification, the bootstrap particle filtering algorithm can be used to sequentially estimate s_t from the observations y_t. Specifically, the weights belonging to candidate samples $\left\{ \tilde{s}_t^{(i)} \right\}_{i=1}^{N}$ can be found via

$$\tilde{w}_t^{(i)} = p(y_t | \tilde{s}_t^{(i)}) = N(y_t; \Phi z_t(\tilde{s}_t^{(i)}), \Sigma) \tag{6.46}$$

and rescaling to normalize across all i. These important weights can be calculated at each time step without the need to perform CS decoding on y. In some sense, the filter is acting purely on compressive measurements and hence the name "compressive particle filter" [54].

6.9 Surveillance Video Processing Using Compressive Sensing

A massive number of cameras are deployed for surveillance, some with wireless connections. The cameras transmit surveillance videos to a processing centre where the videos are processed and analyzed. Of particular interest in surveillance video processing is the ability to detect anomalies and moving objects in a scene automatically and quickly. Detection of moving objects is traditionally achieved by background subtraction methods [55] which segment background and moving objects in a sequence of surveillance video frames. The mixture of Gaussians [56] technique assumes that each pixel has a distribution that is a sum of Gaussians and the background and foreground are modeled by the size of the Gaussians. In low rank and sparse decomposition, the background is modeled by a low-rank matrix, and the moving objects are identified by a sparse component. These traditional background subtraction techniques require all pixels of a surveillance video to be captured, transmitted, and analyzed. A challenge in the network of cameras is the bandwidth. Since traditional background subtraction requires all pixels of video to be acquired, an enormous amount of data is transported in the network due to a large number of cameras. At the same time, most of the data are uninteresting due to inactivity. There is a high risk of the network being overwhelmed by the mostly uninteresting data to prevent timely detection of anomalies and moving objects. Therefore, it is highly desirable to have a network of cameras in which each camera transmits a small amount of data with enough information for reliable detection and tracking of moving objects or anomalies. CS allows us to achieve this goal.

In compressive sensing, the surveillance cameras make compressive measurements of video and transmit measurements in the network. Since the number of measurements is much smaller than the total number of pixels, transmission of measurements, instead of pixels, helps to prevent network congestion. Furthermore, the lower data rate of compressed measurements helps wireless cameras to reduce power consumption. When a surveillance video is acquired by compressive measurements, the pixel values of the video frames are unknown, and consequently, the traditional background subtraction techniques such as [55] cannot be applied directly. A straight forward approach is to recover the video from the compressive measurements [57] and, then, after the pixel values are estimated, to apply one of the known background subtraction techniques. Such an approach is undesirable for two reasons. First, a generic video reconstruction algorithm does not take advantage of special characteristics of surveillance video in which a well-defined, relatively static background exists. The existence of a background provides prior information that helps to reduce the number of measurements. Second, in the straight forward approach, additional processing is needed to perform background subtraction after the video is recovered from the measurements [58].

6.10 Performance Metrics

The performance metrics employed to evaluate the compressive video sensing applications are discussed below [59]:

Qualitative evaluation
To evaluate the performance of any system, it is essential to define the metrics to be used. These qualitative evaluation metrics are different depending on the final application. For measuring accuracy, different frame-based accuracy metrics, namely precision, recall, F1, similarity, false alarm rate, and tracking accuracy, are used to evaluate the performance.

Recall (or) detection rate
Recall gives the percentage of the detected true positives as compared to the total number of true positives in the ground truth. Detection rate is calculated as follows:

$$\text{Recall } (R) = \frac{T_p}{T_p + F_n} \tag{6.47}$$

Recall acts as a frame-based metric.

Precision
Precision also known as positive prediction gives the percentage of detected true positives as compared to the total number of items detected by the method.

$$\text{Precision } (P) = \frac{T_p}{T_p + F_p} \tag{6.48}$$

Using the above-mentioned metrics, generally, a method is considered good if it reaches high recall values, without sacrificing precision.

Figure of merit or F-measure
F1 metric also known as figure of merit is the weighted harmonic mean of precision and recall

$$F_1 \text{ score} = 2 * \frac{\text{Recall} * \text{Precision}}{\text{Recall} + \text{Precision}} \tag{6.49}$$

Figure of merit indicates a good accuracy of the architecture used.

False alarm rate
False alarm rate is calculated as follows:

$$\text{False alarm rate (FAR)} = \frac{\text{FP}}{\text{TP} + \text{FP}} \tag{6.50}$$

Tracking accuracy
Tracking accuracy is also calculated for detected objects frame by frame.

$$\text{Tracker detection rate} = \frac{\text{TP}}{\text{TG}} \tag{6.51}$$

Starting with the first frame of the test sequence, frame-based metrics are computed for every frame in the sequence. From each frame in the video sequence, a few true and false detection and tracking quantities are first computed.

True Negative (TN)
Number of frames where both ground truth and system results agree on the absence of any object.

True Positive (TP)
Number of frames where both ground truth and system results agree on the presence of one or more objects, and the bounding box of at least one or more objects coincides among ground truth and tracker results.

False Negative (FN)
Number of frames where the ground truth contains at least one object, while system either does not contain any object or none of the system's objects fall within the bounding box of any ground truth object.

False Positive (FP)
Number of frames where the system results contain at least one object, while ground truth either does not contain any object or none of the ground truth's objects fall within the bounding box of any system object.

Object-based Metrics
Object-based evaluation computes the metrics based on the complete trajectory and lifespan of the individual system and ground truth tracks. Since a given ground truth track could correspond to more than one system tracks, and likewise, a correspondence mapping has to be established first. Based on this mapping between object tracks, the frame-based and object-based metrics are computed. The first sets of metrics are based on simple threshold-based correspondence. The Euclidean distance between their centroid is computed for each common frame between a system track and ground truth track. The cumulative Euclidean distance is then normalized by the total number of overlapping frames between the ground truth–system track pair under investigation. Finally, two ground truth–system track pairs are declared corresponding if their total normalized distance is within a threshold. Once the correspondence is established, compute the true positive (TP), false positive (FP), and total ground truth (TG) as explained previously in the context of frame-based metrics [59].

The tracker detection rate (TRDR) and false alarm rate (FAR) are then computed as before. We also compute the object-tracking error which is the average discrepancy between the ground truth bounding box centroid and the centroid of the system result:

$$\text{Object} - \text{tracking error (OTE)}$$

$$= \frac{1}{N_{rg}} \sum_{ieg(t_i) \wedge r(t_j)} \sqrt{(x_i^g - x_i^r)^2 + (y_i^g - y_i^r)^2} \qquad (6.52)$$

where N_{rg} represents the total number of overlapping frames between ground truth and system results, x_i^g represents the x-coordinate of the centroid of object in i^{th} frame of ground truth, and x_i^r represents the x-coordinate of the centroid of object in i^{th} frame of tracking system result.

6.11 Summary

In this chapter, compressed sensing for computer vision applications is discussed in detail. The conventional object detection techniques and object-tracking techniques are also discussed in detail. This chapter gives an overview of the video compressed sensing framework and compressed sensing for background subtraction framework. Compressed sensing-based object detection and tracking have been discussed along with the performance metrics that are used for evaluation of the performance of the object detection and tracking framework. Frame-based metrics and object-based metrics have been discussed for evaluating the performance of the framework.

References

[1] Szeliski, R. (2010). *Computer Vision: Algorithms and Applications.* (Springer-Verlag New York Inc.).

[2] Paragios, N., and Deriche, R. (2000). "Geodesic active contours and level sets for the detection and tracking of moving objects," *IEEE Trans. Patt. Analy. Mach. Intell.* 22, 3, 266–280.

[3] Watanabe. M., Takeda. N., Onoguchi. K. (1996). "A moving object recognition method by optical flow analysis", *Proceedings of the 13th International Conference on Pattern Recognition*, vol. 1, 528–533.

[4] Patel, M. P., and Parmar, S.K. (2014). "Moving object detection with moving background using optic flow", *Recent Advances and Innovations in Engineering (ICRAIE)*, 1– 6.

[5] Zinbi, Y., and Chahir, Y.S. (2008). "Moving object segmentation using optical flow with active contour model". *IEEE Conference on ICTTA*, pp. 1–5.

[6] Lucas, B., and Kanade, T. (1981). "An iterative image registration technique with an application to stereo vision," *Proceedings of Imaging understanding workshop*, pp. 121–130.

[7] Stauffer, C., and Grimson, W. E. L. (1999). "Adaptive background mixture models for real-time tracking," *Proc. IEEE CVPR 1999*, pp. 24 & 25. 2 June.

[8] Koller, D., Weber, J., Huang, T., Malik, J., Ogasawara, G., Rao, B., and Russell, S. (1994). "Towards Robust Automatic Traffic Scene Analysis in Real-time", *Proc. ICPR'94*, pp. 126–131, Nov.

[9] Zhu, S., and Yuille, A. (1996). Region competition: unifying snakes, region growing, and bayes/mdl for multiband image segmentation. *IEEE Trans. Patt. Analy. Mach. Intell.* 18 (9), 884–900.

[10] Shaikh, S. H., Saeed, K., and Chaki. N. (2014). *Moving Object Detection Using Background Subtraction.* (Berlin: Springer).

[11] Isard, M. and Maccormick, J. (2001). Bramble: A Bayesian multiple-blob tracker. In *IEEE International Conference on Computer Vision (ICCV).* pp. 34–41.

[12] YIilmaz, A., Javed, O., and Shah, M. (2006). Object tracking: A survey. *ACM Comput. Surv.* 38 (4), 13, Dec.

[13] Kim, I. S. Choi, H. S., Yi, K. M., Choi, J. Y., and Kong, S. G. (2010). "Intelligent Visual Surveillance A Survey." *Intl. J. Control, Auto. Syst.* 8 (5), 926–939.

[14] Elgammal, A. Duraiswami, R., Hairwood, D., and Anddavis, L. (2002). Background and foreground modeling using nonparametric kernel density estimation for visual surveillance. *Proceedings of IEEE* 90, 7, 1151–1163.

[15] Stauffer, C., and Grimson, E. (2000). "Learning patterns of activity using real time tracking," *IEEE Trans. Pattern Anal. Machine Intell.* 22 (8), 747–757, August.

[16] Christopher, R. W., Azarbayejani, A. J. Darrell, T., and Pentland, A. P. (1997). "Pfinder: Real–Time Tracking of the Human Body" in *IEEE Trans. Pattern Anal. and Machine Intell.* 19 (7), 780–785.

[17] Wu, U. Z., and Leahy, R. (1993). "An optimal graph theoretic approach to data clustering: Theory and its applications to image segmentation". *IEEE Trans. Patt. Analy. Mach. Intell.*

[18] Drori, I. (2008). "Compressed video sensing." In *BMVA Symposium on 3D Video-Analysis, Display, and Applications*.

[19] Candès Emmanuel, J., and Wakin Michael, B. (2008). An introduction to compressive sampling. *IEEE Signal Process Mag.* 25 (2), 21–30.

[20] Mohammad Bagher Akbari, H., Aghagolzadeh, A., and Hadi, S. (2011). Multi-focus image fusion for visual sensor networks in DCT domain. *Comput. Electr. Eng.* 37 5, 789–797.

[21] Dua, S., Acharya, R., and Ng, E. Y. K. (2011). *Computational analysis of the human eye with applications.* (Singapore: World Scientific).

[22] Elgammal, A., Harwood, D., and Davis, L. (1999). Non-parametric model for background subtraction. In: *IEEE FRAME-RATE Workshop*, (Berlin: Springer).

[23] Piccardi, M. (2004). Background subtraction techniques: a review. In: *IEEE International Conference on Systems, Man and Cybernetics.* 4.

[24] Cheung, G. K. M., Kanade, T., Bouguet, J. Y., and Holler, M. (2000). Real time system *for robust 3 D voxel reconstruction of human motions. In: CVPR.* 714–720.

[25] Wakin, M. B., Laska, J. N., Duarte, M. F., Baron, D., Sarvotham, S., Takhar, D., Kelly, K. F., and Baraniuk, R. G. (2006). *An architecture for compressive imaging.* In: ICIP, Atlanta, GA, Oct. pp. 273–1276.

[26] Mallat, S., and Zhang, S. (1993). Matching pursuits with time-frequency dictionaries. *IEEE Trans. Signal Process.* 41 (12), 3397–3415, Dec.

[27] Mallat, S. (1999). *A Wavelet Tour of Signal Processing.* (New York: Academic Press).

[28] Cevher, V., Chellappa, R., and McClellan, J. H. (2007). Gaussian approximations for energy-based detection and localization in sensor networks. In: *IEEE Statistical Signal Processing Workshop*, Madison, WI (26–29 August).

[29] Van Trees, H. L. (1968) Detection, Estimation, and Modulation Theory, Part I. (New York: John Wiley & Sons, Inc.)

[30] Cevher, V., Sankaranarayanan, A., Duarte, M. F., Reddy, D., Baraniuk, R. G. and Chellappa, R. (*2008*). "Compressive sensing for background subtraction." *In Computer Vision–ECCV*, pp. 155–168. Springer Berlin Heidelberg, 2008.

[31] Nagabhushana, S. *Computer vision and image processing.* New Age International, 2005.

[32] Patel, V. M., and Rama, C. (2013). *Sparse representations and compressive sensing for imaging and vision.* (New York: Springer).

[33] Wright, J., Yang, A. Y., Ganesh, A., Sastry, S. S., and Ma. Y. (2009). Robust face recognition via sparse representation. *IEEE Trans. Pattern Analysis and Machine Intell.* 31 (2), 210–227.

[34] Black, M., and Jepson, A. (1998). Eigentracking: Robust matching and tracking of articulated objects using a view-based representation. *IJCV* 38, 63–84.

[35] Jepson, A., Fleet, D., and Maraghi, T. (2003). Robust online appearance models for visual tracking. *PAMI* 25, 1296–1311.

[36] Avidan, S. (2004). Support vector tracking. *PAMI* 26, 1064–1072.

[37] Ross, D., Lim, J., Lin, R., and Yang, M.-H. (2008). Incremental learning for robust visual tracking. *IJCV* 77, 125–141.

[38] Mei, X., and Ling, H. (2011). Robust visual tracking and vehicle classification via sparse representation. *PAMI* 33, 2259–2272.

[39] Li, H., Shen, C., and Shi, Q. (2011) Real-time visual tracking using compressive sensing. *CVPR*, pp. 1305–1312.

[40] Babenko, B., Yang, M.-H., and Belongie, S. (2011). Robust object tracking with online multiple instance learning. *PAMI* 33, 1619–1632.

[41] Collins, R., Liu, Y., and Leordeanu, M. (2005). Online selection of discriminative tracking features. *PAMI* 27, 1631–1643.

[42] Grabner, H., Grabner, M., and Bischof, H. (2006). Real-time tracking via online boosting. *BMVC*, pp. 47–56.

[43] Grabner, H., Leistner, C., and Bischof, H. (2008). Semi-supervised online boosting for robust tracking. In: Forsyth, D., Torr, P., Zisserman, A. (eds.) *ECCV 2008, Part I.* LNCS, vol. 5302, pp. 234–247. (Heidelberg: Springer).

[44] Kalal, Z., Matas, J., and Mikolajczyk, K. (2010). P-n learning: bootstrapping binary classifier by structural constraints. *CVPR*, pp. 49–56.

[45] Vaswani, N. (2008). "Kalman filtered compressed sensing." In *ICIP 2008. 15th IEEE International Conference on Image Processing.* pp. 893–896. IEEE.

[46] Cossalter, M., Valenzise, G., Tagliasacchi, M., and Tubaro. S. (2010). "Joint compressive video coding and analysis." *IEEE Trans. Multimedia* 12 (3), 168–183.

[47] Candes, E. J., and Tao, T. (2006). "Near-optimal signal recovery from random projections: Universal encoding strategies?" *IEEE Transactions on Information Theory*, 52 (12), 5406–5425.

[48] Candes, E. J., Romberg, J., and Tao, T. (2006). "Stable signal recovery from incomplete and inaccurate measurements," *Communications on Pure and Applied Mathematics*, 59 (8), 1207–1223.

[49] Khan, S. M., and Shah, M. (2006). "A multi-view approach to tracking people in crowded scenes using a planar homography constraint," in *European Conference on Computer Vision*, vol. 4, pp. 133–146.

[50] Cheung, G. K. M., Kanade, T., Bouget, J. Y., and Holler, M. (2000). "Real time system for robust 3D voxel reconstruction of human motions," *CVPR*, 714–720.

[51] Weinland, D., Boyer, E., and Ronfard, R. (2007). "Action recognition from arbitrary views using 3D exemplars," in *International Conference on Computer Vision*, pp. 1–7.

[52] Sundaresan, A., and Chellappa, R. (2006). "Segmentation and probabilistic registration of articulated body models," in *International Conference on Pattern Recognition*, 2, 92–96.

[53] Reddy, D., and Sankaranarayanan, A. C., Cevher, V. and Chellappa, R. (2008). "Compressed sensing for multi-view tracking and 3-D voxel reconstruction." In *ICIP 2008. 15th IEEE International Conference on Image Processing, 2008*, pp. 221–224. IEEE.

[54] Eric, W., Silva, J., and Carin, L. (2009). "Compressive particle filtering for target tracking." In *SSP'09. IEEE/SP 15th Workshop on Statistical Signal Processing, 2009*, pp. 233–236. IEEE.

[55] Benezeth, Y., Jodoin, P.M., Emile, B., Laurent, H., and Rosenberger, C. (2010). Comparative study of background subtraction algorithms. *J. Electron. Imag.* 19, 033003.

[56] Staucr, C., and Grimson, W. E. L. (1999). Adaptive background mixture models for real-time tracking. *Comput. Vision Patt. Recog.* 2, 252–258.

[57] Jiang, H., Li, C., Haimi-Cohen, R., Wilford, P., and Zhang, Y. (2012). Scalable Video Coding using Compressive Sensing, *Bell Labs Techn. J.* 16 (4).

[58] Jiang, H., Deng, W., and Shen, Z. (2013). "Surveillance video processing using compressive sensing." *arXiv preprint arXiv:1302.1942*.

[59] Bashir, F., and Porikli, F. (2006). "Performance evaluation of object detection and tracking systems." In *Proceedings 9th IEEE International Workshop on PETS*, pp. 7–14.

7

Compressed Sensing for Wireless Networks

7.1 Wireless Networks

A wireless local area network (LAN) uses radio waves to connect devices such as laptops to the Internet and business network. There was a widespread belief in the past that wired networks were faster and more secure than wireless networks. But continued enhancements to wireless networking standards and technologies have eroded those speed and security differences. When laptops are connected to WiFi hot spots in public places, the connection is established to that business's wireless network.

There are four main types of wireless networks:

- Wireless local area network (WLAN): Links two or more devices using the wireless distribution method, providing the connection through access points to the wider Internet.
- Wireless metropolitan area network (WMAN): Connects several wireless LANs.
- Wireless wide area network (WWAN): Covers large areas such as neighboring towns and cities.
- Wireless personal area network (WPAN): Interconnects devices in a short span, generally within a person's reach.

Small business firms can experience many benefits from a wireless network, including the following:

- **Convenience**: Access the network resources from any location within the wireless network's coverage area or from any WiFi hot spot.
- **Mobility**: People can go online during movement, for example, from a car.
- **Easy setup**: There is no need for string cables, so installation can be quick and cost effective.
- **Expandable**: Wireless network can easily be expanded with existing equipment, while a wired network might require additional wiring.

- **Security**: Advances in wireless networks provide robust security protections.
- **Cost**: Operating costs are less owing to elimination of or reduction in wiring.

7.1.1 Categories of Wireless Networks

Many types of wireless communication systems exist, but in network communication, wireless network takes place between computer devices. These devices include personal digital assistants (PDAs), laptops, personal computers (PCs), servers, and printers. Computer devices have processors, memory, and a means of interfacing with a particular type of network. Traditional cell phones do not fall within the definition of a computer device. However, newer phones and even audio headsets are beginning to incorporate computing power and network adapters. Eventually, most electronics will offer wireless network connections. An explicit pictorial representation of evolution of 2G networks to 3G networks is given by Figure 7.1.

7.1.1.1 3G cellular networks

Universal deployment of wideband code division multiple access (WCDMA) and CDMA2000 radio access technologies-based 3G mobile communication is now becoming common. The applications supported by these commercial systems range from circuit-switch services such as voice and video telephony to packet-switched services such as video streaming, email, and file transfer. For 3G cellular systems, there are two camps: 3G partnership project (3GPP) [1] and 3G partnership project 2 (3GPP2) [2], which is based on different 2G technologies. The development of 3G follows a few key trends which include the following: voice services will continue to be important in the future, which would ultimately keep capacity optimization for voice services in continuity. Along with increasing use of IP-based applications, the importance of data as well as simultaneous voice and data will increase. With increasing need for data, improvement in efficiency of data becomes a matter of great necessity. When more and more attractive multimedia terminals emerge in the markets, the usage of such terminals proliferates into offices, homes, and airports to roads, and finally everywhere. This means that high-quality high-data-rate applications will be needed everywhere as well. When the volume of data increases, the cost per transmitted bit needs reduction in order to make new services and applications affordable for everybody. The other current trend is that in the 3G evolution path very high data rates are achieved in hot spots with WLAN rather than cellular-based standards.

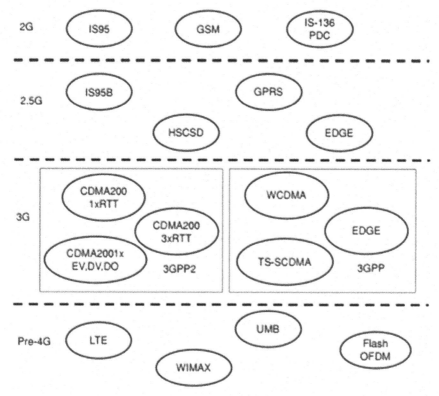

Figure 7.1 Evolution of 2G to 3G networks.

CDMA2000

CDMA uses different spreading codes, enabling simultaneous occupation of the same channel by a large number of users, which could provide a multiple-access scheme for cellular telecommunications. The first standard was the IS-95, and the first network was launched in Hong Kong in 1996 under the brand name CDMA One. The CDMA system also has the following standards in its developmental stages as follows: IS-95, IS-95A, IS 95B, and CDMA2000 (1x/EV-DO, 1xEV-DV, 1xRTT, and 3xRTT). The first version of system standard IS-95 has never been launched for commercial purposes due to its prematurity. The IS-95A has been applied for businesses since then and has extensive applications even today. IS-95B was a short version since the CDMA2000 standard was announced six months after its emergence. The original IS-95A standard allowed only for circuit-switched data at 14.4 kbit/s, and IS-95B provided up to 64 kbit/s data rates as well as a number of additional services. A major step improvement came later with the development of 3G services. The first 3G standard was known as CDMA2000 1x, which initially provided data rates up to 144 kbit/s. With further developments, the systems hold the promise of allowing a maximum data rate of 307 kbits/s.

WCDMA/UMTS
WCDMA (wideband code division multiple access) was developed by NTT DoCoMo as the air interface for their 3G network FOMA. Later, NTT DoCoMo submitted the specification to the international telecommunication union (ITU) as a candidate for the international 3G standard known as IMT-2000. The ITU eventually accepted WCDMA as part of the IMT-2000 family of 3G standards, as an alternative to CDMA2000, EDGE, and the short-range DECT system. Later, WCDMA was selected as the air interface for the universal mobile telecommunications system (UMTS), the 3G successor to GSM. The basic difference between the CDMA and WCDMA lies in bandwidth: CDMA uses 1.25 MHz frequency bandwidth, while WCDMA uses 5 MHz bandwidth. Again, CDMA is 2G telecommunication standard providing smaller data rates compared to WCDMA which is third-generation technology. CDMA being 2G standard provides mainly circuit-switched services while WCDMA being used in UMTS system is used in both circuit-switched and packet switched networks. The connection procedure carried out in WCBMA is shown in Figure 7.2.

TD–SCDMA
Transmit diversity (TD) is one of the key contributing technologies to defining the ITU endorsed 3G systems WCDMA and CDMA2000. Spatial diversity is introduced into the signal by transmission through multiple antennas. The antennas are spaced far enough apart that the signals emanating from them can be assumed to undergo independent fading. In addition to diversity gain, antenna gain can also be incorporated through channel state feedback. This leads to the categorization of TD methods into open-loop and closed-loop methods. Several methods of transmit diversity in the forward link have been either under consideration or adopted for the various 3G standards. An example scenario of TD–SDMA is given in Figure 7.3. Here, many applications such as voice, messaging, and cellular transmission are carried out effectively.

7.1.1.2 WiMAX network
Wireless metropolitan area network (WMAN) technology is a relatively new field that began in 1998. This standard has helped to pave the way for WMAN

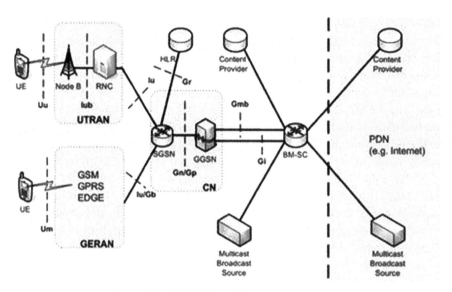

Figure 7.2   WCDMA and its connection process.

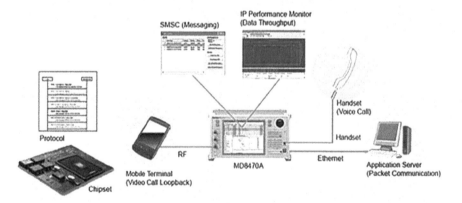

Figure 7.3   TD–SDMA example scenario.

technology globally, and since its first inception, it has received six expansions onto the standards. WMAN differs from other wireless technologies in that it is designed for a broader audience, such as a large corporation or an entire city. The 802.16 MAC uses a scheduling algorithm for which the subscriber station needs to compete once (for initial entry into the network). After the competition, the subscribe station is allocated an access slot by the base station. The time-slot can enlarge and contract, but it remains assigned to the subscriber stations, which means that other subscribers cannot use it. The 802.16 scheduling algorithm is stable under overload and oversubscription (unlike 802.11). It can also be more bandwidth efficient. The scheduling algorithm also allows the base station to control QoS parameters by balancing the time-slot assignments among the application needs of the subscriber stations. The MAC layer is also in charge of protocol data unit (PDU) assembly and disassembly. The operation standards for WMANs are regulated under IEEE standard 802.16 [3], and the WMANs are allowed the operating frequency range of 10–66 GHz. With such a broad spectrum to work with, WMANs have the ability to transmit over previous wireless frequencies such as IEEE 802.11b/g, causing less interference with other wireless products. A simple WiMAX network is shown in Figure 7.4. The only drawback of using such high frequencies is that WMAN needs a line of sight between the transmitters and receivers, much like a directional antenna. Line of sight, however, decreases multi-path distortion, allowing higher bandwidths to be achieved, and it can attain up to 75 Mbps for both uplink and downlink on a single channel [4].

IEEE 802.16a: The IEEE has developed 802.16a for use in licensed and license-exempt frequencies from 2 to 11 GHz. Most commercial interest in IEEE 802.16 is in these lower frequency ranges. In the lower ranges, the signals

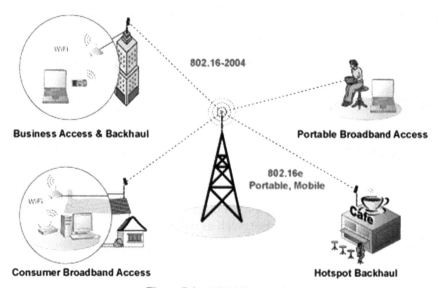

Figure 7.4 WiMAX network.

can penetrate barriers and, hence, do not require a line of sight between the transceiver and the antenna. This enables more flexible WiMax implementations while maintaining the data rate of the technology and transmission range. IEEE 802.16a supports mesh deployment, in which transceivers can pass a single communication on to other transceivers, thereby extending the basic 802.16s transmission range.

IEEE 802.16b: This extension increases the spectrum that the technology can use in the 5- and 6-GHz frequency bands and improves quality of service. WiMax provides QoS to ensure priority transmission for real-time voice and video and to offer differentiated service levels for different traffic types.

IEEE 802.16c: IEEE 802.16c represents a 10- to 66-GHz system profile that standardizes large volume of details of the technology. This encourages highly consistent implementation and, therefore, interoperability.

IEEE 802.16d: IEEE 802.16d includes minor improvements and is based on 802.16a. This extension also creates system profiles for compliance testing of 802.16a devices.

IEEE 802.16e: This technology standardizes networking between base stations of the carriers and mobile devices, rather than just between base stations and fixed recipients. IEEE 802.16e would enable the high-speed signal handoffs necessary for communications with users moving at vehicular speeds. In addition to IEEE 802.16, IEEE 802.20 (IEEE802.20), and the mobile broadband wireless access (MBWA), working group aim to prepare a formal specification for a packet-based air interface designed for IP-based services. The goal is to create an interface that allows the creation of low cost, always on, and truly mobile broadband wireless networks, nicknamed Mobile-Fi. IEEE 802.20 will be specified according to a layered architecture, which is consistent with other IEEE 802 specifications. The working group consists of the following layers: the physical (PHY), medium access control (MAC), and logical link control (LLC) layers. The air interface will operate in bands below 3.5 GHz and with a peak data rate of more than 1 Mbit/s. 802.20 and 802.16e, the so-called mobile WiMAX, have identical goals. WiMAX can be viewed as "last mile" connectivity at high data rates. This could result in lower pricing for both home and business customers as competition lowers prices. In areas without preexisting physical cable or telephone networks, WiMAX may be a viable alternative for broadband access that has been economically unavailable. Prior to WiMAX, many operators have been using proprietary, fixed wireless technologies for broadband services. For this reason, WiMAX has its significant markets in rural areas and developing countries.

7.1.1.3 WiFi networks

Figure 7.5 shows the general wireless connectivity including WiFi and ethernet. IEEE 802.11 denotes a set of wireless local area network (WLAN) standards developed by working group 11 of the IEEE LAN/MAN standards committee (IEEE 802). WiFi is a brand originally licensed by the WiFi alliance to describe the underlying technology of WLAN based on the IEEE 802.11 specifications. It was developed for use in mobile computing devices, such as laptops; in LANs, but it is now increasingly used for more services, including Internet and VoIP phone access, and gaming; and basic connectivity of consumer electronics such as televisions, DVD players, and digital cameras. In the physical layer, 802.11b operates within the 2.4-GHz industrial, scientific, and medical (ISM) band. The original 802.11b defines data rates of 1 Mbps and 2 Mbps via radio waves using frequency hopping spread spectrum (FHSS) or direct sequence spread spectrum (DSSS). For FHSS, the 2.4-GHz band is divided into 75 1-MHz sub-channels. The sender and the receiver agree on a hopping pattern, and data are sent over a sequence of the sub-channels. Each conversation within the 802.11 network occurs over a different hopping pattern. Because of federal communications commission (FCC) regulations that restrict sub-channel bandwidth to 1 MHz, FHSS

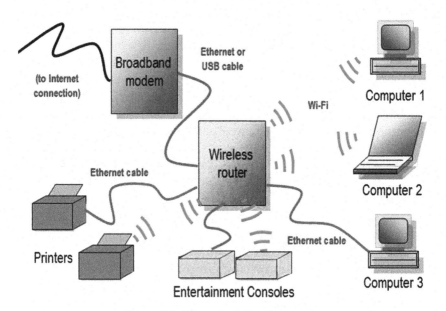

Figure 7.5   General wireless framework.

techniques are limited to speeds of no higher than 2 Mbps. DSSS divides the 2.4-GHz band into 14 22-MHz channels. Adjacent channels overlap one another partially, with three of the 14 being completely non-overlapping. The spreading code is an 11-bit barker sequence. Binary-phase shift keying (BPSK) and quadrature-phase shift keying (QPSK) are used for providing different rates. For increasing the data rate to 5.5 Mbps and 11 Mpbs in the 802.11b standard, an advanced coding technique, complementary code keying (CCK), is employed. A complementary code contains a pair of finite bit sequences of equal length, in which the number of pairs of identical elements (1 or 0) with any given separation in one sequence is equal to the number of pairs of unlike elements having the same separation in the other sequence. A network using CCK can transfer more data per unit time for a given signal bandwidth than a network using the barker code, because CCK makes more efficient use of the bit sequences. CCK consists of a set of 64 8-bit code words. The 5.5 Mbps rate uses CCK to encode 4 bits per carrier, while the 11 Mbps rate encodes 8 bits per carrier. Both speeds use QPSK as the modulation technique and signal at 1.375 MSps. Standard 802.11a adopts orthogonal frequency division multiplexing (OFDM) at 5.15–5.25 GHz, 5.25–5.35 GHz, and 5.725–5.825 GHz to support multiple data rates up to 54 Mbps. Standard 802.11g utilizes the 2.4-GHz band with OFDM modulation and is also backward compatible with 802.11b. For OFDM, the FFT has 64 subcarriers. There are 48 data subcarriers and 4 carrier pilot subcarriers, for a total of 52 nonzero subcarriers defined in IEEE 802.11a, plus 12 guard subcarriers. The IEEE 802.11a/g physical layer provides eight PHY modes with different modulation schemes and different convolutional coding rates, and it can offer various data rates. In January 2014, IEEE announced the formation of a new 802.11 group for the development of a new amendment to the 802.11 standard for wireless local area networks. This was meant for achieving higher data rates in the PHY layer. 802.11n builds upon previous 802.11 standards by adding MIMO (multiple-input multiple-output). MIMO uses multiple transmitter and receiver antennas to allow for increased data throughput through spatial multiplexing and increased range by exploiting the spatial diversity. There are several proposal groups named TGnSync, WWiSE (short for "world-wide spectrum efficiency"), and MIT-MOT ("Mac and MIMO technologies for more throughput"). All proposals occupy frequency band 2.5 GHz with 20-MHz or 40-MHz bandwidth so as to support the communication speed of more than 200 Mbps. 802.11n is backward compatible with 802.11b and 802.11g. Recently, the 802.11ac standard has been under development. Several key challenges have to be overcome for taking full advantage of future marketing opportunity for WiFi. In the following, we list some near-future design topics and their possible solutions.

7.1.1.4 Wireless Ad hoc networks

An ad hoc network is an autonomous collection of mobile users that communicate over bandwidth-constrained wireless links. The network is decentralized, so that all network activity that includes discovering the topology and delivering messages require execution by the nodes themselves. Ad hoc networks need efficient distributed algorithms to determine network organization, link scheduling, and routing. For a special case of ad hoc network, mobile ad hoc networks (MANETs), due to the mobility of the nodes, the network topology may change rapidly and unpredictably over time. The first generation of ad hoc networks was initiated in the early 1970s, when packet radio networks (PRNET) were proposed from the defense advanced research projects agency (DARPA) for multihop networks in a combat environment and areal locations of hazardous atmospheres (Aloha) was proposed in Hawai'i for distributed channel access management. The second generation of ad hoc networks emerged in the 1980s, when the ad hoc network systems were further enhanced and implemented as a part of the survivable adaptive radio networks (SURAN) program. SURAN provided a packet-switched network to the mobile battlefield in an environment without infrastructure, so as to be beneficial in improving the performance of radios by making them smaller, cheaper, and resilient to electronic attacks. In the 1990s, the concept of commercial ad hoc networks arrived with notebook computers and other viable communications equipment. For example, the IEEE 802.11 subcommittee had adopted the term "ad hoc networks".

The advantages of ad hoc networks are the ease and speed of deployment, which are important requirements for military applications. Ad hoc networks decrease dependence on expensive infrastructure for civilian applications. The set of applications for ad hoc networks is diverse, ranging from small, static networks that are constrained by power sources, to large-scale, mobile, highly dynamic networks. Some typical applications are personal area networks, emergency operations such as policing and fire-fighting, civilian environments such as taxi networks, and military use on the battlefields. In contrast to the traditional wireless network with infrastructure, an ad hoc network needs its own design requirements so as to be functional. We list some important aspects as follows: distributed operation and self-organization. No node in the ad hoc network can depend on a network in the background to support the basic functions such as routing. Instead, these functions must be implemented and operated efficiently in a distributed manner.

Moreover, if the event topology changes due to mobility, the network can be self-organized to adapt to the changes. In addition, because the ad hoc nodes might belong to different authorities, they might not be willing to cooperate to fulfill the network functions due to ad hoc nodes belonging to different authorities. However, this non-cooperation can cause severe network breakdown. Motivation of the distributed autonomous users is an important research and design topic. Traditionally, pricing anarchy is employed using the distributed control theory. Exploration of other methods such as game theory for motivating users' cooperative behavior is dealt with later in this book. Dynamic routing for MANET, the routing problem between any pair of nodes, is challenging due to the mobility of nodes. The optimal source-to-destination route is time variant. Moreover, compared to the traditional network in which the routing protocols are proactive, the ad hoc dynamic routing protocols are reactive. The routes are determined only when the source requests the transmission to the destination.

There are two types of ad hoc dynamic routing protocols: Table-driven routing protocols and source-initiated on-demand routing protocols. The table-driven routing protocols require each node to maintain one or more tables to store routing information. The protocols rely on an underlying routing table update mechanism that involves the constant propagation of routing information. Packets can be forwarded immediately since the routes are always available. However, this type of protocol causes substantial signaling traffic and power consumption problems. Some protocols existing in literature are destination-sequenced distance-vector routing [5], cluster head gateway switch routing [6], and wireless routing protocol [7]. Source-initiated on-demand routing creates routing only when desired by the source node. The disadvantage is that the packet at the source node must wait until a route can be discovered. But the advantage is that periodic route updates are not required. Some of the available routing protocols in the literature are ad hoc on-demand distance vector routing [8], dynamic source routing [9], temporally ordered routing algorithm [10], associativity-based routing [11], and signal stability-based adaptive routing protocol [12].

7.1.1.5 Wireless sensor networks

A wireless sensor network (WSN) as in Figure 7.6 is a wireless network consisting of spatially distributed autonomous devices using sensors for cooperative monitoring of physical or environmental conditions, such as

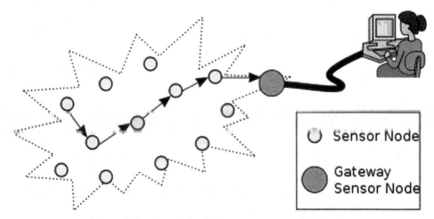

Figure 7.6   A typical wireless sensor network (WSN).

temperature, sound, vibration, pressure, motion, or pollutants, at different locations. The goals and tasks of sensor networks are to determine the value of some parameter at a given location, to detect the occurrence of events of interest and estimate parameters of the detected events, to classify a detected object, or to track an object. The development of wireless sensor networks was originally motivated by military applications such as battlefield surveillance. However, wireless sensor networks are now used in many civilian application areas too, including environment and habitat monitoring, healthcare applications, home automation, and traffic control. Some examples are listed as follows: military sensor networks to detect and gain as much information as possible about enemy movements, explosions, and other phenomena of interest. Sensor networks to detect and monitor environmental changes such as those seen in plains, forests, and oceans. Wireless traffic sensor networks to monitor vehicle traffic on highways or in congested parts of a city. Wireless surveillance sensor networks provide security in shopping malls, parking garages, and other civil facilities. Manufacturing sensors can facilitate the monitoring and control process, which can reduce the cost, improve the flexibility, and enhance the accuracy.

Sensors for supermarkets can speed products to shelves and provide customers with better quality products and the benefits of easy location of products of their choice. Medical sensors, especially those implanted sensors, need to constantly monitor the patients, have sufficient battery life, and be able to transmit the sensed information out of the body via wireless channels. In addition to one or more sensors, each node in a sensor network is typically equipped with a radio transceiver or other wireless communications device, a small microprocessor, some memory, and an energy source, usually a battery. For the sensors, different sensing applications can include temperature, light, humidity, pressure, accelerometers, magnetometers, chemical, acoustics, and image/video. The microprocessor has a significant constraint in computational power. Currently, the devices typically have the component-based embedded operating system. The available memory is also very limited. The current radio transceivers for sensor networks are of low rate and of short range. Some sensors can be powered by a wired power source, while most of the widely deployed sensors are powered by battery. Exchanging the energy-depleting sensors is a challenging job, so power saving is critical in the design of such wireless sensor networks. The size of a single sensor node can vary from shoe-box-sized nodes down to devices the size of a grain of dust. The cost of sensor nodes is similarly variable, ranging from hundreds of dollars to a few cents, depending on the size of the sensor network and the complexity required of individual sensor nodes. Size and cost constraints on sensor nodes result in corresponding constraints on resources such as energy, memory, computational speed, and bandwidth. Basically, to design a wireless sensor network, the following requirements should be considered [13]:

1. Large number of (mostly stationary) sensors;
2. Low energy use to extend sensor network lifetime;
3. Network self-organization;
4. Collaborative signal processing;
5. Querying ability.

In practice, there are some other design issues [14] such as those relating to deployment of sensor networks, location of specific sensors, shutting down and reactivation of a sensor for saving energy, routing of information back to data-collecting points, reduction in packet forwarding loads by data fusion, and finally having secure sensor networks. Sensor network deployment problem is to select the locations to place the sensor networks, given a particular application context, an operational region, and a set of wireless sensor devices. The sensors can be deployed in a structured sense or in a randomly scattering manner. The density of sensors is judged by the robustness and cost of the networks. Usually, the sensor networks have two types of sensors. The first type is the low capability sensor that is in charge of collecting data, while the other type is the cluster head or data sink that is more powerful in computation and data transmission. The network topology between these two types of sensors can be star-connected single-hop, multi-hop mesh/grid, or a multiple-tier hierarchical cluster.

Because the transmit power is bounded, the sensor can reach the other sensor with limited distance. This fact causes connectivity problems for sensor networks. Moreover, the sensing range is also restrained. The coverage of the sensing area is a function of the density of the sensors. There are different types of coverage metrics such as k-coverage, minimum coverage, and maximal breach distance [15]. Localization means determining the location of a certain event sensed by the sensor. This location information can provide a location stamp over the event, track the monitored object, determine the coverage, form the cluster, facilitate routing, and perform efficient querying. Despite the capability of GPS in obtaining the information, the cost and indoor environments prohibit the sensors from being equipped with GPS. Nevertheless, the task of localization captures multiple aspects of sensor networks: The physical layer imposes measurement challenges, due to multi-path, shadowing, sensor imperfections, and changes in propagation properties. Extensive computation is necessary for many formulations of localization problems.

Moreover, the problems sometimes should be solved in a distributed manner or on a memory-constrained processor. Next, for networking and coordination issues, sensor nodes have to collaborate and communicate with each other to know the topology of the whole network. Finally, for system integration issues, integration of location services with other applications is a challenging task. There are several types of localization mechanisms. First, an active localization system sends signals to localize a target. The examples are RADAR or LIDAR (LADAR). Second, in cooperative localization, the target cooperates with the system. For example, the target emits a signal with known characteristics, and then, the system deduces location by detecting a signal. Third, a passive localization system deduces location from observation of signals that are already present. The example technique is to use the geometric methods to calculate the location by measuring the signal strength over the receivers in different locations.

Finally, a blind localization system deduces the location of target without a priori knowledge of its characteristics. Time synchronization localization provides spatial information to the sensor networks, while accurate time synchronization is also essential, as shown in the following examples: first, since the delay for the information to the sink is unpredictable, each sensor needs to have a consistent time stamp for the message. This is more important for some types of data such as tsunami alarming, since the time information provides many scientific clues. For localization, the transmitter and receiver need to have synchronized time to enable calculation of the time of the flight. For multiple access such as TDMA-based schemes, each sensor needs to transmit at the exact time-slots. For sleeping scheduling, the energy is saved by turning on and turning off of sensors at certain times. The accurate time can be obtained by a GPS signal. But this approach is very expensive. Quartz crystal oscillators can provide accuracy of about several μs. For better accuracy, some techniques such as phase-locked loops need to be implemented to synchronize the clocks.

Sleeping Mechanism

In most sensors, the primary source of power consumption is the radio for transmitting, receiving, and listening. If the sensors wake up only during the time when the radio is active and sleep in the remaining time, the limited energy can be conserved and the lifetime of sensor networks can be prolonged. However, the sleep-and-wake-up mechanism causes other design problems. First, there is a trade-off between the delay of information and energy consumption. Moreover, the design of the MAC layer multiple access protocol needs consideration of the wake-up time. Further, the transmitter and receiver should be synchronized for simultaneous waking up. Finally, the fairness issue needs to be considered so that some sensors do not get overloaded to be energy depleted too early. Energy-efficient routing since the energy is a major concern for the design of wireless sensor networks, to select energy-efficient routes from the data-collecting sensors to the data sink, can significantly improve the network lifetime.

In addition, when multiple routes are considered, individual optimal energy-efficient routes are not optimal, in the sense that some sensors on the critical paths might be depleted first. So, joint optimization is necessary. In addition to the energy concern, the routing protocols also should take consideration of latency due to the sleeping mechanism of sensors. The routing protocols should also consider the data fusion/aggregation. Finally, scalability is an important issue for large sensor networks. For the situations in which the sensors are mobile or can join/leave the network frequently, the adaptive ability is also a design challenge. Fusion/aggregation is a process dealing with the association, correlation, and combination of data and information from single and multiple sources to achieve refined position and identity estimates for observed entities, and to achieve complete and timely assessments of situations and threats, and their significance. In wireless sensor networks especially large ones, it is impossible and energy inefficient to gather the information to make the decision. Instead, along the path to the data sink, a data fusion node collects the results from multiple nodes, fuses the results with its own based on a decision criterion, and then sends the fused data to another node/base station. By doing this, the traffic load can be greatly reduced and energy can be conserved. There are two forms of data fusion/aggregation. In the first form, data from different node measurements are combined together to form larger packets. It is simple to implement, but requires higher computational burden, higher communication burden, and larger training data requirement. The second type is decision fusion, in which the decisions (hard or soft) based on node measurements are combined. The decision fusion solves the problems seen in the first form. The decision can be made by mechanisms such as voting. For example, a fusion node arrives at a consensus by a voting scheme: majority voting, complete agreement, and weighted voting. Other fusion decision algorithms include the probability-based Bayesian model [16] and stacked generalization [17].

For sensor networks, there are different fusion architectures from the data-collecting sensors to the data sink. In the centralized architecture, a central processor fuses the reports collected by all other sensing nodes. The centralized one has the advantages that an erroneous report can be easily detected and it is simple to implement. On the other hand, it has the disadvantage that it is inflexible to sensor changes and the workload is concentrated at a single point. In the decentralized architecture, data fusion occurs locally at each node on the basis of local observations and the information obtained from neighboring nodes. There is no central processor node. The advantages are scalable and tolerant to the addition or loss of sensing nodes or dynamic changes in the network. In the hierarchical architecture, nodes are partitioned into hierarchical levels. The sensing nodes are at level 0 and the data sink at the highest level. Reports move from the lower levels to higher ones. This architecture has the advantage that workload is balanced among nodes.

Security
Since sensor networks may interact with sensitive data and/or operate in hostile unattended environments such as military sensors, it is important for the security to be addressed from the system design. Moreover, due to inherent resource and computing constraints, security in sensor networks poses more challenges than traditional network/computer security. The possible security attacks include denial of service attack, Sybil attack, traffic analysis attacks, node replica attack, privacy attack, physical attack, and collusion attack. There are some defensive mechanisms in literature such as key cryptography and

trust management. The security issue is beyond the scope of this book. A good survey of security issues in wireless sensor networks can be found in [18].

7.1.2 Advanced Wireless Technologies

7.1.2.1 OFDM technology

Orthogonal frequency division multiplexing (OFDM) is a technique of transmitting multiple digital signals simultaneously over a large number of orthogonal subcarriers. Based on the fast Fourier transform algorithm for generation and detection of the signal, data transmission can be performed over a large number of carriers that are spaced apart at precise frequencies. The frequencies (or tones) are orthogonal to each other. Therefore, the spacing between the subcarriers can be reduced and hence high spectral efficiency can be achieved. Figure 7.7 portrays the OFTM transmitter and receiver pair that has modulator, inverse fast Fourier transform, and D/A converter in the transmitter side and the vice versa at the receiver. OFDM transmission is also resilient to interference and multi-path distortion, which causes inter-symbol interference (ISI). OFDM transmitter and receiver block diagrams are shown in Figures 2.14 and 2.15, respectively. $s[n]$ is a serial stream of binary digits to transmit. After the serial-to-parallel converter, the data are split into N streams. Each stream is then coded to $X_0, \ldots X_{N-1}$ with possible different modulation methods (such as PSK and QAM), depending on the sub-channel condition. An inverse FFT is computed on each set of symbols, giving a set of complex time-domain samples. These samples are then quadrature-mixed to passband in the standard way: the real and imaginary components are first

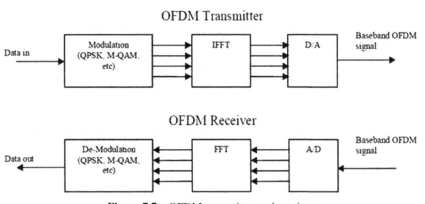

Figure 7.7 OFDM transmitter and receiver.

converted to the analog domain using digital-to-analog converters (DACs); the analog signals are then used to modulate cosine and sine waves at the carrier frequency, f_c, respectively. These signals are then summed up to yield the transmission signal, $s(t)$. The receiver picks up the signal $z(t)$, which is $s(t)$ transmitted through radio channels and contaminated by noise. Then, $r(t)$ is quadrature-mixed down to baseband using cosine and sine waves at the carrier frequency. The baseband signals are then sampled and digitized using analog-to-digital converters (ADCs), and a forward FFT is used to convert back to the frequency domain. This returns N parallel streams, each of which is converted to a binary stream using an appropriate symbol detector. These streams are then recombined into a serial stream, $\hat{s}[n]$, which is an estimate of the original binary stream at the transmitter.

For a mathematical description, the low-pass equivalent OFDM signal is expressed as follows:

$$V(t) = \sum_{k=0}^{N-1} X_k e^{j2\Pi kt/T}, 0 \le t < T,$$

where X_k are the modulated data symbol for the kth data stream and T is the OFDM symbol time. A guard interval of length T_g is inserted prior to the OFDM block for avoiding inter-symbol interference in multi-path fading channels. During this interval, a cyclic prefix is transmitted such that the signal in the interval $-T_g \le t < 0$ equals the signal in the interval $(T - T_g) \le t < T$. Cyclic prefix is often used in conjunction with modulation for retaining the sinusoids' properties in multi-path channels. It is well known that sinusoidal signals are eigen functions of linear, and time-invariant, systems. Therefore, if the channel is assumed to be linear and time invariant, a sinusoid of infinite duration would be an eigen function. However, this cannot be achieved in practice, as real signals have chronological limitations. So, to mimic the infinite behavior, prefixing the end of the symbol to the beginning makes the linear convolution of the channel appear as though it were circular convolution and thus preserves this property in the part of the symbol after the cyclic prefix. Orthogonal frequency division multiple access (OFDMA) is a multiple access scheme implemented based on OFDM. In OFDMA, different users are allocated with different subcarriers, and hence, multiple users can transmit their data simultaneously. QoS can be achieved in OFDMA by allocating different numbers of subcarriers to the users with different QoS requirements. In OFDMA, different users can occupy different time frequency slots to enable full utilization of the diversity. OFDM can be combined with the

CDMA scheme (i.e., multicarrier code division multiple access (MC-CDMA) or OFDM-CDMA). In this case, different codes are assigned to different users for concurrent transmissions. Dynamic spectral management for OFDMA can be categorized into two groups based on the network control architecture.

1. Centralized control: A spectral management center (SMC) monitors and controls the transmit spectra of all the users in the system. A considerable amount of coordination and communication between the users and the SMC is required. This greatly increases the control signaling overhead but leads to a satisfactory performance when compared to the distributed control. Centralized control can be further categorized as follows [19]: Level 1— The data rate and transmit power of the user are reported to a central controller and corresponding control signals are generated to control the rate and transmit power. Level 2—The noise spectra and the received signal spectra are monitored, and the transmit power is controlled by the central controller. Level 3—It allows complete coordination in real-time control of transmit power while monitoring the signals and noise spectra.

2. Distributed control: The users have the capability of sensing the channel conditions and adjust their transmit spectra accordingly. This may be efficient from the perspective of signaling and coordination but has the drawback of converging to a suboptimal point. It is also referred to as Level 0 [19], where there is no dynamic spectrum management (DSM) involved and the control is fully distributed. The control exhibited by the SMC over the users determines the computational complexity of the overall system. A centralized control is efficient but consumes a lot of bandwidth in control messaging between the users and the base station. On the other hand, distributed algorithms increase the complexity of the receiver of the user. The resource allocation of OFDMA can be broadly categorized into three areas: subcarrier assignment—The subcarriers with the best channel gains as seen by the user are allocated to the particular user. Rate allocation—The data rate is allocated depending on the user application requirements. Power control—Optimal transmit power is to be allocated to the user, in order to meet its rate requirements without interfering with the other users. OFDM and OFDMA are used in the emerging standards that include IEEE 802.11a/Wi-Fi, IEEE 802.15/WiPAN, IEEE 802.16/WiMAX, IEEE 802.20/MobileFi, IEEE 802.22/WiRAN, digital audio broadcasting (DAB), terrestrial broadcasting of digital television (DVB-T, DVB-H), fast low latency access with seamless handoff orthogonal frequency division multiplexing (Flash-OFDM), SDARS for satellite radio, G.DMT (ITU G.992.1) for ADSL, and ITU-T G.hn for power line communication.

7.1.2.2 Multiple antenna systems

Transceivers employ antenna arrays for spatial diversity and adjust their beam patterns to enable having good channel gain toward the desired directions, while the aggregate interference power is minimized at its output. Antenna array-processing techniques such as beam forming, MIMO technology, and space–time coding can be applied to receive and transmit multiple signals that are separated in space. Hence, multiple cochannel users can be supported in each cell to increase the capacity by exploring the spatial diversity. In this subsection, we briefly discuss various multiple antenna technologies.

Beam-Forming Technique

Beam forming is a technique in signal processing used for directional data transmission and reception. A typical beam forming is shown in Figure 7.8 using which the signal is received/transmitted in a particular direction to improve the receive/transmit gain. When transmitting, a beam former controls the phase and amplitude of the signal to create a pattern of constructive and destructive interference in the wave front. When receiving, these signals are combined to ensure preferential observation of the expected pattern of

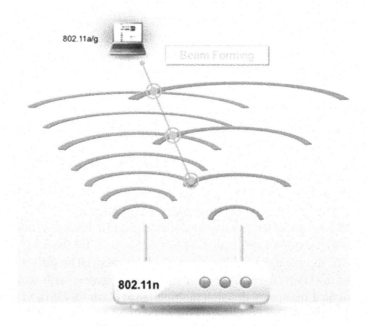

Figure 7.8   Transmit beam-forming.

radiation. Different weights can be assigned to the different signals so that the decoded information has the smallest error probability. In a conventional beam-forming system, these weights are fixed and can be obtained according to the location of the receive antenna and the direction of the signal. Alternatively, the weights can be adaptively adjusted by considering the characteristics of the received signal to mitigate the interference from unwanted sources. In Figure 7.7, we show an illustration of beam-forming technology.

MIMO Technology and Space–Time Coding
Multiple-input multiple-output (MIMO), or multiple antennas, can be used to transmit and receive the radio signals. Data transmitted from multiple antennas will experience different multi-path fading, and at the receiver, these different multi-path signals are received by multiple antennas. By using advanced signal-processing techniques, multi-path signals at the receiver can be combined to reconstruct original data. MIMO systems take advantage of this spatial diversity to achieve higher data rates or lower bit error rates. There are two basic types of MIMO systems, namely the space–time coding MIMO system (for diversity maximization) and spatial multiplexing MIMO system (for data rate maximization). In a space–time coding MIMO system, a single data stream is redundantly transmitted over multiple antennas by using suitably designed transmit signals. Multiple copies of the signal are received at the receiver end, which are used to construct the original data.

Space–time coding is shown in Figure 7.9. Space–time coding can be categorized into space–time trellis coding (STTC), in which trellis code is transmitted over multiple antennas and multiple time-slots, and space–time block-coding (STBC) in which a block of data is transmitted over multiple antennas. Both STTC and STBC can achieve diversity gain, which improves

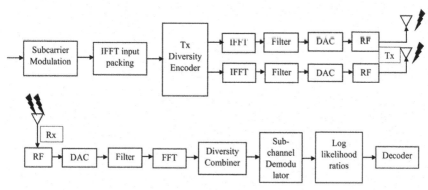

Figure 7.9 Space–time coding.

the error performance. While STTC can also achieve coding gain (i.e., results in lower error rate), STBC can be implemented with less complexity. Instead of transmitting the same data over multiple antennas, a spatial multiplexing MIMO system transmits different data streams over multiple antennas. In this case, the number of transmit antennas is equal to or larger than that of receive antennas and the data rate can increase by a factor of the number of transmit antennas [20]. There is a fundamental trade-off between diversity gain and multiplexing gain. In simple words, the diversity gain improves the BER performance, while multiplexing gain will increase the rate, but on the other hand, the BER performance might be reduced. Detailed analysis can be found in the literature [21].

One simple example of an STBC is the Alamouti code [22], which was designed for a two transmit antenna system and has the coding matrix:

$$C = \begin{bmatrix} s_1 & s_2 \\ -s_2^* & s_1^* \end{bmatrix}$$

where s_1 and s_2 are two symbols, $*$ denotes complex conjugate, and components of the matrix are sent on two antennas and two time-slots, respectively. The coding rate is 1, and using the optimal linear decoding scheme, the BER of this STBC is equivalent to maximal ratio combining (MRC). Multiuser MIMO (MU-MIMO) was proposed to support data transmission from multiple users simultaneously. In this case, data from different users can be transmitted over different antennas. MIMO broadcast channels and MIMO multiple access channels are used in this MU-MIMO, for downlink and uplink transmissions, respectively. Alternatively, space division multiple access (SDMA) can exploit the information of users' locations to adjust the transmission and reception parameters to achieve the best path gain in the direction of each user. Phased array antenna techniques are generally used for SDMA. MIMO is an optional feature in the IEEE 802.16/WiMAX standard, while it is being used in the IEEE 802.11n standard.

7.2 CS-based Wireless Communication

The CS theory [23–25] is a new technology that has emerged in the area of signal processing, statistics, and wireless communication. By utilizing the sparsity of signal or compressibility in some transform domain, CS can powerfully acquire a signal from a small set of randomly projected measurements with a sampling rate much lower than the Nyquist sampling rate. The word "sparse" means that a signal vector contains only very few nonzero entries. As more and more experimental evidences suggest the presence of an inherent sparse multi-path structure in broadband or multi-antenna communication channels, CS has recently become a promising tool to deal with the channel estimation problem [26–31]. In this chapter, we present several new approaches to estimating sparse (or effectively sparse) multi-path channels that are based on CS for different scenarios. The traditional channel estimation method is also mentioned and compared. We highlight channel estimation via CS and explain the problem formulations for different kinds of wireless channels.

7.2.1 Multi-Path Channel Estimation

In this section, based on [26], we discuss the multi-path channel model first, and then, we illustrate the traditional least-squares method for channel estimation. Finally, we discuss the CS-based channel sensing and study four different scenarios: sparse frequency selective channel, sparse doubly selective channel, sparse non-selective MIMO channel, and sparse frequency-selective channel.

7.2.1.1 Channel model and training-based model

For the MIMO multi-path channel, the channel response can be written as

$$H(t, f) = \sum_{i=1}^{N_R} \sum_{k=1}^{N_T} \sum_{i=0}^{L-1} \sum_{m=-M}^{M} H_v(i, k, l, m) \alpha R \left(\frac{i}{N_R} \right)$$

$$\alpha T \left(\frac{k}{N_T} \right) e^{-j2\pi \frac{l}{W} f} e^{j2\pi \frac{W}{t} t}, \tag{7.1}$$

where N_R is the dimension of the channel output, N_T is the dimension of the channel input, L is the maximal number of resolvable delays, M is the maximal number of resolvable one-side Doppler shifts, $H_v(i, k, l, m)$ is the sampled channel response, αR is the receiver array response vector, αT is the transmitter array steering vector, W is the bandwidth, and T is the duration. In the traditional training-based solution, the transmitter sends a known sequence X to the receiver. The received signal can be written as

$$Y = \sqrt{\frac{\varepsilon}{N_T}} X H_v + Z, \tag{7.2}$$

where ε is the transmit power, H_v is the channel response comprising of the channel coefficient $H_v(i, k, l, m)$, and Z is the thermal noise. The goal is to estimate the least-squares channel response as the following problem:

$$H_v^{LS} = \arg \min_H \left\| Y - \sqrt{\frac{\varepsilon}{N_T}} X H \right\|^2. \tag{7.3}$$

The well-known solution is given by

$$H_v^{LS} = \arg \min_H \left\| Y - \sqrt{\frac{\varepsilon}{N_T}} X H \right\|^2 \tag{7.4}$$

The reconstruction error can be expressed and bounded by

$$E[\Delta(H_v^{LS})] = \frac{\text{trace}((X^H X)^{-1}) N_R N_T}{\varepsilon} \geq \frac{N_R N_T^2 L(2M+1)}{\varepsilon} \tag{7.5}$$

7.2.1.2 Compressed channel sensing

The maximal total number of coefficients to estimate is given by

$$D = N_R N_T L(2M+1). \tag{7.6}$$

However, most of the coefficients are below the noise level, and there are only very few components that are large. This sparsity motivates the CS-based approaches. From the perspective of CS, in (7.2), Y is the compressed vector, X is the random projection matrix, and H_v is the unknown sparse signal. In this subsection, we discuss four special cases and study how to write the CS problem formulations.

Sparse Frequency-Selective Channel

For the single-antenna case, the frequency-selective channel in (7.1) is reduced to

$$H(f) = \sum_{l=0}^{L=1} H_v(l) e^{-j2\pi \frac{1}{W} f} \tag{7.7}$$

The question is how to construct the random projection matrix X so that the CS can be employed. We discuss two cases for CDMA and OFDM, respectively. In the CDMA case, the training sequence can be written as

$$x(t) = \sqrt{\varepsilon} \sum_{n=o}^{N_0-1} x_n g(t - nTc), \; 0 \leq t \leq T, \tag{7.8}$$

where $g(t)$ is the unit energy chop waveform and xn is the $N0$-dimensional spreading code with unit energy $n\left[xn\right]^{n} = 1$. The received signal can be expressed by

$$y = \sqrt{\varepsilon}Xh_{v} + z, \tag{7.9}$$

where X is an $N0 \times L$ Toeplitz matrix. For the OFDM case, the training sequence takes the form as

$$x(t) = \sqrt{\frac{\varepsilon}{N_{tr}}} \sum_{n\in S_{tr}} g(t)e^{j2\pi\frac{n}{j}t}, 0 \leq t \leq T, \tag{7.10}$$

where $g(t)$ is a prototype pulse having unit energy, S_{tr} is the set of indices of pilot tones used for training, and N_{tr} is the number of receive training dimensions or the number of pilot tones ($N_{tr} = [S_{tr}]$). The received signal can be expressed similarly as $y = \sqrt{\varepsilon}Xh_{v} + z$. But here X is the $N_{tr} \times L$ sensing matrix that comprises $\{(1/\sqrt{N_{tr}})[1wn\,N_0 \ldots \ldots wn\,N_{(L-1)}] : N \in S_{tr}\}$ as its row with $wN_0 = e^{-j2\pi/N0}$.

Sparse Doubly Selective Channel
For the single-antenna doubly selective case [32–34], the channel in (7.1) is reduced to

$$H(t, f) = \sum_{l=0}^{L-1} \sum_{m=-M}^{M} H_{v}(l, m)e^{-j2\pi l_{w}f}e^{j2\pi\frac{m}{T}l}. \tag{7.11}$$

For the CDMA case, X is an $N_0 \times L(2M + 1)$ block matrix of the form $X = [X - M \ldots X_0 \ldots X_M]$. Here, X_m is an ($N_0 \times L$)-dimension matrix as

$$X_m = W_m T, \tag{7.12}$$

where $W_m = \mathrm{diag}\left(w_{N_0}^{-mo}, w_{N_0}^{-m1}, \ldots, w_{N_0}^{-m(N_0-1)}\right)$ and T is an $N_0 \times L$ Toeplitz matrix whose first row and first column are given by $\lfloor x_0 0_{L-1}^{l} \rfloor$ and $[X^T 0_{L-1}^T]^T$, respectively. The multicarrier short-time Fourier (STF) waveforms are a generalization of OFDM waveforms for doubly selective channels. The training signal can be written as

$$x(t) = \sqrt{\frac{\varepsilon}{N_{tr}}} \sum_{(n,m)\in S_{tr}} g(t - nT_0)e^{j2\pi mW0t}, 0 \leq t \leq T \tag{7.13}$$

where $g(t)$ is a prototype pulse having unit energy, $S_{tr} \subset \{0, \ldots . N_{t-1}\} \times \{0, \ldots N_{f-1}\}$, $N_f = N_0/Nt$, T_0 is the time separation of STF basis waveform, $g(t - nT_0)ej^{2\pi m W_{0t}}$, W_0 is the frequency separation of the STF basis waveform, $T_0 W_0 = 1$, $Nt = T/T_0$, and $N_f = W/W_0$. Now, the $N_{tr} \times L(2M+1)$ random projection matrix (or training sequence) matrix $\mathbf{X}$ has the following as its row

$$\left\{ \frac{1}{\sqrt{N_{tr}}} \left[w_{Nt}^{nM} w_{Nt}^{n(M-1)} w_{Nt}^{-nM} \right] \otimes \left[1 w_{Nf}^1 \ldots w_{Nf}^{m(L-1)} \right] : (n, m) \in S_{tr} \right\}$$

(7.14)

Sparse Non-Selective MIMO Channel

For a non-selective MIMO channel, the channel in (7.1) is reduced to (7.14)

The training sequence for MIMO can be written as

$$H = \sum_{i=1}^{N_R} \sum_{k=1}^{N_r} H_v(i, k) \alpha_R \left(\frac{i}{N_R} \right) \alpha_T \left(\frac{k}{N_T} \right) = A_R H_v^T A_T^H$$

(7.15)

where $g(t)$ is a prototype pulse having unit energy, $xn \in C\ NT$ is the training sequence with energy $n \times n^2 = NT$, and M_{tr} is the number of temporal training dimensions (the total number of time-slots dedicated to training). For MIMO, the received signal can be written as

$$y = \sqrt{\frac{\varepsilon}{N_T}} X H_v + Z,$$

(7.16)

where $\mathbf{X}$ is an $M_{tr} \times N_T$ matrix of form $X = [X_0 \ldots X_{M_{tr}-1}]^T A_T^*$

Sparse Frequency-Selective MIMO Channel

For a frequency-selective MIMO channel, we have

$$H(f) = \sum_{l=0}^{L-1} A_R H_v^T(l) A_T^H e^{-j2\pi \frac{1}{W} f}$$

(7.17)

The training sequence for this case can be written as

$$x(t) = \sqrt{\frac{\varepsilon}{N_T}} \sum_{(n,m) \in S_{tr}} x_{n,m} g(t - nT_0) e^{j2\pi m W_{0t}},$$

(7.18)

where $S_{\text{tr}} \subset \{0, 1, \ldots \ldots N_t - 1\} \times \{0, 1, \ldots \ldots N_f - 1\}$ and $x_{n,m} \in C^{N_1}$ is the training sequence with energy $\sum_{S_{\text{tr}}} \|x_{n,m}\|^2 = N_T$. The random projection matrix X is an $M_{\text{tr}} \times N_T L$ matrix with the following as its row

$$\left\{ \left[1 w_{N_f}^m \ldots w_{N_f}^{m(L-1)} \right] \otimes X_{n,m}^T A_T^* : (n, m) \in S_{\text{tr}} \right\}$$

In summary, four different types of wireless channels are formulated as the CS problem in this section.

7.2.2 Random Field Estimation

The task of identifying the unoccupied spectrum is crucial for the efficiency and reliability of future wireless networks, especially for cognitive radio networks. A quantitative and precise understanding of current spectrum occupancy can help the secondary users to make more effective dynamic spectrum access decisions. Moreover, knowledge of the spectrum occupancy can help analysis of connectivity, capacity, and other properties of cognitive radio networks. Extensive measurement campaigns are needed for obtaining spectrum usage data. An illustration of the spectrum occupancy sampling is shown in Figure 7.9. Spectrum occupancy data can be obtained at the spots where cognitive radio nodes are present. However, due to hardware and geographical limitations, only limited samples can be collected at specific locations during a certain period of time, which is impractical and inefficient for the longtime measurements collection. Other approaches [35, 36] of spectrum occupancy prediction are to apply the model of primary users' activity or the experience of the secondary users, and few works take the correlations of the spectrum usage data into account.

Existing works [37] show correlation of the spectrum occupancy in space, time, and frequency. For example, adjacent secondary users may share very similar spectrum occupancy, and the spectrum usage is usually not independent of time, thus incurring spatial–temporal correlation. Many existing works [38–42] have exploited mathematically tractable models for describing the correlations. A point to be noted is that the spectrum occupancy data reconstruction problem is very similar to the matrix completion [23, 43]: only a portion of cognitive radio nodes report the spectrum usage samplings to the fusion center, where the spectrum occupancy data for the whole network are correlated in nature and can be reconstructed. By leveraging the low-rank nature of the spectrum usage data matrix, we can reconstruct the original spectrum occupancy from a limited number of measurements. Some other existing work for CS-based random fields can be found in [44] for wireless sensor networks and in [45] for image processing.

7.2.2.1 Random field model

Topology of Cognitive Radio Network
The plenary grid topology is adopted for the cognitive radio networks. The geographic area is divided into many grids, and for simplicity, it is assumed each gird has only two states: busy or idle. Each plane stands for the spectrum occupancy at a specific time, and thus, a series of the planes detail the spatial–temporal spectrum usage in cognitive radio networks. For facilitating the modeling using the random fields, the neighborhood and clique are defined in the network topology. For any grid that is not on the boundary of the network, the four immediately adjacent grids are considered as its neighbors. Also, a clique is defined as a set of points that are all neighbors of each other.

Ising Model
The Ising model defines a probability measure on the set of all possible configurations. We define the sample space as a collection of all possible realizations of random variables. Here, we assume the random variable X stands for the spectrum occupancy in the cognitive radio networks.

The 2D Ising model is given in Figure 7.10 where the dipole moments represented by the states +1 or −1. This contributes to the energy of the entire spectrum. The energy function corresponding to a specific spectrum occupancy $X = \{x_0, \ldots, x_i, \ldots, x_n\} \in$ is defined as

$$U(X) = \alpha \sum_i x_i + \beta \sum_{ij} x_i x_j \tag{7.19}$$

Here, the first term describes the effect of the exogenous inputs and the second sum represents the endogenous effect. Each random variable x_i

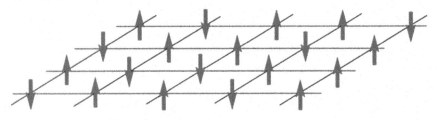

Figure 7.10　2D Ising model.

represents a grid in the topology and has two possible values, -1 and 1. $x_i = -1$ means busy and $x_i = 1$ means idle. α and β are two parameters of the Ising model. Intuitively, the Ising model can address the fact that the spectrum occupancy in the cognitive radio network depends on the natural (exogenous) and manufactured (endogenous) inputs. Note that the second sum is taken over all pairs i, j of points that are neighbors. The Ising model makes the simplifying assumption that only interactions between neighboring points need to be taken into account. This is called a first-order Markovian property. The case $\beta > 0$ is called the attractive case, and the $\beta < 0$ is called the repulsive case. In cognitive radio networks, the first-order Markovian property holds rigorously; then, the interruption ranges of primary users are small. As the interruption range increases, this may no longer be valid. However, some useful properties can still be quantified by the Ising model [40]. In the Ising model, specific spectrum occupancy is assigned to a probability according to the probability distribution:

$$P(\mathrm{X}) = Z^{-1}e^{-\frac{1}{KT}U(\mathrm{X})}, \tag{7.20}$$

where T is the temperature and K is a universal constant. For simplicity of analysis, we assume $T = 1$ and $K = 1$. The normalizing constant Z is called the partition function, which is given by

$$Z = \sum_X e^{-\frac{1}{kt}U(X)} \tag{7.21}$$

Formula (7.20) is called the Gibbs distribution, and it defines the probability distribution of the random variables X over the sample space.

Metropolis–Hastings Algorithm
The Metropolis–Hastings algorithm can be described as follows: Suppose the initial state we sampled is X_t, we draw a new proposal state X with probability $0(X/X_t)$. Then, we calculate a ratio

$$r = \frac{P(X')Q(X'|X')}{P(X')Q(X'|X')}, \tag{7.22}$$

where $P(X)|P(X_t)$ is the likelihood between the proposed sample X and the previous sample X_t, $Q(X_t|X)/Q\,X|X_t)$ is the ratio of the proposal density in two directions. Each time, the new state $X_t + 1$ is chosen according to the following rules: When $\gamma \geq 1$, we assign X to X_t+1; when $\gamma < 1$, $X_t+1 = X$ with probability γ, and $X_t + 1 = X_t$ with probability $1 - \gamma$. Thus, once the α

and β for the Ising model are specified, we can obtain a sequence of random samples from the given Gibbs distribution, which represents the spectrum occupancy during a period of time.

7.2.2.2 Matrix completion algorithm

In this subsection, we discuss the reconstruction of the spectrum occupancy in cognitive radio networks from a limited number of measurements. Assume $X = \{X_{t0}, \ldots, X_{tn}\}$; X represents the spectrum occupancy of the entire network from $t0$ state to tn state. Each $X_{ti} \in X$ is the spectrum occupancy matrix corresponds to the network topology at that time. For simplicity, we just consider the spectrum usage in a single channel. Usually, X is a 3D data matrix that is inconvenient for algebraic operations and optimization. Without loss of generality, for every matrix $X_{ti} \in X$, we can stack the columns of X_{ti} to form a vector that stands for the spectrum utilization of ti state. Then, we can combine these vectors into columns of a larger matrix that represents the spectrum occupancy over a period of time. For consistency, here we still use the same notation X to denote the overall larger matrix. At each time instant, only a limited number of cognitive radio nodes in the network measure the spectrum occupancy data and report their measurements to the fusion center.

The sampled measurements can be expressed as follows:

$$Y = M^0 X, \tag{7.23}$$

where $M(i, j) = 1\,0$, if otherwise, $X(i, j)$ is collected, and $\circ\!\!\leftarrow$ is the notion of a Hadamard product operator. $Y = M^0 X$, means $Y(i, j) = M(i, j)X(i, j)$. Our goal is to recover spectrum occupancy of cognitive radio networks from the partial measurements. It is usually difficult to unambiguously determine the original X for this kind of underdetermined linear inverse problems. However, because of the spatial–temporal correlations of the spectrum occupancy in the cognitive radio network, matrix X inherits the property of low rank. We can use the matrix completion techniques to recover it:

$$\min\ Rank(X)\,\text{s.t.}\quad M(i, j) = X(i, j). \tag{7.24}$$

Generally, (7.25) is NP-hard due to the combinational natural of the *rank* function. Instead of solving (7.25), we can replace $Rank(X)$ by its convex envelop to get a convex and more computationally tractable approximation [46]:

$$\min\ \|X\|_*\quad \text{s.t.}\quad M(i, j) = X(i, j), \tag{7.25}$$

where $X*$ is the nuclear norm that is defined as the sum of singular values of X,

$$\|X\|_* = \sum_{i=1}^{r} \sigma_i(X), \tag{7.26}$$

where $\sigma_i(X)$ are the positive singular values of matrix X, and r is the rank of matrix X. The original data matrix X represents the spectrum occupancy data that can be factorized by the singular value decomposition:

$$X = U \sum V^T = LR^T, \tag{7.27}$$

where U and V are unitary matrices and $\sum$ is a diagonal matrix containing the singular values of X. V^T is the transpose of $V.L = U \sum 1/2$ and $R = V \sum 1/2$. Note that the singular value decomposition cannot reconstruct the original matrix, and typical operations for singular value decomposition of the matrix assume the matrix is completely known. Hence, we adopt an algorithm to first estimate the L and R, and then, we reconstruct the original data matrix by calculating $X = LRT$. The optimization problem (7.26) can thus be solved by the following optimization [47, 48]:

$$\min_{L.R} \|L\|_F + \|R\|_F \quad \text{s.t.} \quad M(i,j) = X(i,j), \tag{7.28}$$

where the L_F and R_F denote the Frobenius norm of the matrix L, R, respectively,

$$\|L\|_F = \left(\sum_{i,j} L_{i,j}^2 \right)^{1/2}, \quad \|R\|_F = \left(\sum_{i,j} R_{i,j}^2 \right)^{1/2} \tag{7.29}$$

In order to reduce the recovery error as well as guarantee a low-rank solution of the optimization function, we rewrite the formula (7.29) as

$$\min \|Mo(Y - X)\|_F + \lambda(\|L\|_F + \|R\|_F), \tag{7.30}$$

where λ is a tunable parameter that balances between the precise fit of the measurement data and the goal of achieving low rank.

7.2.3 Other Channel Estimation Models

This section deals with several other channel estimation methods, showing the broad scope of the CS channel estimation approaches.

7.2.3.1 Blind channel estimation

In [49, 50], the concept of blind CS is introduced so as to avoid the need to know the sparsity basis in both the sampling and the recovery process. In other words, the only prior is that there exists some basis in which the signal is sparse. This is very different from CS, in which the basis representation should be known. The goals are to investigate the basic conditions under which blind recovery from compressed measurements is possible theoretically, to propose concrete algorithms for this task, and to study the applications in the spectrum sensing. Since the sparsity basis is unknown, the uncertainty about the signal is larger in the blind case than in the traditional CS case. A straightforward solution is to increase the number of measurements, which is shown to have multiple solutions. Additional constraints are needed for determination of the signal by the measurements. In [49], three constraints are listed together with the condition for uniqueness and algorithms as follows:

1. Finite set: The basis is in a given finite set of possible bases.
2. Sparse basis: The columns of the basis are sparse under some known dictionary.
3. Structure: The basis is block diagonal and orthogonal.

In [50], the above ideas are employed in multi-channel systems, in which the constraint is implied from the system structure. For a system such as microphone arrays or antenna arrays, the ensemble of signals can be constructed by dividing the signals from each channel into time intervals, and constructing each column of the signal as a concatenation of the signals from all the channels over the same time interval. In other words, we can show that the basis is block diagonal and orthogonal. Under certain conditions, it is shown that the blind CS algorithms can achieve results similar to those of standard CS, which rely on prior knowledge of the sparsity basis.

7.2.3.2 Adaptive algorithm

Many signal-processing applications require adaptive channel estimation with minimal complexity and small memory requirements. Most of the CS algorithms have to compute the results for each time instance and do not consider the adaptive algorithms. In [51], a conversion procedure that turns greedy algorithms into adaptive schemes is established for sparse system identification. A sparse adaptive orthogonal matching pursuit (SpAdOMP) algorithm of linear complexity is developed for providing optimal performance guarantees. An analytical study of the steady-state mean-square error (MSE) of the SpAdOMP algorithm has been attempted. The developed algorithm is used for estimating autoregressive moving-average (ARMA) and nonlinear ARMA channels. In addition, in [52], sparse adaptive regularized algorithms based on Kalman filtering and expectation maximization are reported.

7.2.3.3 Group sparsity method

The methodology of group sparse CS (GSCS) [53] allows efficient reconstruction of signals whose support is contained in the union of a small number of groups (sets) from a collection of predefined disjoint groups. Conventional CS is a special case of GSCS, with each group containing only a single element. Several CS recovery algorithms such as basis pursuit denoising and orthogonal matching pursuit have been extended to the group sparse case [54]. GSCS techniques are interesting in channel estimation, because the propagation paths of typical wireless channels are often structured in clusters. This is an additional physical reason that explains the occurrence of the channel delay Doppler components in groups. The performance of compressive channel estimation tends to be limited by leakage effects that are caused by the finite bandwidth and block length and that impair the effective delay Doppler sparsity of doubly dispersive channels. In [55], the authors demonstrate that the leakage components in the channel's delay Doppler representation exhibit a group sparse structure, which can be exploited by the use of GSCS recovery techniques.

7.3 Multiple Access

In this chapter, we discuss multiple access using CS in the context of wireless communications. Essentially, multiple access means multiple users sending their data to a single receiver that needs to distinguish and reconstruct the data from these users, as illustrated in Figure 9.1. It exists in many wireless communication systems, such as cellular systems, wherein the mobile users within a cell are served by a base station. The users may need to transmit their data during the same time period; e.g., multiple cellular phone users make phone calls simultaneously. Ad hoc networks: In such systems, there is no centralized base station. However, one node may need to receive data from multiple neighbors simultaneously; e.g., a relay node in a sensor network receives the reports from multiple sensors and forwards them to a data sink. The key challenge of multiple access is how to distinguish the information from different transmitters. Hence, there are basically two types of multiple-access schemes: orthogonal multiple access—This type of multiple access includes time division multiple access (TDMA), orthogonal frequency division multiple access (OFDMA), and carrier sense multiple access (CSMA). In TDMA, each transmitter is assigned a timeslot and it can only transmit over the assigned time-slot. Hence, the signals from different transmitters are distinguished in the time. OFDMA has a similar scheme to TDMA. The only difference is that OFDMA allocates different frequency channels to different transmitters and thus separates the transmitters in the frequency domain. CSMA also separates the transmitters in the time domain. However, there is no dedicated time-slot for each transmitter. Each transmitter needs to sense the environment first. If there is no transmission, it will go ahead to transmit; otherwise, it backs off for a random period of time and then senses the environment again, until it sends out the packet.

Non-orthogonal multiple access: A typical non-orthogonal multiple-access scheme is the code division multiple access (CDMA) [56], which is the fundamental signaling technique in 3G systems. In CDMA, the signals of all transmitters are mixed in both time and frequency domains. Each transmitter is assigned a code based on which the receiver can distinguish the signals from different transmitters. As seen already, it is very easy to distinguish the signals for different transmitters in the orthogonal multiple access. The main challenge of these multiple-access schemes is how to keep the data channels (time-slot or frequency channel) orthogonal (e.g., how to keep the transmitters synchronized in the time domain). Hence, we study the non-orthogonal multiple access scheme since the CS technique may play an important role in the signal separation/reconstruction. When we mention CS, the first question is where the sparsity is from. In the context of multiple access, the sparsity may come from the following: Data sparsity—At each time, only a small fraction of the transmitters have data to transmit; the set of active transmitters is unknown in advance since the packet arrivals at different transmitters could be random. Observation sparsity—When the packets at the transmitters are mutually correlated (e.g., they are the observations on the temperatures in a region), the received signal may also be written as a linear transformation of a sparse vector in a certain domain. We will consider both types of sparsity in this chapter. Note that the CS-based multiple access has a tight relationship to the technique of multiple-user detection in CDMA systems. Hence, we will provide a brief introduction on multiuser detection and then discuss the CS-based multiple access in both cellular and sensor networks.

7.3.1 Multiuser Detection

7.3.1.1 Comparison between multiuser detection and compressive sensing

A comparison of the expression in (9.2) with the standard CS problem $y=Hx$, reveals similarity between them. In both cases, the known signal is the linear transformation of the unknown signal (b in multiuser detection and x in CS), perhaps contaminated with noise. Hence, in [57], Donoho et al., pointed out the similarity between multiuser detection and CS: "There are strong connections to stagewise/stepwise regression in statistical model building, successive interference cancelation multiuser detectors in digital communications and iterative decoders in error-control coding."

However, there are also significant differences between multiuser detection and CS: The elements in b are binary, while x is a continuous real or complex vector. The nonzero elements in x are sparse, while there is no sparsity assumption in multiuser detection. In multiuser detection, the dimension of the observation N is usually comparable to the dimension of unknown vector K (or equivalently the number of users). However, in CS, the observation has a much lower dimension than the original signal. Hence, although looking similar to each other, multiuser detection and CS have different challenges. The difficulty of multiuser detection is how to handle the curve of dimensions incurred by the binary unknowns, while the CS is mainly challenged by the low-dimensional observation. Meanwhile, they can also share many ideas with each other, as we will see in the next subsection.

7.3.1.2 Algorithm for multiuser detection

In this subsection, a brief introduction to various algorithms for multiuser detection is provided, explaining its similarity to algorithms in CS. Optimal multiuser detector maximizes the likelihood of the received signal x. There are two types of optimal multiuser detectors: Joint optimal multiuser detector (JO-MUD)—In this case, the unknown signal b is chosen to maximize the joint conditional probability of y given b, on assuming that the noise n is white Gaussian with variance $\sigma^2 n$; i.e.,

$$\hat{b} = \arg\max_b \exp\left[-\frac{1}{2\alpha_n^2} \|r - SAb\|^2 \right]. \tag{7.31}$$

Individual optimal multiuser detector (IO-MUD)—Different from the JO-MUD, the IO-MUD considers the detection of each individual bit and maximizes the marginal distribution; i.e.,

$$\hat{b} = \arg \max_{b_k} P(r\,|b_k), \tag{7.32}$$

where

$$P(r\,|b_k = b) \propto \sum_{b, b_t = b} \exp\left[-\frac{1}{2\alpha_n^2}\|r - SAb\|^2\right] \tag{7.33}$$

The optimization problems in both (7.32) and (7.33) are discrete, thus requiring exponentially complex computations and being infeasible when K and N are large. Note that there are no counterparts in the context of CS since the unknown vector x in CS is real.

Linear Multiuser Detector

In this family of multiuser detectors, the unknown vector b is first considered as a real vector and then is transformed by a linear vector. The final decision is obtained from the output of the linear transform. Take user k, for example, we first carry out the linear transformation using a vector v_k and the output z_k is given by

$$z_k = V_k^T r, \tag{7.34}$$

and the decision is given by

$$\hat{b}_k = sgn(z_k) \tag{7.35}$$

where $sgn(z_k)$ is the sign of number (z_k). The following choices are available for the vector vk Matched filter (MF): In MF, we set $v_k = s_k$. Hence, the output of the linear multiuser detector is the projection of the received signal on the spreading code of user k.

Decorrelator: vk is set as the kth column vector of STS. It is easy to verify the possibility of the decorrelator's perfect reconstruction of the original bits in the absence of any noise and when $K \leq N$.

Minimum Mean-Square Error (MMSE) Detector

We set

$$v = \left(\alpha_n^2 I + \sum_{n=1}^{K} g_n s_n s_n^T\right)^{-1} s_K. \tag{7.36}$$

The possibility of the MMSE detector measuring the signal-to-interference-and-noise (SINR) ratio is shown. There is no counterpart in CS corresponding to the above algorithms. However, they are similar to some steps in MP or OMP algorithms in CS. For example, the MF detector can be considered as the step of finding the dictionary element hk and the corresponding unknown xk. The MMSE detector is similar to the one of finding the unknown xk in OMP.

7.3.2 Multiuser Access in Cellular Systems

In this section, we study the CS-based multiple access in cellular systems; i.e., there is a base station that carries out centralized processing and each transmitter sends data to the base station in a single-hop manner. The uplink (transmitters to base station) and downlink (base station to transmitters) are discussed separately.

7.3.2.1 Uplink

The uplink in cellular systems is discussed first. The framework in [58] is followed, which considers a random on–off access channel. In this framework, the transmitters are allowed to transmit simultaneously, similarly to the CDMA systems introduced in the previous section. However, the number of active users is known in traditional multiple-access systems (e.g., the number of cellular phones in call is known according to the management procedure). In sharp contrast, the set of active transmitters in the scenario considered in [58] is random and is not known in advance. The base station has to determine the set of active users in a manner similar to determining the nonzero elements in CS. Moreover, in [58], it is assumed that each user transmits at most one bit, and thus, the channel coding techniques are irrelevant.

Problem Formulation

The presence of K transmitters sharing a single wireless channel is assumed. Each transmitter is assigned an N-dimensional code word sk (for transmitter k). Note that this code word is equivalent to the spreading codes in the CDMA systems discussed in the previous section. Only a fraction of the transmitters transmit their data and the ratio, coined activity ratio, is denoted by λ. $M = \lambda K$ is assumed as an integer and is known in advance. The set of active users is denoted by *Itrue*. The received signal at the base station is then given by

$$\begin{aligned} r &= \sum_{k=1}^{K} s_k x_k + w \\ &= Sx + w, \end{aligned} \tag{7.37}$$

where x_k is the complex symbol transmitted by transmitter k and w is the noise. When a transmitter is not active, $x_k = 0$. Similarly to the CDMA system, S

contains the code words $s1, \ldots, sK$, while X is the vector containing the symbols of the transmitters. A point to note is that an implicit assumption here is the perfect time synchronization among the transmitters. For simplicity of analysis, we assume that the noise w is Gaussian with expectation zero and covariance matrix such that the total noise power of w is one. Then, the SNR is given by X_2. We also define the minimum-to-average ratio (MAR), which is given by

$$MAR = \frac{\min_{j \in I_{true}} |x_j|^2}{\|X\|^2 / \lambda K}, \qquad (7.38)$$

where the numerator is the minimum power of nonzero elements, while the denominator is the average power in s. Note that the metric MAR is used to measure the near–far effect among the transmitters.

Performance Analysis of Traditional Approaches

The performance of reconstructing the information symbols $x_1, \ldots, x_K$ for various receiver policies is now taken for discussion. Definition of asymptotic reliable detection is necessary prior to evaluation of the performance of different algorithms.

Definition: Consider deterministic sequences $N = N(K)$ and $X = X(K)$ with respect to K[58]. The probability of error of a given detection algorithm is given by

$$p_{\text{err}} = \Pr(\hat{I} \neq I_{\text{true}}), \qquad (7.39)$$

where $\hat{I}$ is the estimated set of active transmitters, while I_{true} is the true set. The probability is over the randomness of the code matrix S. Then, the algorithm is said to achieve an asymptotic reliable detection if $p_{\text{err}}(K) \to 0$ if $K \to \infty$. The measurement of success of the detection by the set of active transmitters is seen. Even if the demodulations for some active transmitters are wrong (actually the probability is nonzero if the SNR is bounded).

Optimal detection without noise: When there is no noise, the optimal detection algorithm is to exhaustively search for the nonzero elements in x, or equivalently the active transmitters. This requires $N \geq M$. A point to note is the total absence if any assumption on the modulation scheme of xk and which is considered just as an arbitrary complex scalar. If the modulation of xk is known, e.g., BPSK, the condition $N \geq M$. M may no longer be necessary since the prior knowledge about x can be exploited. Maximum likelihood (ML) detection with noise: x is considered as an unknown but deterministic vector. Hence, the ML criterion can be used for estimating x. Intuitively, the

ML detector looks for the subspace that is spanned by the codes of active transmitters and has the larger power. Again, no assumption is made of any knowledge about the modulation of x. It has been shown by Wainwright [59] that there exists a constant C such that, if

$$N \geq C \max \left\{ \frac{1}{SNR_{\min}} \log(K(1-\lambda)), \lambda K \log \left(\frac{1}{\lambda} \right) \right\}, \qquad (7.40)$$

the ML detector can detect the correct set of active users asymptotically, where $SNR_{\min}$ is called the minimum component SNR, which is defined as

$$SNR_{\min} = \min_{j \in I_{\text{true}}} |x_j|^2, \qquad (7.41)$$

namely the minimum power of nonzero elements. Note that (7.40) is a sufficient condition for the perfect detection of nonzero elements in x. A necessary condition for the perfect detection is found in [58], which is given by

$$N \geq \frac{1-\delta}{SNR_{\min}} \log(K(1-\lambda)) + \lambda K, \qquad (7.42)$$

for any $\delta > 0$.

Single-user detection: This is very similar to the MF detection in multiuser detection in CDMA. In such a single-user detection, a threshold μ is set such that the estimated set of active transmitters is given by

$$\hat{I} = \left\{ k \Big| \frac{|s_k^T y|^2}{\|s_k\|^2 \|y\|^2} > \mu \right\}, \qquad (7.43)$$

i.e., a claim is made of transmitter k as active when the correlation coefficient between the code of transmitter k, namely s_k, and the received signal y is larger than the threshold μ. It has been shown in [58] that, when the following inequality holds, we have

$$N > \frac{(1+\delta)L(\lambda, K)(1 + SNR)}{SNR_{\min}}, \qquad (7.44)$$

where $L(\lambda, K) = \left(\sqrt{\log(K(1-\lambda))} + \sqrt{\log(K\lambda)} \right)^2$, and then, there exists a sequence of thresholds $\mu(n)$ such that the single-user detection can detect the true set of active users asymptotically. A disadvantage of the single-user detection is the saturated performance in the high *SNR* regime; i.e., when $SNR_{\min} \to \infty$, *perr* does not converge to 0. This performance saturation is due to the MAI that exists only when *SNR* is sufficiently high. Another feature of the single-user detection scheme is the constant offset $L(\lambda, K)$. Lasso and OMP: The Lasso estimator [60] can also be used for reconstructing the sparse signal by solving the following optimization problem:

$$\hat{X} = \arg\min_{X} \left(\|y - AX\|_2^2 + \mu \|x\|_1 \right), \tag{7.45}$$

where μ is a weighting factor. It has been shown in [58] that, if K, N and λK, all tend to infinity, and $SNR_{\min} \to \infty$ (the minimum *SNR* also tends to infinity), the following inequality is necessary and sufficient for asymptotic reconstruction of the signal:

$$m > \lambda K \log(K(1 - \lambda)) + \lambda K + 1. \tag{7.46}$$

When the greedy OMP algorithm is applied, a sufficient condition for the asymptotically reliable reconstruction is given by

$$N > 2\lambda K \log(K) + C\lambda K, \tag{7.47}$$

where $C > 0$ is a constant, when S has Gaussian entries. Compared with the single-user detection, we find that the constant offset $L(\lambda, K)$ is removed by the Lasso and greedy OMP. The performance is also limited in the high-*SNR* regime due to the redundant MAI.

Sequential Orthogonal Matching Pursuit

As shown in the previous discussion, there is a significant gap between the ML detector and the existing traditional CS approaches in the context of multiple access. In [60], a sequential orthogonal matching pursuit algorithm is proposed for significant bridging of this gap. Moreover, it does not saturate in the high-*SNR* regime. This algorithm is similar to the OMP algorithm in CS. However, it is carried out in a sequential manner. The detailed algorithm of the sequential OMP algorithm is given by the following:

1. Initialize the iteration index as $l = 1$ and make the set of active transmitters $\hat{I}(0)$ as an empty set.
2. Compute $P(j)Sj$, where $P(j)$ is an operator that projects Sj onto the orthogonal complement of the subspace spanned by $\{sl, l \in \hat{I}(j-1)\}$.

3. Compute the correlation, which is given by

$$p(j) = \frac{\left| s_j^T P(j) y \right|^2}{\| P(j) s_j \|^2 \, \| P(j) y \|^2} \tag{7.48}$$

4. If $p(j) > \mu$, where μ is a threshold, add the index j to the set of active transmitters $\hat{I}(j-1)$.
5. Increase $j = j + 1$. If $j \leq K$, go to Step 2.
6. Output the set of active transmitters $\hat{I}(K)$.

The sequential OMP algorithm is similar to a single-user detector (or MF detector in CDMA), since it uses the correlation to detect the existence of nonzero elements. The difference is that, in the sequential OMP algorithm, it is $P(j)Sj$, instead of sj, that is used for computing the correlation and detect the nonzero elements. Unfortunately, numerical simulations show the performance of the sequential OMP algorithm as generally much worse than the standard OMP. However, the value of the sequential OMP algorithm is that it can achieve a better scaling in the high-*SNR* regime. Here, the high-*SNR* regime is represented by the definition of minimum signal-to-interference-and-noise ratio (SINR), which is given by

$$\gamma = \min_{l=1,\dots,K} \frac{pt}{\hat{\sigma}^2(l)},$$

Where

$$\hat{\sigma}^2(l) = 1 + \lambda \sum_{j=l+1}^{K} p_j \tag{7.49}$$

An intuitive explanation for $\hat{\sigma}^2(l)$ is the expected power of interference plus noise when the signals of transmitters 1 to l have been perfectly detected and canceled, where the power of transmitter j is scaled by λ because the probability that it is active is λ. Then, γ is the corresponding SINR.

7.3.2.2 Downlink
In the downlink of cellular systems, the information is sent from the base station to different users. Assuming there are totally K users served by a base station. However, the base station can serve M users simultaneously. A special case is $M = 1$, i.e., time division multiplexing (TDM). When code division multiplexing (CDM) or orthogonal frequency division multiplexing

(OFDM) is used, M could be larger than one. When K is very large and the number of corresponding channels (e.g., codes in CDM or subcarriers in OFDM) is limited, M could be much less than K. Then, the base station needs to schedule the traffic of the users. An efficient scheduling algorithm is the opportunistic scheduling [61], in which users with good channel qualities are scheduled. This scheduling algorithm can improve the throughput of data traffic. A challenge for the opportunistic scheduling is that the base station needs to know the channel qualities of different users. However, the downlink channel qualities must be reported by the users; e.g., the base station can broadcast pilot signals from which the users can estimate the channel qualities. If all users report their channel qualities in each scheduling period, the large number of K may cause a high overhead for the control channel or uplink system.

CS-based scheme exists for the channel quality report mechanism in the downlink scheduling. In this scheme, only users having good channel qualities will report to the base station; e.g., there could exist a threshold and, only when the channel quality is better than the threshold, can the user send its report. K_0 is denoted by the number of users that send channel-quality reports. When K is sufficiently large and the channel quality is an i.i.d. random variable for different users, K_0 is almost a constant with some random fluctuations. The base station can also have a good estimate for K_0 due to the channel-quality distribution and the threshold. K_0 K if K can be assumed as very large and M is much smaller. However, the base station is unable to locate the set of users which have good channel qualities and send their reports. This is very similar to the unknown nonzero elements in CS. A reporting period can be set before the downlink scheduling, considering the similarity to CS. Similarly to CDMA, each user is assigned a code s_k (say, for user k). In the reporting period, if user k has a channel quality better than the threshold, it transmits $s_k x_k$, in which x_k is an information symbol containing the channel quality. Then, the received signal at the base station is given by

$$r = \sum_{k \in I_{\text{act}}} S_k x_k + w, \tag{7.50}$$

where I_{act} is the set of users that plan to report the channel qualities.

Thus, this chapter gives an idea about the various wireless technologies based on compressed sensing. Also, multiple access methods and uplink and downlink of wireless channels are discussed with respect to compressed sensing.

7.4 Summary

This chapter deals with various wireless networks that can effectively use compressed sensing procedure for wireless communications with expected efficiency. The general wireless networks, advanced wireless networks, and cellular networks are explained in the first part of the chapter. The CS-based wireless networks then follow where field and channel estimation models are discussed in depth. This includes the CS-based multi-path channel estimation, the random field estimation, and other estimation models. The multiple access and multiuser detection using CS are then discussed. This covers the uplink and downlink processes of multiuser access in cellular networks using CS.

References

[1] [Online] available: http://www.umtsworld.com
[2] [Online] available: http://www.cdg.org
[3] "IEEE standard for local and metropolitan area networks part 16: Air interface for fixed broadband wireless access systems IEEE Std. 802.16-2004" (Revision of IEEE Std. 802.16-2001), pp. 27–28, 2004.
[4] "Achieving wireless broadband with WiMAX", IEEE Computer Society, Tech. Rep., 2004.
[5] Perkins, C. (1994). "Highly dynamic destination-sequenced distance-vector routing (DSDV) for mobile computers", In *Proc. of ACM Conference on Communications Architectures, Protocols and Applications*, London, UK, September.
[6] Chiang, C.-C., Wu, H.-K., Liu, W., and Gerla, M. (1997). "Routing in clustered multihop, mobile wireless networks with fading channel", In *Proc. of IEEE Singapore International Conference on Networks*, Singapore, April.
[7] Murthy, S., and Garcia-Luna-Aceves, J. J. (1996). "An efficient routing protocol for wireless networks", *ACM Mobile Networks App. J.*, (Special Issue on Routing in Mobile Communication Networks), 1 (2), 183–197, Oct.
[8] Perkins, C., and Royer, E. (1999). "Ad-hoc on-demand distance vector routing", In *Proc. of the 2nd IEEE Workshop on Mobile Computing Systems and Applications*, New Orleans, LA, USA, February.
[9] Johnson, D. B. (1994). "Routing in ad hoc networks of mobile hosts", In *Proc. of the Workshop on Mobile Computing Systems and Applications*, Santa Cruz, CA, USA, December.

[10] Park, V. D., and Corson, M. S. (1997). "A highly adaptive distributed routing algorithm for mobile wireless networks", In *Proc. IEEE Conf. on Comp. Comm.*, Kobe, Japan, April.

[11] Toh, C. K. (1997). "Associativity-based routing for ad hoc mobile networks", *Int. J. Wireless Personal Commun.* 4 (1), 103–139, Mar.

[12] Dube, R., Rais, C. D., Wang, K. Y., and Tripathi, S. K. (1997). "Signal stability-based adaptive routing (SSA) for ad-hoc mobile networks", *IEEE Pers. Commun. Mag.*, 4 (1), 36–45, Feb.

[13] [Online] available: http://www.antd.nist.gov/wahn ssn.shtml.

[14] Krishnamachari. B. (2005). *Networking Wireless Sensors.* Cambridge, UK: Cambridge University Press.

[15] Meguerdichian, S., Koushanfar, F., Potkonjak, M., and Srivastava, M. B. (2001). "Coverage problems in wireless ad hoc sensor networks", In *Proc. IEEE Conf. on Comp. Comm.* Anchorage, AK, USA, April.

[16] Chen, B., and Varshney, P. K. (2002). "A Bayesian sampling approach to decision fusion using hierarchical models", *IEEE Trans. Signal Process.*, 50 (8), 1809–1818, Aug.

[17] Ozay, M., and Vural, F. T. Y. (2009). "A theoretical analysis of feature fusion in stacked generalization", In *Proc. of Signal Processing and Communications Applications Conference*, Antalya, Turkey, April.

[18] Walters, J. P., Liang, Z., Shi, W., and Chaudhary, V. (2006). *Wireless Sensor Network Security: A Survey Security in Distributed, Grid, and Pervasive Computing.* (CRC Press: Auerbach Publications).

[19] Kerpez, K. J., Waring, D. L., Galli, S., Dixon, J., and Madon, P. (2003). "Advanced DSL management", *IEEE Commun. Mag.*, 41 (9), 116–123, Sept.

[20] Tse, D., and Viswanath, P. (2005). *Fundamentals of Wireless Communication.* Cambridge, UK: Cambridge University Press.

[21] Zheng, L., and Tse, D. N. C. (2003). "Diversity and multiplexing: A fundamental tradeoff in multiple antenna channels", *IEEE Trans. Inform. Theory*, 49 (5), 1073–1096, May.

[22] Alamouti, S. M. (1998). "A simple transmit diversity technique for wireless communications", *IEEE J. Select. Areas Commun.*, 16 (8), 1451–1458, Oct.

[23] Candes, E., and Tao, T. (2006). "Near optimal signal recovery from random projections: Universal encoding strategies", *IEEE Trans. Inform. Theory*, 52 (1), 5406–5425, Dec.

[24] Donoho, D. (2006). "Compressed sensing", *IEEE Trans. Inform. Theory*, 52 (4), 1289–1306, Apr.

[25] Baraniuk, R. (2007). "Compressive sensing", *IEEE Signal Process. Mag.*, 24 (4), 118–121, July.

[26] Bajwa, W. U., Haupt, J., Sayeed, A. M., and Nowak, R. (2010). "Compressed channel sensing: A new approach to estimating sparse multipath channels", *Proc. IEEE*, 98 (6), 1058–1076, June.

[27] Paredes, J. L., Arce, G. R., and Wang, Z. (2007). "Ultra-wideband compressed sensing channel estimation", *IEEE J. Select. Topics Signal Process.*, 1 (3), 383–395, Oct.

[28] Berger, C. R., Wang, Z., Huang, Z., and Zhou, S. (2010). "Application of compressive sensing to sparse channel estimation", *IEEE Commun. Mag.*, 48 (11), 164–174, Nov.

[29] Zhang, P., Hu, Z., Qiu, R. C., and Sadler, B. M. (2009). "Compressive sensing based ultra-wideband communication system", In *Proc. of IEEE International Conference on Communications (ICC)*, Kyoto, Japan, June.

[30] Romberg, J. (2009). "Multiple channel estimation using spectrally random probes", In *Proc. SPIE Wavelets XIII*, San Diego, CA, USA.

[31] Bajwa, W. U., Sayeed, A., and Nowak, R. (2008). "Compressed sensing of wireless channels in time, frequency and space", In *Proc. of 42nd Asilomar Conf. Signals, Systems, and Computers*, Pacific Grove, CA.

[32] Taubock, G., Hlawatsch, F., Eiwen, D., and Rauhut, H. (2010). "Compressive estimation of doubly selective channels in multicarrier systems: leakage effects and sparsity-enhancing processing", *IEEE J. Select. Topics Signal Process.*, 4 (2), 255–271, Apr.

[33] Taubock, G., and Hlawatsch, F. (2008). "Compressed sensing based estimation of doubly selective channels using a sparsity-optimized basis expansion", In *Proceedings of the 16th European Signal Processing Conference (EUSIPCO 2008)*, Lousanne, Switzerland, August.

[34] Bajwa, W. U., Sayeed, A., and Nowak, R. (2008). "Learning sparse doubly-selective channels", In *Proc. of 46th Allerton conf. Communication, Control, and Computing*, Monticello, IL, USA, October.

[35] Lai, L., Gamal, H. E., Jiang, H., and Poor, H. V. (2011). "Cognitive medium access: Exploration, exploitation and competition", *IEEE Trans. Mobile Comput.*, 10 (2), 239–253, Feb.

[36] Zhao, Q., and Swami, A. (2007). "A decision-theoretic framework for opportunistic spectrum access", *IEEE Wireless Commun. Mag.*, 4 (4), 14–20, Aug.

[37] Chen, D., Yin, S., Zhang, Q., Liu, M., and Li, S. (2009). "Mining spectrum usage data: a large-scale spectrum measurement study", In *Proceedings*

of the 15th Annual International Conference on Mobile Computing and Networking*, Beijing, China, September.

[38] Riihijarvi, J., Mähönen, P., Wellens, M., and Gordziel, M. (2008). "Characterization and modelling of spectrum for dynamic spectrum access with spatial statistics and random fields", In *Proc. of IEEE 19th International Symposium on Personal, Indoor and Mobile Radio Communications (PIMRC)*, Cannes, France.

[39] Wellens, M., Riihijarvi, J., Gordziel, M., and Mähönen, P. (2009). "Spatial statistics of spectrum usage: from measurements to spectrum models", In *Proc. of IEEE International Conference on Communications (ICC)*, Bresden, Germany, June.

[40] Li, H., Han, Z., and Zhang, Z. (2011). "Communication over random fields: A statistical framework for cognitive radio networks", In *Proc. of IEEE Globe Communication Conference (Globecom)*, Houston, TX, USA, December.

[41] Wellens, M., Riihijarvi, J., and Mähönen, P. (2009). "Spatial statistics and models of spectrum use", *Elsevier Comp. Commun.*, 32(18), 1998–2011, Aug.

[42] Li, H. (2010). "Reconstructing geographical-spectral pattern in cognitive radio networks", In *Proc. of the Fifth International Conference on Cognitive Radio Oriented Wireless Networks Communications (CROWNCOM)*, Cannes, France, June.

[43] Candes, E. J., and Recht, B. (2012). "Exact matrix completion via convex optimization", *Mag. Commun. ACM*, 55 (6), 111–119, June.

[44] Oka, A., and Lampe, L. (2008). "Compressed sensing of Gauss-Markov random field with wireless sensor networks", In *Proc. of Sensor Array and Multichannel Signal Processing Workshop, 2008. SAM 2008. 5th IEEE*, Vancouver, Canada, July.

[45] Cevher, V., Duarte, M. F., Hegde, C., and Baraniuk, R. G. (2008). "Sparse signal recovery using Markov random fields", In *Proc. of the Workshop on Neural Information Processing Systems*, Vancouver, Canada, Dec.

[46] Ma, S., Goldfarb, D., and Chen, L. (2009). "Fixed point and Bregman iterative methods for matrix rank minimization", *Math. Program. Ser. A and B*, 128 (1–2), 321–353, June.

[47] Yin, W., Wen, Z., Li, S., Meng, J., and Han, Z. (2011). "Dynamic compressive spectrum sensing for cognitive radio networks", In *Proc. of Conference on Information Sciences and Systems (CISS)*, Baltimore, MD, USA, March.

[48] Burer, S., and Monteiro, R. (2005). "Local mimima and convergence in low-rank semi definite programming", *Math. Prog.*, 103 (3), 427–444.

[49] Gleichman, S., and Eldar, Y. C. (2011). "Blind compressed sensing", *IEEE Trans. Inf. Theory*, 57 (10), 6958–6975.

[50] Gleichman, S., and Eldar, Y. C. (2010). "Multichannel blind compressed sensing", In *IEEE Sensor Array and Multichannel Signal Processing Workshop*, Hoboken, NJ, June.

[51] Mileounis, G., Babadi, B., Kalouptsidis, N., and Tarokh, V. (2010). "An adaptive greedy algorithm with application to nonlinear communications", *IEEE Trans. Signal Process.*, 58 (6), 2998–3007, June.

[52] Kalouptsidis, N., Mileounis, G., Babadi, B., and Tarokh, V. (2009). "Adaptive algorithms for sparse nonlinear channel estimation", In *IEEE/SP 15th Workshop on Statistical Signal Processing*, Cardiff, United Kingdom, Sept.

[53] van den Berg, E., Schmidt, M., Friedlander, M., and Murphy, K. (2008). "Group sparsity via linear time projection", Technical Report TR-2008-09, University of British Columbia.

[54] Eldar, Y., and Mishali, M. (2009). "Robust recovery of signals from a structured union of subspaces", *IEEE Trans. Inform. Theory*, 55 (11), 5302–5316, Nov.

[55] Eiwena, D., Taubock, G., Hlawatsch, F., and Feichtinger, H. G. (2010). "Group sparsity methods for compressive channel estimation in doubly dispersive multicarrier systems", In *2010 IEEE Eleventh International Workshop on Signal Processing Advances in Wireless Communications (SPAWC)*, Vienna, Austria, June.

[56] Viterbi, A. J. (1995). *CDMA: Principles of Spread Spectrum Communication*. Addison-Wesley, Boston, MA, USA.

[57] Donoho, D. L., Tsaig, Y., Drori, I., and Starck, J. (2012). "Sparse solution of underdetermined systems of linear equations by stagewise orthogonal matching pursuit", *IEEE Trans. Inform. Theory*, 58 (2), 1094–1121, Feb.

[58] Fletcher, A. K., Rangan, S., and Goyal, V. K. (2009). "On-off random access channels: A compressed sensing framework", *preprint*.

[59] Wainwright, M. J. (2008). *Sharp thresholds for high-dimensional and noisy recovery of sparsity*. Univ. of California, Berkeley, Dept. of Statistics, Tech. Rep.

[60] Wang, H., and Leng, C. (2007). "Unified lasso estimation via least square approximation", *J. Am. Statist. Assoc.*, 102 (479), 1039–1048, Sept.

[61] Liu, X., Chong, E. K. P., and Shroff, N. B. (2003). "A framework for opportunistic scheduling in wireless networks", *Comput. Networks*, 41 (4), 451–474, Mar.

8

Compressive Spectrum Sensing for Cognitive Radio Networks

8.1 Introduction

The radio frequency portion of the electromagnetic spectrum is one of the most valuable natural resources for wireless systems, which have become ubiquitous with many applications (Wi-Fi and cellular) and devices (laptops, tablets, smart phones). The demand for wireless spectrum is increasing with both the competing wireless applications and the number of users. New wireless applications such as smart home appliances, telemedicine, wireless sensor networks, and many others are emerging as concrete systems. This dramatic increase in the number of wireless systems and applications has severely limited the availability of wireless spectrum. Thus, new wireless applications find the wireless spectrum overcrowded leading to scarcity in spectrum. Nevertheless, the usage of spectrum is coordinated by regulatory bodies such as the federal communications commission (FCC) and office of communications (OfCom). Spectral bands are assigned to various communication services on a long-term basis. Usually, the spectral resources are allocated in three ways:

i. Frequencies reserved for radio astronomy in which no one can transmit.
ii. Frequencies are open such as the industrial, scientific, and medical (ISM) bands in which any one can transmit.
iii. Frequencies are licensed to a particular service (such as cellular and TV) in which only the licensed users can transmit.

Recent studies performed on the usage of the licensed spectrum reveal most of the assigned spectrum as underutilized with respect to time and geographical location. According to the recent statistics study, the spectrum utilization in licensed bands is approximately 5–50%. Figure 8.1 shows the spectrum measurement pattern indicating sporadic use of the spectrum. This inefficient use of spectrum is mainly due to static spectrum allocations, rigid policies, and fixed

Maximum Amplitudes

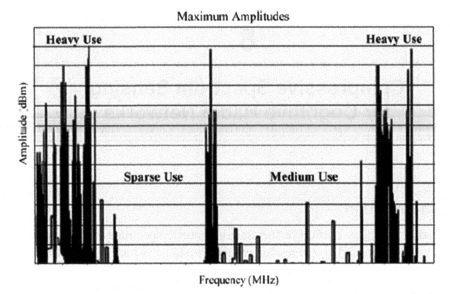

Figure 8.1   Spectrum utilization.

radio functions. However, spectrum usage can be improved by allowing the unlicensed users to access the licensed spectrum in an opportunistic manner. Cognitive radio is an emerging technology envisaged for the improvement of the spectrum utilization through dynamic spectrum access (DSA) techniques and is enabled by software radios. Cognitive radio technology allows the opportunistic use of spectral resources in licensed bands. This innovative idea of cognitive radio is apparent from its definition adopted by the FCC which is reiterated as follows:

"Cognitive radio: a radio or system that senses its operational electromagnetic environment and can dynamically and autonomously adjust its radio operating parameters to modify interference, facilitate interoperability, and access secondary markets."

Despite the likelihood of DSA enabled cognitive radio techniques improving the spectrum efficiency, they impose several research challenges due to the wide range of spectrum as well as diverse quality of service (QoS) requirements. The following sections explain in detail about the architectures of cognitive radio, DSA principle, cognitive cycle, and the functions of cognitive radio.

Despite cognitive radio concept being a promising solution to the spectral scarcity problem, efficient methods for detecting spectrum holes in wideband wireless spectrum remain challenged in which cognitive radio users reliably detect spectral opportunities across a wide range of spectrum. The conventional methods available for detection use a very high sampling rate as it is a primary requirement of Nyquist criterion. Thus, there is a limitation on their operational bandwidth by available hardware resources and many extensive theoretical works on spectrum sensing are not possible to implement in real time over a wide frequency band. Researchers have started to use a technique called compressive sensing, for reducing the bottleneck in which the acquisition of sparse signals is permitted at sub-Nyquist rates, in conjunction with cognitive radios. In this chapter, a brief introduction to cognitive radio technology is provided and various spectrum sensing algorithms are discussed along with their advantages and disadvantages with their future challenges. Specifically, the compressive sensing technique based on sub-Nyquist sampling and multi-channel sub-Nyquist sampling techniques are concentrated in detail.

8.2 Cognitive Radio and Dynamic Spectrum Access

Dynamic spectrum access is a new spectrum sharing paradigm that enables secondary users to access the spectrum holes in the licensed spectrum bands. It is a promising technology for alleviating the spectrum scarcity problem and increasing the spectrum utilization. DSA using cognitive radio paves the way for effective spectrum utilization. This section gives an introduction to the DSA principle and cognitive radio.

8.2.1 Dynamic Spectrum Access

Dynamic as opposed to static spectrum access involves the opportunistic utilization of licensed spectrum by unlicensed users, during their idle time. It aims at maximizing the usage of the finite spectrum which would otherwise remain idle. Thus, DSA models are necessary for efficient allocations of spectral resources. Cognitive radios are best suited for incorporating the DSA techniques. Figure 8.2 shows the DSA paradigm in cognitive radio networks [1].

The main objective of the cognitive radio is to capture and access the spectral holes without interfering with the licensed users. Spectrum holes are the vacant portions of the licensed spectrum and are otherwise called white spaces. If a cognitive radio is occupying a particular spectrum hole, it should move to another spectrum hole on the appearance of the incumbent. This can

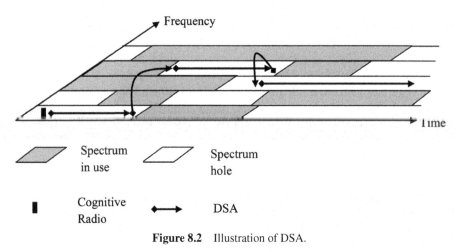

Figure 8.2   Illustration of DSA.

be accomplished by the cognitive radio through alteration of its modulation type, power level, etc.

8.2.2 Cognitive Radio

Cognitive radio is a novel concept defined as "an intelligent radio which is aware of its surrounding environment and capable of changing its behavior to optimize the user experience" [2]. It is the key enabler for DSA and has various definitions adopted by different regulatory agencies.

In the existing cognitive radio terminology, licensed users are called primary users and unlicensed users are called secondary users or cognitive radio users. Secondary users equipped with cognitive capabilities are allowed to share the licensed spectrum only when unoccupied. This is referred to as opportunistic access of the primary spectrum by the secondary cognitive users and is called DSA. A secondary user can utilize the primary spectrum of different primary users on the basis of the availability and its own require-ments. However, the secondary users should take utmost care to avoid causing interference to the primary users. The essential characteristics of a cognitive radio are awareness, cognition, and reconfigurability. These characteristics are explained below:

Awareness
The ability of cognitive radio to sense its outside environment is known as awareness. The cognitive radio should be aware of its environment with respect

to time, space, and frequency to enable capturing the best possible spectrum available to them.

Cognition

The ability of cognitive radio to learn the environment, process the available information, and make decisions to achieve the predefined goals is known as cognition. Upon monitoring the power in some frequency band, use of sophisticated techniques to locate the unused portions of the spectrum at a specific time and location is needed. This is known as the cognitive capability of the secondary user and results in the best available spectrum.

Reconfigurability

Reconfigurability is the ability to adjust the operating parameters without changing the underlying hardware. This enables dynamic programming of the cognitive radio on the basis of the environment.

8.2.3 Cognitive Radio Architectures

The primary network and the secondary network coexist in a cognitive radio system irrespective of the type of architecture. The basic design principle is that the primary users are as unaffected as possible. In order to fulfill such coexistence, there are three possible architectural solutions, namely

- Overlay architecture
- Underlay architecture
- Interweave architecture

Overlay Architecture

Concurrent transmissions between primary and secondary users are allowed in the *overlay* architecture. Secondary users have the ability to sense primary users' message and then use advanced coding schemes (e.g., dirty paper coding) for interference cancelation, so that primary users' transmissions remain unaffected.

Underlay Architecture

The *underlay* architecture also allows simultaneous primary and secondary transmissions. The secondary users spread their signals over a wide bandwidth, which is large enough to ensure the amount of interference caused to primary transmissions being under the tolerable thresholds. Such interference constraint restricts the usage of the underlay architecture to short range communications.

Interweave Architecture

The *interweave* architecture is proposed based on the opportunistic communication. Secondary users have the intelligence to periodically monitor the spectrum, detect the activities of primary users in time and frequency domain, and then opportunistically interweave secondary transmissions through the sensed spectrum holes. In this scenario, accurate spectrum sensing is critical to the performance, especially when signal-to-noise ratio (SNR) is low.

This chapter focuses on the interweave configuration of cognitive radio in which spectrum sensing is the primary function. The physical architecture of cognitive radio and the various functions of cognitive radio in a cognitive cycle are explained in detail in the following section.

8.2.4 Physical Architecture of Cognitive Radio

The generic architecture of a cognitive radio transceiver is shown in Figure 8.3. The important functional blocks of a cognitive radio transceiver system are the RF front end and the baseband processing unit. Each component of the architecture should be controlled, to ensure adaptation to the changing environment.

The received signal is amplified, mixed, and A/D converted in the RF front-end section. In the baseband processing unit, the received signal is modulated/demodulated and encoded/decoded which process is similar to that conventional transceiver sections. However, the RF front end is the important building block of the cognitive radio. The RF front end possesses

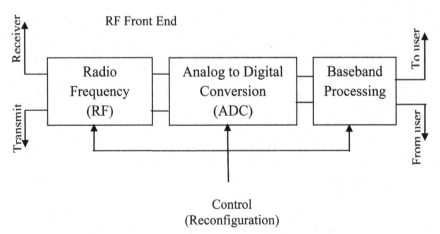

Figure 8.3   Cognitive radio transceiver [3].

the novel characteristic of cognitive radio transceiver viz its wideband sensing capability. This function is directly related to that of RF hardware technologies such as wideband antenna, adaptive filter, and power amplifier. RF hardware section for the cognitive radio unit should be capable of tuning to any portion of a wide range of frequency spectrum. Such spectrum sensing also enables real-time measurements of spectrum information from radio environment. Figure 8.3 shows the general structure of a wideband RF front-end architecture for the cognitive radio. The components of a cognitive radio RF front-end are shown in Figure 8.4.

1. RF filter: The RF filter selects the desired frequency band using a band pass filter on the received RF signal.
2. Low noise amplifier (LNA): The LNA amplifies the desired signal while minimizing noise signal.
3. Mixer: In the mixer, the received signal is mixed with locally generated frequency and converted to the intermediate frequency (IF) or the baseband signal.
4. Voltage-controlled oscillator (VCO): The VCO is a voltage-controlled oscillator and generates a signal at a specific frequency for a given voltage to combine with the incoming signal. This process converts the incoming signal to an IF or baseband frequency.
5. Phase-locked loop (PLL): The PLL ensures that a signal is locked on a particular frequency. It can be also be used for generating precise frequencies with fine resolution.

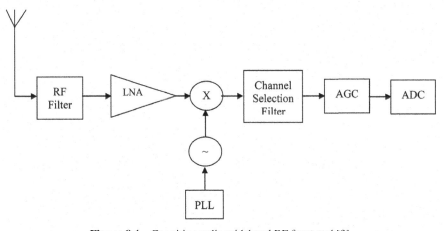

Figure 8.4 Cognitive radio wideband RF front end [3].

6. Channel selection filter: The channel selection filter is used for selecting only the desired channel and to reject the adjacent or unwanted channels. It can be a low-pass filter as used in a direct conversion receiver or a band pass filter as used in a super-heterodyne receiver.
7. Automatic gain control (AGC): The AGC maintains the constant gain or output power level of an amplifier over a wide range of input signal levels.

In this architecture, the RF front-end section receives a wideband signal, which is sampled by the high-speed analog-to-digital converter (ADC) and measurements are performed for the detection of the primary user signal. However, there exist some limitations on developing the cognitive radio RF front end. The wideband RF antenna receives signals from various transmitters operating at different bandwidths, power levels, and locations. As a result, the RF front end should have the capability to detect a weak signal in a large dynamic range. However, this capability requires a multi-GHz speed with high-resolution ADC, which might not be feasible.

The requirement of a multi-GHz high-speed ADC is reduction in the dynamic range of the signal before ADC conversion. This reduction can be achieved by filtering strong signals. Since strong signals can be located anywhere in the wide spectrum range, tunable notch filters are required for the reduction. An alternative approach is to use multiple antennas such that signal filtering is performed in the spatial domain rather than in the frequency domain. Multiple antennas can receive signals selectively using beam-forming techniques. Thus, the key challenge of the physical architecture of the cognitive radio is an accurate detection of weak signals of primary users over a wide range of spectrum. Hence, the implementation of RF wideband front end and A/D converter is a critical issue in cognitive radio networks.

Cognitive Radio Functions
The important functions of a cognitive radio include spectrum sensing, spectrum management, spectrum sharing, and spectrum mobility. Spectrum sensing is the fundamental task of obtaining the awareness about the idle portions of the primary spectrum. Spectrum sensing can be performed individually by a secondary user called local/single-user spectrum sensing or can be performed collaboratively by a group of secondary users called cooperative spectrum sensing. Spectrum management is the task of identifying the best available spectrum based on the cognitive radio requirements.

Spectrum sharing refers to the allocation of spectrum among multiple secondary users in a fair manner. Spectrum mobility refers to shifting to another free spectrum upon the arrival of the primary user. These functions of cognitive radio networks enable spectrum-aware communication protocols. However, the opportunistic use of the spectrum causes adverse effects on the performance of conventional communication protocols, which were developed considering a fixed frequency band for communication. Hence, in this chapter, the intrinsic challenges in cognitive radio networks are captured and guidelines for further research are laid out in this area. More specifically, the recent proposals for spectrum sensing are overviewed in cognitive radio networks as well as the challenges for compressive sensing.

Cognitive radio technology is extremely multi-disciplinary involving research in many fields such as signal processing, communication systems, antenna designs, medium access control (MAC) protocols, and cognitive radio hardware architectures. Cognition can be built in all layers of the wireless network protocol stack. However, this thesis concentrates on analysis and design of spectrum sensing algorithms and DSA techniques from the physical and MAC layer perspective.

The Cognitive Cycle
The cognitive cycle begins with spectrum sensing, by which the secondary users detect the available white spaces. The spectrum access mechanism is initiated on the basis of the availability of spectrum holes. Spectrum sharing is also conducted for sharing the spectrum among multiple secondary users. Finally, on the reappearance of the primary user, the secondary user should vacate to another available spectrum band which is referred to as spectrum mobility.

The cognitive capability of a cognitive radio enables real-time interaction with its environment for the determination of appropriate communication parameters and adapt to the changing radio environment. The functions required for this dynamic operation in an open spectrum are shown in Figure 8.5, which is referred to as the cognitive cycle. In this section, we provide an overview of the three main steps of the cognitive cycle: spectrum sensing, spectrum analysis, and spectrum decision. The steps of the cognitive cycle are as follows:

1. Spectrum sensing: A cognitive radio monitors the available spectrum bands, captures information relating to them, and then detects the spectrum holes.

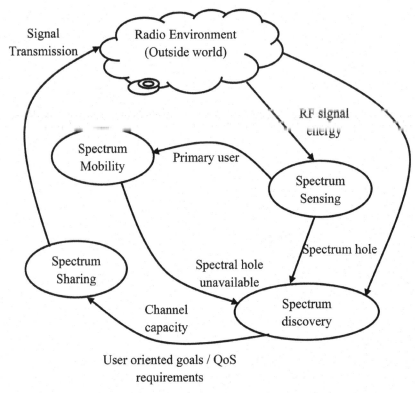

Figure 8.5   Cognitive cycle.

2. Spectrum analysis: The characteristics of the spectrum holes that are detected through spectrum sensing are estimated.

3. Spectrum decision: A cognitive radio determines the data rate, the transmission mode, and the bandwidth of the transmission. Then, the appropriate spectrum band is selected according to the spectrum characteristics and user requirements. Once the operating spectrum band is chosen, the communication can be performed over this spectrum band. However, since the radio environment changes over time and location, the cognitive radio should keep track of the changes periodically.

When the current spectrum band in use becomes unavailable, the spectrum mobility function should be performed to provide a seamless communication. Any change in the RF environment during the transmission such as reappearance of primary user, user movement, or traffic variation can trigger this adjustment.

8.3 Spectrum Sensing for Cognitive Radio

Spectrum sensing is an important requirement of the cognitive radio network which is employed for detecting the availability of spectrum holes. A cognitive radio should be designed to ensure its awareness of and sensitivity to the changes in its environment. The spectrum sensing function enables the cognitive radio to adapt to its environment. It is very important to detect the presence of primary users within the communication range of the secondary user. In reality, however, it is difficult for a cognitive radio to have direct measurement of a channel between a primary receiver and a transmitter. Thus, the cognitive radio network focuses on primary transmitter detection based on local observations of the secondary users. Generally, the spectrum sensing techniques can be classified as transmitter detection, cooperative detection, and interference-based detection, as shown in detail in Figure 8.6. In the following sections, these spectrum sensing methods and the open research issues on spectrum sensing are described in detail.

The cognitive radio should be able to distinguish between occupied and unoccupied portions of the spectrum bands. Thus, the cognitive radio should have the capability to determine the presence of a signal from primary transmitter in a certain spectrum band. Transmitter detection approach is based on the detection of the weak signal from a primary transmitter through the local observations of the cognitive radio users.

Basic hypothesis model for transmitter detection can be defined as follows:

$$y(n) = \begin{cases} v(n); & \text{under } H_0 \\ hs(n) + v(n); & \text{under } H_1 \end{cases} \tag{8.1}$$

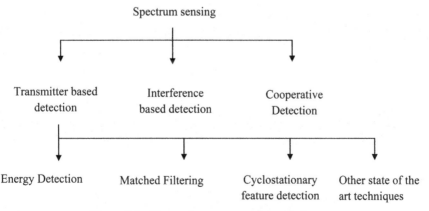

Figure 8.6   Types of spectrum sensing techniques.

where $y(n)$ is the signal received by the secondary user, $s(n)$ is the transmitted signal of the primary user, $v(n)$ is the AWGN, and h is the amplitude gain of the channel. H_0 refers to null hypothesis, which denies the presence of a primary user in a specific spectrum band. On the other hand, H_1 is an alternative hypothesis, which indicates the presence of primary user signal. There are three popular schemes, generally used for transmitter detection according to this hypothesis model. They are energy detection, matched filter detection, and cyclostationary feature detection techniques proposed. Apart from these, there are many other detection schemes proposed for spectrum sensing, namely eigenvalue detection, autocorrelation based detection, and wavelet detection compressed sensing-based detection. In this chapter, we focus briefly on different types of spectrum sensing techniques and elaborate in detail the compressed sensing-based detection technique.

8.3.1 Spectrum Sensing Techniques

The important spectrum sensing functions of a cognitive radio are explained in detail below:

Matched Filtering
A matched filter is a linear, coherent, and optimum filter which maximizes the output SNR under AWGN channel conditions. The received signal is correlated with a known sequence using a matched filter. Matched filtering can be employed for cognitive radio spectrum sensing if the primary transmitted signal is known. A decision statistic is formulated for detecting the presence or absence of licensed user and takes the form

$$T(y) = s^{\mathrm{H}} C \mathbf{y} \tag{8.2}$$

where s is vector of the known signal to be detected $s(n)$, C is the covariance matrix of the AWGN $v(n)$, and y is vector of the received signal $y(n)$. The inference from Equation (8.2) is that the decision statistic depends on s, which is a known signal. A simple block diagram of the spectrum sensing technique using a matched filter is shown in Figure 8.7.

The advantages of matching filtering are linearity, ease of implementation and less sensing time compared to other techniques. The major limitations are that the filter implementation requires prior knowledge about the primary user signal and also its inability to detect different primary user signals. In order to use matched filtering to detect different primary user signals,

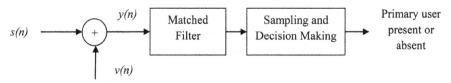

Figure 8.7 Spectrum sensing using a matched filter.

a bank of such filters is required. This certainly increases the hardware complexity and becomes difficult to modify when any new primary user signals are introduced. Matched filtering also requires perfect synchronization for coherent processing.

Matched filtering technique-based spectrum sensing is proposed for cognitive radio in the literature for various primary user signals. In [4, 5], the authors proposed matched filtering techniques for digital video broadcasting terrestrial (DVB-T) signal which exploits the pilot structure present in the DVB-T signal. In [6], a novel reconfigurable matched filter-based spectrum sensing for cognitive radios which is adaptable to primary users of multiple standards is presented. Matched filter detection was proposed in [7] for primary user detection in the global system for mobile (GSM) band. The authors have carried out the spectrum sensing in GSM 900 for identifying spectrum opportunities for the cognitive users. An adaptive matched filter constant false alarm rate detector has been proposed for spectrum sensing [8]. The authors have used a reduced rank multistage Wiener filter detector and compared the results with those of a full rank filter. The proposed filter achieved better performance with reduced complexity.

Energy Detection
The conventional energy detection algorithm compares the received signal energy over a finite interval with a predetermined threshold. The decision statistic of the energy detector takes the form

$$T(y) = \frac{1}{M} \sum_{m=1}^{M} |y(n)|^2 \qquad (8.3)$$

where $y(n)$ is the received signal and M is the number of samples collected over the observation interval. The decision statistic depends only on the received signal dispensing with the knowledge of the primary user signal. They are very popular because of their simplicity and low computational complexity.

In a fading environment, the SNR is a random variable. In the typical case of Rayleigh fading, the SNR is exponentially distributed with parameter μ, the average SNR. In such a situation, the detection probability depends on the exponentially distributed average SNR. The block diagram of the energy detector is shown in Figure 8.8.

Similar to time-domain implementation, the energy detector can also be conveniently implemented in the frequency domain. An important advantage of energy detection lies in the ease of implementation. The SNR must be known to set the detection threshold and characterize its performance. Significant weaknesses are its lack of robustness when the variances are unknown or varying with time. When the noise variance is known only within a range, the threshold must be conservatively set to protect the primary user. At low SNR, this means that performance degradation is dramatic when the uncertainty range is comparable to the SNR. This phenomenon has been called the *SNR wall*.

Originally, the energy detector is proposed for the detection of an unknown deterministic signal considering a flat band-limited Gaussian channel [9]. In the last decade, the energy detector is being used for spectrum sensing to a large extent due to its simplicity and low computational complexity [10]. The expressions for the probability of false alarm and detection were derived under AWGN and fading channel models for energy detection algorithm [11, 12]. The authors have also analyzed the receiver operating characteristic (ROC) performance. The secondary user spectrum sensing throughput problem was analyzed using energy detection sensing [13]. An optimal sensing time which maximized the secondary user throughput has been identified. The authors proved that for a 6 MHz channel, when the duration of the frame is 100 ms at 90% detection probability, the optimum sensing time is 14.2 ms. The authors of [14] proposed a blindly combined energy detection technique which does not require any information about the primary signal. The authors validated the proposed technique using wireless microphone signals and randomly generated signals and proved that their method outperforms energy detection for highly correlated signals.

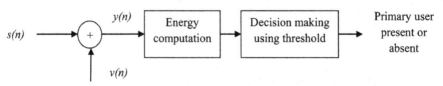

Figure 8.8　Spectrum sensing using an energy detector.

An adaptive threshold-based energy detection suitable for time-varying nature of the wireless channel and primary user activities was proposed in [15]. The authors of [16] put forward the Barlett's estimate as the decision statistic for energy detection. The authors investigated the performance for unknown signals under Rayleigh and Rician fading channels. The accuracy of their method was also compared with that of periodogram technique and found to achieve low miss detection probability. But their technique was able to achieve low false alarm only for a higher detection threshold. An energy detection-based spectrum sensing was performed using Welch periodogram technique in [17]. The authors observed improved performance when the parameters of the Welch periodogram are included in the distribution of the decision statistic. They also observed that improved detection performance has been achieved at the expense of increased false alarm probability under noise uncertainty. An improved version of energy detection algorithm was proposed for spectrum sensing in [18]. The improved detection scheme initially employs the conventional energy detection algorithm and confirms with additional verifications to avoid any missed detection due to instantaneous energy drops and improves the detection performance. The authors have analyzed the computational complexity of the improved energy detection algorithm and found to be similar to that of the conventional energy detection algorithm. Another approach to improve the traditional energy detection algorithm was proposed in [19, 20]. The algorithm computes an arbitrary positive power operation on the received signal to compute the decision statistic instead of the squaring operation showing a better performance. In [21], a formal measure for the utilization of spectrum holes is defined and a new adaptive sensing duration is proposed for energy detection-based spectrum sensing. By dynamically changing the sensing duration, the authors proved that more transmission time is available for the secondary users, thereby improving their throughput.

There are several contributions in which the energy detection-based spectrum sensing algorithms are analyzed under noise uncertainty. In [22], the authors demonstrate the existence of SNR wall below which effective detection performance is not possible. The performance of energy detection under log normal approximation of noise uncertainty was analyzed in [23]. In [24], a uniform distribution for noise uncertainty is assumed and they have analyzed the performance of energy detection. A generalized energy detector was analyzed under noise uncertainty and the TED was seen to be best suited for spectrum sensing under noise uncertainty [25]. One of the important issues of the energy detector is its inability to differentiate between signals from primary and secondary users. These motivate us to consider detectors that can exploit additional signal features.

Cyclostationary Detection

Cyclostationary features are unique to modulated signals. A signal is said to be cyclostationary if its statistical parameters such as the mean and autocorrelation are periodic as a function of time [26, 27]. Wireless communication signals exhibit cyclostationary features at multiple cyclic frequencies. These features typically depend on parameters such as the carrier frequency, code rate, symbol rate, and chip rates. In sharp contrast, noise signals do not possess cyclostationary features as they are less correlated. These features present in the primary user signals are often specified in standards from which the knowledge about the cyclic frequencies of various types of primary users is known.

A feature detector should be able to distinguish different types of primary user signals which require management of the cyclostationary signatures of all possible types of primary users. Hence, the design of a feature detector is highly complex for multiple primary user signals. Feature detectors are popular for their accuracy in signal detection and have received considerable attention in the literature. The basic block diagram of a cyclostationary feature detection is shown in Figure 8.9.

Cyclostationary features are susceptible to channel fading and clock offset [28]. The performance of a feature detector is found to be poor when the noise is stationary and possesses an SNR wall phenomenon similar to that of the energy detector [22]. Detection of weak primary signals in the presence of interference and noise using cyclostationary features was addressed [29]. Spectrum sensing of orthogonal frequency division multiplexing (OFDM) signals using their embedded pilots was proposed [30]. They have presented a correlated aided cyclostationary detection for detecting OFDM signals under asynchronous conditions. Cyclostationary feature detection with the joint use of cyclic frequencies and lags to form a reliable decision statistic was proposed [31].

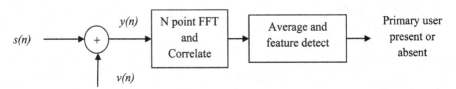

Figure 8.9   Spectrum sensing using a cyclostationary feature detector.

Other State-of-the-Art Detection Techniques

In recent literature, alternative techniques for spectrum sensing have been proposed and have gained much attention. Some of them include autocorrelation detection, eigenvalue detection, and wavelet detection. In autocorrelation-based detection, the sample autocorrelation of the received signal is compared with a predefined threshold and the presence or absence of the incumbent is identified. In [32], autocorrelation-based primary user detection algorithm under AWGN and Rayleigh fading channels has been proposed and the performance compared with a covariance detector, correlation detector, and an energy detector. In [33], an efficient spectrum sensing scheme using an autocorrelation-based sequential detection for the detection of OFDM primary user has been proposed. Waveform-based detectors have been designed for the detection of a known primary user signal through correlation detection [34]. The received signal is correlated with the pre-known primary user patterns such as preambles and pilots. The advantages include less sensing time and good detection performance. However, the performance is severely degraded due to synchronization errors. The authors of [35] have proposed a new detection scheme based on the eigenvalues of the covariance matrix of received signals. The performance of the eigenvalue detector is found to be superior when the received signals are highly correlated. A wavelet-based spectrum sensing approach for cognitive radios is introduced for primary user detection [36]. In [37], a blind moment-based detector which exploits the primary signal constellation is proposed for unknown noise power and signal power. Apart from these specific techniques, many hybrid detectors are also proposed which combine the advantages of two or more sensing techniques discussed above, but at the expense of increased complexity [38–40].

8.3.2 Cooperative Spectrum Sensing

The hidden terminal problem cannot be avoided when the spectrum sensing is conducted independently by secondary users. This is because the secondary user may suffer from the problem of shadowing or receiver uncertainty [3]. Therefore, the sensing results from the other secondary users are required for improving accuracy. Cooperative spectrum sensing schemes are thus introduced for incorporation of the sensing information from multiple secondary users and arrive at a final decision. The final decision is usually accurate compared to the single user sensing as it is less likely that the channels between the primary user and the secondary users will experience deep fade or shadowing simultaneously. Figure 8.10 depicts the typical architecture for cooperative spectrum sensing [41].

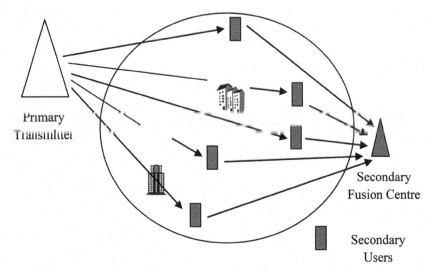

Figure 8.10 Cooperative spectrum sensing.

Cooperative schemes can be either centralized or distributed. Centralized architecture for cooperative sensing employs a base station or fusion center for collecting the sensing data from the secondary users and identifying the white spaces. In a distributed architecture, the sensing outcomes are exchanged between the secondary users and the users arrive at a decision independently. Considerable attention has been drawn for cooperative spectrum sensing in the cognitive radio literature [3, 34, 42]. The main drawback of cooperative sensing is the overhead involved in the collection of sensing data, additional operations, and the time required for data fusion.

Cooperative detection proposed by [43, 44] helps in mitigating the channel impairments through multi-path diversity. Cooperative sensing based on energy detection has been proposed by [45] in which the energy values from multiple secondary users are combined using soft combination techniques such as maximum ratio combination. A cluster-based collaborate spectrum sensing technique was proposed by [46]. Clustering which reduces the bandwidth needed for the control channel is regarded as one of the challenges of the cognitive radio network when number of users increases. The performance of cooperative spectrum sensing was optimized with the energy detection technique [41, 47]. In [48], the authors have presented a comprehensive analysis and comparison of hard decision and soft decision fusion rules for cooperative spectrum sensing under imperfect reporting channels. An adaptive cooperative sensing scheme with random access which deals with the problem of sensing data collection has been presented in [49].

With the development of numerous algorithms for spectrum sensing, there is still a tradeoff between the complexity and accuracy of the detector. Recently, compressed sensing-based detectors have been designed for cognitive radio spectrum sensing. Compressed sensing is employed for reducing the complexity of the sensing algorithm. It is used for both narrowband and wideband spectrum sensing. Both cooperative and distributed detection based on compressed sensing are also becoming popular. In the following sections, the techniques for cognitive radio primary user detection based on compressed sensing are discussed in detail.

Cognitive Radio Challenges
The primary users or the licensed spectrum owners are not receptive of the cognitive radios and the spectrum trade policies as they are likely to cause disruption to their access in the radio spectrum and resources. The fundamental challenges of cognitive radio lie in effective identification of vacant portions over a wide range of spectrum and also over spatio-temporal domain while maintaining signal-to-noise ratio (SNR) of secondary users' communications [3]. Secondary users should be able to dynamically sense the radio spectrum and adapt to the dynamically changing radio environment by altering its transmitter parameters accordingly. Wideband spectrum sensing requires a very fast high-resolution digital sampler for meeting the dynamic and wide spectrum. In addition, the radio sensitivity of the cognitive radios should be very precise with very short sensing duration as possible to meet the agility requirements. The implementation challenges in wideband spectrum sensing for cognitive radios can be broadly viewed under the following aspects.

1. **Detection capability**: Primary users are not receptive of the cognitive radio users as they cause harmful effects to them. A secondary radio, sometimes, may not be able to correctly sense the occupancy of the spectrum and may start using it when primary user uses it which results in interference with the primary user communications. Thus, the function of any cognitive radio is not only to sense the spectrum before beginning its transmission but also have to continuously sense during its transmission period so as to detect the reappearance of primary users, if any, and vacate the spectrum for them. The receiver sensitivity plays a major role in cognitive radio network in achieving this function. Moreover, the noisy and dynamic fading characteristics of the channel make the problem

of finding the threshold for spectrum detection and decision process complex. Sensing time duration is also an important factor as cognitive radios as it cannot spend a long time in spectrum sensing resulting in decrease in the opportunities for its own communication purposes. The SNR margin for cognitive radio communication creates a trade-off between the receiver sensitivity and power allocation at the transmitter side. This requires adaptation of the cognitive radios to transmit power based upon the communications environment.

2. **Wideband sampling circuitry**: Wideband spectrum sensing is a challenging task as it requires dynamic sensing of wide range of frequency bands. It requires either grouping of wideband spectrum into narrowbands and performs multiple narrowband sensing or using digital signal processors to sample the wideband analog signal and then performs spectrum sensing. Practically, the multiple narrowband sensing is not feasible in practice and less attractive due to its complexity in wideband model, whereas the wideband sensing using DSP necessitates high-speed ADCs.

A new sampling paradigm based upon sparse sampling named compressive sensing provides a sampling mechanism at rates lower than the Nyquist rates. The signal reconstruction scheme in compressive sensing is a norm optimization problem. The main objective of this chapter is to present a novel spectrum sensing mechanism for cognitive radio based upon compressive sensing.

8.4 Compressed Sensing in Cognitive Radio

Compressive sensing-based technique is used for accurate and fast spectrum sensing in cognitive radio technology-based systems and standards. The first standard utilizes the concept of cognitive radio, providing an air interface for wireless communication in the TV spectrum band is the IEEE 802.22 standard. Specific spectrum sensing methods are not explicitly defined in the standard. It has, however, to be fast and accurate. Fast Fourier sampling (FFS) is an algorithm based on compressed sensing which can be used for detecting wireless signals. According to the FFS algorithm, only very few with highly energetic frequencies of the spectrum are detected and the whole spectrum band is approximated from these values using non-uniform inverse fast Fourier transform (IFFT). Thus, with just fewer samples, FFS algorithm results in faster sensing, which enables more spectrum to be sensed in the same time window.

Wideband spectrum sensing based on compressive sampling uses multiple secondary receivers for sensing the same wideband signal through analog-to-information converters (AICs) and produces the autocorrelation vectors of the compressed signal and sends them to a centralized fusion center for decision on spectrum occupancy. This exploits joint sparsity and spatial diversity of signals for getting performance gains over a non-distributed CS system. In the past, work has also been done with the application of compressive sensing in parallel to time-windowed analog signals. Load of samples on the digital signal processor gets reduced through the use of compressed sensing but the ADC still has to digitize analog signal to digital signal. In order to overcome this problem, compressed sensing is applied directly on analog signals. This is done on segmented pieces of signal; each block is compressively sensed independent of the other. At the receiver, however, a joint reconstruction algorithm is implemented for recovering the signal. The sensing rate is greatly reduced through use of "parallel" compressed sensing while reconstruction quality improves. A cyclic feature detection framework based on compressed sensing for wideband spectrum sensing utilizes second-order statistics for coping with high-rate sampling requirement of conventional cyclic spectrum sensing. Signal sparsity level has temporal variation in cognitive radio networks, and thus, optimal compressive sensing rate is not static. A framework to dynamically track optimal sampling rate and determine unoccupied channels in a unified way is also introduced for cognitive radios.

Compressed Sensing

Let N-dimensional space real-valued signal $z \in R^{N \times 1}$ can be represented as a linear combination of orthogonal basis vectors $\{\psi_i\}_i^N$,

$$z = \sum_{i=1}^{N} \theta_i \psi_i \text{ or } z = \psi \theta$$

where θ is the $N \times 1$ vector of weighting coefficients, and $\psi = [\psi_1, \psi_2, \dots \psi_N] \in R^{N \times N}$ is the orthogonal dictionary basis. Let θ is K-sparse if the number of nonzero coefficients $K \leq N$; that is, only K of the θ_i coefficients are nonzero and $(N - K)$ are zero. The signal z is compressible only if the representation has just a few large coefficients and many small coefficients. The measurement matrix ϕ: $M \times N(M)$ which is irrelevant to ψ is adopted. Consider a general compressed process for signal z,

$$y = \phi z = \phi \psi_z = A^{CS} z$$

where $A^{CS} = \phi\psi$ is an $M \times N$ matrix. The measurement process is not adaptive, which means that ϕ is fixed and not depending on the signal z. Usually, the following aspects are of main focus:

    a) A stable measurement matrix ϕ to ensure the compressible signal is not harmed by the dimensionality reduction from $z \in R^{N \times 1}$ to $y \in R^{M \times 1}$;

    b) A proper reconstruction algorithm to recover y from z.

There is a requirement of non-reconstruction methods which can directly determine the presence or absence of primary user from the measurement matrix y without recovering the original signal z. The choice of the measurement matrix ϕ has a great impact on the resolving reconstruction problem. The restricted isometry property (RIP) is a necessary and sufficient condition for this problem. This property is described as follows: If signal z is *K-sparse*, for some δ, ϕ can satisfy the formula (10), then ϕ possess the K level RIP property.

$$(1 - \delta)\frac{M}{N} \leq \frac{\|\phi_z\|_2^2}{\|z\|_2^2} \leq (1 + \delta)\frac{M}{N}$$

For instance, let the matrix ϕ's elements be independent and identically distributed random variables from a Gaussian probability density with mean zero and variance $1/N$. Then, the measurements y are merely M different randomly weighted linear combinations of the elements of z. The Gaussian measurement matrix ϕ is be used to compress the received signal.

8.5 Collaborative Compressed Spectrum Sensing

Figure 8.11 shows the system model of collaborative compressed spectrum sensing. In local spectrum sensing, each secondary user $1, 2, \ldots N$ acquires the received signal data with compressed sampling; that is, they utilize the Gaussian measurement matrix for obtaining the compressed sampling data.

    The presence of the primary user is then determined with the statistic measurement values after compressed sampling. In a fusion center, the final decision can be obtained according to the "AND" rule, "OR" rule, or "MAJORITY" fusion rule.

    The binary hypothesis model based on the compressed sampling in local spectrum sensing is given by

$$y_i = \begin{cases} \phi_i n_i & H_0 \\ \phi_i(x_i + n_i) & H_1 \end{cases}$$

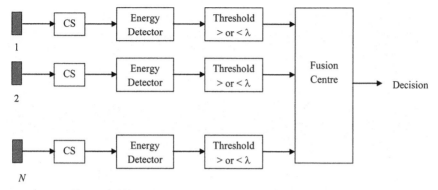

Figure 8.11 Compressed cooperative spectrum sensing model.

where ϕ is $M \times L$ dimension Gaussian measurement matrix, and x_i and n_i are $L \times 1$ dimension vector and the received signal y_i dimension vector. For hypothesis H_0, r_i should follow the Gaussian distribution with mean 0 and variance $\sigma_{n,i}^2 \left| \phi\phi^T \right|_0$.

For hypothesis H_1, r_i should follow the Gaussian distribution with mean 0 and variance $\left(\sigma_{x,i}^2 + \sigma_{n,i}^2 \right) \left| \phi\phi^T \right|_0$. And $\left| \phi\phi^T \right|_0$ denotes the sum of diagonal elements for matrix $\phi\phi^T$.

The decision statistic value T_i of the ith secondary user is calculated using

$$T_i = \frac{1}{M} \sum_{k=1}^{M} |y_i(k)|^2$$

and the decision statistic value T_i should obey the following distribution

$$H_0 : T_i \sim N \left(\frac{\sigma_{n,i}^2 \eta}{M}, \frac{2\eta^* \sigma_{n,i}^4}{M^2} \right)$$

$$H_1 : T_i \sim N \left(\frac{\left(\sigma_{x,i}^2 + \sigma_{n,i}^2 \right) \eta}{M}, \frac{2\eta^* \left(\sigma_{x,i}^2 + \sigma_{n,i}^2 \right)^2}{M^2} \right)$$

where η is equal to $\left| \phi\phi^T \right|_0$ and denotation η^* is the sum of the square of diagonal elements for matrix $\phi\phi^T$. The probability of false alarm P_{fi} and the probability of detection P_{di} in local spectrum sensing can be calculated as

$$P_{fi} = Q \left(\frac{\lambda_i - \sigma_{n,i}^2 \eta/M}{\sigma_{n,i}^2 \sqrt{2\eta^*}/M} \right)$$

$$P_{di} = Q \left(\frac{\lambda_i - \left(\sigma_{x,i}^2 + \sigma_{n,i}^2 \right) \eta/M}{\left(\sigma_{x,i}^2 + \sigma_{n,i}^2 \right) \sqrt{2\eta^*}/M} \right)$$

and the detection threshold λ_i is calculated using

$$\lambda_i = Q^{-1}(P_{fi})\sigma_{ni}^2 \sqrt{2\eta^*} \Big/ M + \sigma_{ni}^2 \eta/M$$

The expression for the probability of detection P_{di} is found using

$$\Gamma_{di} = Q\left(\frac{Q^{-1}(P_{fi})\sigma_{ni}^2\sqrt{2\eta^*} + \sigma_{ni}^2\eta - \left(\sigma_{x,i}^2 + \sigma_{n,i}^2\right)\eta}{\left(\sigma_{x,i}^2 + \sigma_{n,i}^2\right)\sqrt{2\eta^*}}\right)$$

At the fusion center, the fusion rules such as the "AND" rule, "OR" rule, and the "MAJORITY" rules are followed for mitigating the impact of multi-path fading and shadowing on the detection performance. For "AND" rule, the probability of false alarm $P_f^{\mathbf{and}}$ and the probability of detection $P_d^{\mathbf{and}}$ can be calculated as

$$P_f^{\mathbf{and}} = \prod_{i=1}^N P_{fi}$$

$$P_d^{\mathbf{and}} = \prod_{i=1}^N P_{di}$$

For "OR" rule, the probability of false alarm $P_f^{\mathbf{or}}$ and the probability of detection $P_d^{\mathbf{or}}$ can be calculated as

$$P_f^{\mathbf{or}} = 1 - \prod_{i=1}^N (1 - P_{fi})$$

$$P_d^{\mathbf{or}} = 1 - \prod_{i=1}^N (1 - P_{di})$$

For "MAJORITY" rule, the probability of false alarm $P_f^{\mathbf{maj}}$ and the probability of detection $P_d^{\mathbf{maj}}$ can be calculated as

$$P_f^{\mathbf{maj}} = \sum_{l=K}^N \binom{N}{l} (1 - p_f)^l (p_f)^{N-l}$$

$$P_d^{\mathbf{maj}} = \sum_{l=K}^N \binom{N}{l} (p_d)^l (1 - p_d)^{N-l}$$

8.6 Distributed Compressed Spectrum Sensing

The distributed spectrum sensing scheme comprises two steps. The system model for distributed compressed sensing scheme is shown in Figure 8.12.

Step 1: The N secondary users sample the spectrum independently at a very low rate which is equivalent to the bandwidth of a single sub-band and report the aliased spectrum information to the fusion center later.

Step 2: After receiving all the information from the secondary user nodes, the fusion center performs an appropriate compressed sensing recovery algorithm and broadcast the sensing result to all the secondary nodes within the cluster.

Distributed Sampling Based on Spectrum Aliasing

According to Figure 8.13, after the primary signal $x^k(t)$ arrives at the kth secondary node, it is multiplied by a high-frequency mixing function $p^k(t)$.

$p^k(t)$ is a T_s-length periodic spread-spectrum mixing function which aims to alias the spectrum and thus obtain a mixture of signals from all the sub-bands.

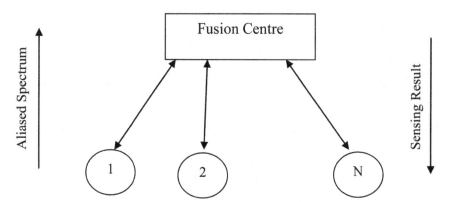

Figure 8.12 Distributed cooperative spectrum sensing model.

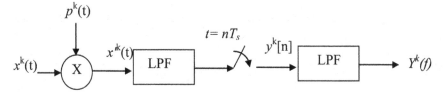

Figure 8.13 Compressed sampling structure based on aliasing.

$$p^k(t) = \alpha_{km}, \; m\frac{T_s}{M} \leq t \leq (m+1)\frac{T_s}{M}, \; 0 \leq m \leq M-1$$

8.7 Compressive Sensing for Wideband Cognitive Radios

Wideband spectrum sensing is the technique in which the frequency spectrum to be sensed has the bandwidth more than the coherent bandwidth of the channel. Spectrum sensing techniques based on narrowband are limited as they make use of the single binary decision and cannot detect every individual spectrum opportunity available over the wideband spectrum. Looking at the prevailing techniques, the wideband spectrum sensing can be classified into two main categories: Nyquist (rate based) spectrum sensing and sub-Nyquist (rate based) spectrum sensing. As the name suggests, the Nyquist (rate based) spectrum sensing uses the sampling rate for spectral estimation equal to or more than the Nyquist rate, while the sub-Nyquist uses the rate of sampling below the Nyquist rate.

Due to the drawbacks of high sampling rate such as the high implementation complexity of Nyquist systems requiring high-speed ADC, sub-Nyquist wideband sensing systems are preferred for acquiring wideband signals at lower sampling rates and determine spectrum opportunities. Sub-Nyquist techniques use partial measurements of the spectrum. Two main techniques are proposed in literature for sub-Nyquist wideband sensing: (i) compressive sensing-based Wideband sensing and (ii) multi-channel sub-Nyquist wideband sensing.

Compressive sensing-based wideband sensing is a technique that can help acquisition of a signal using relatively few measurements. This can help unique representation of the signal can be found based on the signal's sparseness or compressibility in some domain. This technique uses fewer samples closer to the information rate, rather than the inverse of the bandwidth, to perform wideband spectrum sensing. Wavelet-based edge detection was used for detecting spectral opportunities across wideband spectrum [50] after reconstruction of the wideband spectrum. Furthermore, [51] has proposed a cyclic feature detection-based compressive sensing algorithm for wideband spectrum sensing and improving robustness against noise uncertainty. It is the two-dimensional cyclic spectrum (spectral correlation function) of a wideband signal which can be directly reconstructed from the sparse measurements. In [52], a distributed compressive sensing-based wideband sensing algorithm for cooperative multi-hop cognitive radio networks is proposed to further reduce the data acquisition cost. By enforcing consensus among local spectral estimates, such a collaborative approach can benefit from spatial diversity to mitigate the effects of wireless fading. In addition, decentralized consensus optimization approach has been proposed that aims to achieve high sensing performance at a reasonable computational cost.

AIC-based Sub-Nyquist Wideband Sensing

As compressive sensing is applicable to finite-length and discrete-time signals, innovative technologies are required for extending the idea of compressive sensing to continuous time signals, i.e., implementing compressive sensing in analog domain. To realize the analog compressive sensing, an analog-to-information converter (AIC) is proposed in [53], which could be considered as a good basis for the above-mentioned algorithms.

As shown in Figure 8.14, the AIC-based model consists of a pseudorandom number generator, a mixer, an accumulator, and a low-rate sampler. The pseudorandom number generator produces a discrete time sequence that is used for demodulating the signal $x(t)$ using a mixer. The accumulator is used for integrating the demodulated signal for "1/w" seconds, while its output signal is sampled using a low sampling rate, after which, the sparse signal can be directly reconstructed from partial measurements using compressive sensing algorithms.

Multi-Channel Modulated Sub-Nyquist Wideband Sensing

For eliminating model mismatches, Mishali and Eldar have proposed a modulated wideband converter (MWC) model in [54] by modifying the AIC model. The main difference between MWC and AIC is that MWC has multiple

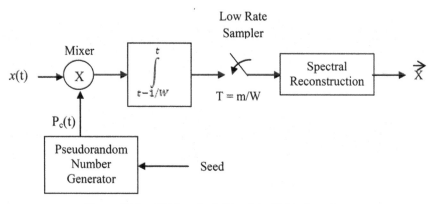

Figure 8.14 AIC-based sub-Nyquist wideband sensing.

sampling channels, with the accumulator in each channel replaced by a general low-pass filter.

One significant advantage of introducing parallel channel structure as shown in Figure 8.15 is that it provides robustness against noise and model mismatches. In addition, the dimension of the measurement matrix is reduced, making the spectral reconstruction more computationally efficient.

An alternative to multi-channel sub-Nyquist sampling approach is the multi-coset sampling as shown in Figure 8.16. The multi-coset sampling is similar to choosing some samples from a uniform grid and can be obtained with a sampling rate of f_s that is slightly higher than the Nyquist rate.

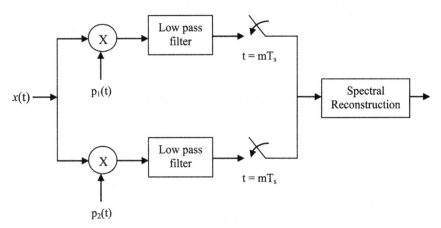

Figure 8.15　Multi-channel modulated sub-Nyquist wideband sensing.

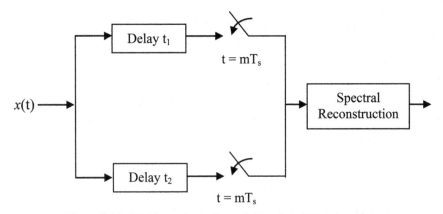

Figure 8.16　Multi-coset sampling sub-Nyquist wideband sensing.

The uniform grid is then divided into blocks of m consecutive samples, with v ($v < m$) samples being retained in each block, while the rest of samples are discarded. Thus, the multi-coset sampling is often implemented by using v sampling channels with sampling rate of f_x/m, with different sampling channels having different time offsets. Sampling patterns should be carefully designed, for obtaining a unique solution for the wideband spectrum from these sparse measurements. In [55], few sampling patterns were proved to be valid for unique signal reconstruction. The advantage of multi-coset approach is that the sampling rate in each channel is m times lower than the Nyquist rate. Moreover, the number of measurements is only v-m^{th} of that in the Nyquist sampling case. One drawback of the multi-coset approach is that the channel synchronization should be met such that accurate time offsets between sampling channels are required to satisfy a specific sampling pattern for a robust spectral reconstruction.

Multi-Rate Sampling Sub-Nyquist Wideband Sensing
This technique is an asynchronous multi-rate wideband sensing approach studied in [56], for relaxing the multi-channel synchronization requirement. Sub-Nyquist sampling was induced in this technique in each sampling channel to wrap the sparse spectrum occupancy map onto itself, the sampling rate can, therefore, be significantly reduced.

The performance of wideband spectrum sensing can be improved through the use of different sampling rates in different sampling channels as shown in Figure 8.17. Specifically, in the same observation period, the numbers of samples in multiple sampling channels are selected as different consecutive

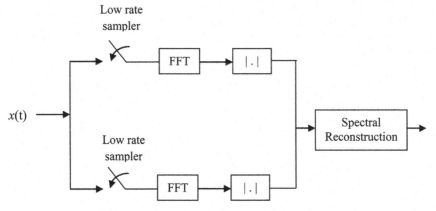

Figure 8.17 Multi-rate sampling sub-Nyquist wideband sensing.

prime numbers. Furthermore, as only the magnitudes of sub-Nyquist spectra are of interest, such a multi-rate wideband sensing approach does not require perfect synchronization between multiple sampling channels, leading to easier implementation. Also, [57] proposes an analog/mixed signal topology for wideband spectrum sensing that replaces the conventional Nyquist ADCs and digital fast Fourier transform (FFT) core with a bank of sample and hold (S/H) circuits, each operating at sub-Nyquist rate, and an all-analog FFT processor.

8.8 Research Challenges

The following are the research challenges that need specific attention for implementing a feasible wideband spectrum sensing device based on compressed sensing for future cognitive radio networks.

8.8.1 Sparse Basis Selection

Mostly, all sub-Nyquist wideband sensing techniques necessitate the sparsity of the wideband signal on a suitable basis. When the utilization of spectrum is low, most of existing wideband sensing techniques assumes that the wideband signal is sparse with respect to the frequency domain, i.e., the sparsity basis is a Fourier matrix. However, with the improvement of spectrum utilization (e.g., the use of cognitive radio techniques for future cellular networks), the wideband signal may not be considered sparse in the spectral domain. Thus, a significant challenge in future cognitive radio networks is to carry out a wideband sensing using partial measurements, if the wideband signal is not sparse in the frequency domain. Thus, it will be indispensable to study suitable wideband sensing techniques that are capable of exploiting sparsity in any known sparsity basis. Moreover, in real time, it may not be easy to acquire enough knowledge about the sparsity basis in cognitive radio networks, e.g., when it is not able to obtain enough prior knowledge about the primary signals. Hence, future cognitive radio networks will be required to perform wideband spectrum sensing even when the sparsity basis is unknown. In this context, a challenging issue is to study "blind" sub-Nyquist wideband sensing algorithms, where prior knowledge about the sparsity basis is not required for the sub-Nyquist sampling or the spectral reconstruction [56].

8.8.2 Adaptive Wideband Sensing

In most of the sub-Nyquist wideband sensing systems, the number of partial measurements will vary prorate to the sparsity level of wideband signal.

Therefore, sparsity-level estimation is often required to choose an appropriate number of measurements in cognitive radio networks. However, in real time, the sparsity level of wideband signal is difficult to estimate and often time-varying, because of either the dynamic activities of primary users or the time-varying fading channels between primary users and secondary users. Because of this uncertainty in the sparsity level, most of sub-Nyquist wideband spectrum sensing systems should pessimistically choose the number of measurements, leading to more energy consumption in cellular networks. Hence, future cognitive radio networks should be capable of performing wideband sensing, given the unknown or changing sparsity level. In such a scenario, it is very challenging to develop adaptive wideband sensing techniques that can cleverly choose an appropriate number of compressive measurements without any prior knowledge of the sparsity level [56].

8.8.3 Cooperative Wideband Sensing

The reliability of wideband spectrum sensing may be compromised due to the fluctuations in the wireless channel and their fading effects. It is also possible to improve the accuracy of spectrum sensing through collaborative techniques significantly. In a multi-path or shadow fading wireless environment, the primary signal received at the secondary users may undergo severe degradation, leading to unreliable wideband sensing results at each cognitive radio node. In this situation, cognitive radio networks of the following generation should employ cooperative strategies for improving the reliability of wideband sensing by exploiting spatial diversity. Actually, in a cluster-based cognitive radio network, the wideband spectrum as observed by different cognitive radios could share some common spectral components, while each cognitive radio may observe some innovative spectral components. Thus, it is possible to fuse compressive measurements from different users and exploit the spectral correlations among cognitive radios in order to save the total number of measurements and thus the energy consumption in wireless networks. Such a data fusion-based cooperative technique, however, leads to intense data transmission load in the common control channels. Development of data fusion-based cooperative wideband sensing techniques subject to reduced data transmission burden is, therefore, a challenging task. An alternative is to develop decision fusion based on wideband sensing techniques, if each secondary user is able to detect wideband spectrum independently. Due to the limited computational resource in cellular networks, the challenge that remains in the decision fusion-based cooperative approach is how to appropriately combine information in practice [56].

Furthermore, one of the problems in cooperation is combining the sensing results of various secondary users which may have different accuracy levels and sensing duration. Some form of weighted combining methods needs to be performed in order to take this into account.

The design of control channels is also a major task in cooperative spectrum sensing. A control channel can be implemented either as a dedicated channel or as an underlay UWB channel. Wideband RF frontend tuners or filters can be shared between the UWB control channel and normal cognitive radio transceivers. Furthermore, when there are multiple cognitive radio groups that are active simultaneously, the control channel bandwidth has to be shared. With a dedicated frequency band, a CSMA scheme may be desirable. For a spread-spectrum UWB control channel, different spreading and sequencing could be allocated to different groups of users, thus imposing the challenge of defining the spreading sequences for an underlay spread-spectrum UWB control channel.

8.9 Summary

Compressed sensing is a popular sub-Nyquist sampling technique which finds application in signal and image processing. The use of compressed sensing is required for applications which are considered to be time bound. Cognitive radio is an intelligent radio which tries to opportunistically utilize the licensed spectrum when the primary user is not using it. The critical function of cognitive radio is spectrum sensing in which the state of the identification of primary user should be done in a very short time. Thus, compressed sensing-based primary user detection is becoming popular in the recent times. Thus, in this chapter, a detailed introduction to cognitive radio, dynamic spectrum access, and its related functionalities is presented. The state-of-the-art literature review related to spectrum sensing is described; specifically, sensing based on compressed sampling is explained in detail.

References

[1] Haykin, S. (2005). "Cognitive Radio: Brain-Empowered Wireless Communication," *IEEE J. Select. Areas Commun.* 23 (2), 201–220.

[2] Mitola, J., Maguire, G. Q. (1999). "Cognitive Radio: Making Software Radios more personal," *IEEE Perss. Commun.* 6 (4), 13–18.

[3] Akyildiz, F., Lee, W. Y., Chen, S., and Varshney, P. (2006). "Next Generation/Dynamic Spectrum Access/Cognitive Radio Wireless Networks: A Survey," *Elsevier Comput. Netw. J.* 50 (13), 2127–2159.

[4] Cabric, D., Tkachenko, A., and Brodersen, R. (2006). "Spectrum Sensing Measurements in Pilot, Energy and Collaborative Detection," *Militiary Commun. Conf.* 1–7.

[5] Danev, D., Axell, E., and Larsson, E. (2010). "Spectrum Sensing Methods of DVB-T Signals in AWGN and Fading Channels," IEEE International Symposium on Personal Indoor and Mobile Radio Communications, pp. 2721–2726.

[6] Gholamipour, A. H., Gorcin, A., Celebi, H., Toreyin, B. U., Saghir, M. A. R., Kurdahi, F., and Eltawil, A. (2011). "Reconfigurable Filter Implementation of a Matched-Filter Based Spectrum Sensor for Cognitive Radio systems," International Symposium on Circuits and Systems, pp. 2457–2460.

[7] Mohamad, M. H., Haw, C. W., and Ismail, M. (2012). "Matched Filter Detection Technique for GSM Band," International Symposium Telecommunication Technologies (ISTT), pp. 271–274.

[8] Fu, K. C., Yang, M. J., and Chen, Y. F. (2012). "Multistage Wiener Filter Based Spectrum Sensing in Cognitive Radio," 12th International Conference on ITS Telecommunications (ITST), pp. 700–705.

[9] Urkowitz, H. (1967). "Energy Detection of Unknown Deterministic Signals," Proceedings of the IEEE, Vol. 55, pp. 523–531.

[10] Kostylev, V. I. (2002). "Energy Detection of a Signal with Random Amplitude," IEEE International Conference on Communications, Vol. 3, pp. 1606–1610.

[11] Digham, F. F., Alouini, M. S., and Simon, M. K. (2003). "On the Energy Detection of Unknown Signals over Fading Channels," IEEE International Conference on Communications, Vol. 5, pp. 3575–3579.

[12] Digham, F. F., Alouini, M. S., Simon, M. K. (2007). "On the Energy Detection of Unknown Signals over Fading Channels," *IEEE Trans. Commun.* 55 (1), 21–24.

[13] Liang, Y. C., Zeng, Y., Peh, E. C. Y., and Hoang, A. T. (2008). "Sensing Throughput Tradeoff for Cognitive Radio Networks," *IEEE Trans. Wireless Commun.* 7 (4), 1326–1337.

[14] Zeng, Y., Liang Y. C., and Zhang, R. (2008). "Blindly Combined Energy Detection for Spectrum Sensing in Cognitive Radio," *IEEE Signal Process. Letts.* 15, 649–652.

[15] Ling, X., Wu, B., Wen, H., Ho, P. H., Bao, Z., and Pan, L. (2012). "Adaptive Threshold Control for Energy Detection Based Spectrum Sensing in Cognitive Radios," *IEEE Wireless Commun. Letts.* 1 (5), 448–451.

[16] Gismalla, E. H., and Alsusa, E. (2012) "On the Performance of Energy Detection Using Bartlett's Estimate for Spectrum Sensing in Cognitive Radio Systems," *IEEE Trans. Signal Process.* 60 (7), 3394–3404.

[17] Martiìnez, D. M., and Andrade, A. G. (2012). "Performance Evaluation of Welch's Periodogram-Based Energy Detection for Spectrum Sensing," *IET Commun.* 7 (11), 1117–1125.

[18] Lopez Benitez, M., and Casadevall, F. (2012). "Improved Energy Detection Spectrum Sensing for Cognitive Radio," *IET Commun.* 6 (8), 785–796.

[19] Chen, Y. (2010). "Improved Energy Detector for Random Signals in Gaussian Noise," *IEEE Trans. Wireless Commun.* 9 (2), 558–563.

[20] Song, J., Feng, Z., and Liu, Z. (2012). "Spectrum Sensing in Cognitive Radios Based on Enhanced Energy Detector," *IET Commun.* 6 (8), 805–809.

[21] Treeumnuk, D., and Popescu, D. C. (2014). "Enhanced Spectrum Utilization in Dynamic Cognitive Radios with Adaptive Sensing," *IET Signal Process.* 8 (4), 339–346.

[22] Tandra, R., and Sahai, A. (2008). "SNR Walls for Signal Detection," *IEEE J. Sel. Top. Signal Process.* 2 (1), 4–17.

[23] Jouini, W. (2011). "Energy Detection Limits Under Log-Normal Approximated Noise Uncertainty," *IEEE Signal Process. Lett.* 18 (7), 423–426.

[24] Yin, W., Ren, P., Cai, J., and Su, Z. (2013). "Performance of Energy Detector in the Presence of Noise Uncertainty in Cognitive Radio Networks," *Wireless Networks J.* 19 (5), 629–638.

[25] Kalamkar, S. S., Banerjee, A., and Gupta, A. K. (2013). "SNR Wall for Generalized Energy Detection Under Noise Uncertainty in Cognitive Radio," 19th Asia-Pacific Conference on Communications (APCC), pp. 375–380.

[26] Ghozzi, M., Marx, F., Dohler, M., and Palicot, J. (2006). "Cyclostatilonarilty-Based Test for Detection of Vacant Frequency Bands," 1st International Conference on Cognitive Radio Oriented Wireless Networks and Communications, pp. 1–5.

[27] Xu, S., Zhao, Z., and Shang, J. (2008). "Spectrum Sensing Based on Cyclostationarity," Workshop on Power Electronics and Intelligent Transportation System, pp. 171–174.

[28] Yucek, T., and Arslan, H. (2009). "A Survey of Spectrum Sensing Algorithms for Cognitive Radio Applications," *IEEE Commun. Surveys Tutor.* 11 (1), 116–130.

[29] Sadeghi, H., Azmi, P., and Arezumand, H. (2012). "Cyclostationarity-Based Soft Cooperative Spectrum Sensing for Cognitive Radio Networks," *IET Commun.* 6 (1), 29–38.

[30] Ghasabeh, Z. A., Tarighat, A., and Daneshrad, B. (2012). "Spectrum Sensing of OFDM Waveforms Using Embedded Pilots in the Presence of Impairments," *IEEE Trans. Veh. Technol.* 61 (3), 1208–1221.

[31] Shen, J., and Alsusa, E. (2013). "Joint Cycle Frequencies and Lags Utilization in Cyclostationary Feature Spectrum Sensing," *IEEE Trans. Signal Process.* 61 (21), 5337–5346.

[32] Pour, M. N., and Ikuma, T. (2010). "Autocorrelation Based Spectrum Sensing for Cognitive Radios," *IEEE Trans. Veh. Technol.* 59 (2), 718–732.

[33] Chaudhari, S., Koivunen, V., and Poor, H. V. (2009). "Autocorrelation-Based Decentralized Sequential Detection of OFDM Signals in Cognitive Radios," *IEEE Trans. Signal Process.* 57 (7), 2690–2700.

[34] Yucek, T., and Arslan, H. (2009). "A Survey of Spectrum Sensing Algorithms for Cognitive Radio Applications," *IEEE Commun. Surveys Tutor,* 11 (1), 116–130.

[35] Zeng, Y., and Liang, Y. C. (2009). "Eigenvalue-Based Spectrum Sensing Algorithms for Cognitive Radio," *IEEE Trans. Commun.* 57 (6), 1784–1793.

[36] Tian, Z., and Giannakis, G. B. (2006). "A Wavelet Approach to Wideband Spectrum Sensing for Cognitive Radios," International Conference on Cognitive Radio Oriented Wireless Networks and Communications, pp. 1–5.

[37] Cui, T., Tang, J., Gao, F., and Tellambura, C. (2011). "Moment-Based Parameter Estimation and Blind Spectrum Sensing for Quadrature Amplitude Modulation," *IEEE Trans. Commun.* 59 (2), 613–623.

[38] Khalaf, Z., Nafkha, A., Palicot, J., and Ghozzi, M. (2010). "Hybrid Spectrum Sensing Architecture for Cognitive Radio Equipment," Sixth Advanced International Conference on Telecommunications, pp. 46–51.

[39] Lu, L., Zhou, X., Onunkwo, U., and Li, G. Y. (2012). "Ten Years of Research in Spectrum Sensing and Sharing in Cognitive Radio," *Eurasip J. Wireless Commun. Networks,* 2012 (28), 1–16.

[40] Ejaz, W., Hasan, N., Lee, S., and Kim, H. S. (2013). "I3S: Intelligent Spectrum Sensing Scheme for Cognitive Radio Networks," *Eurasip J. Wireless Commun. Networks,* 2013 (26), 1–12.

[41] Zhang, W., Mallik, R. K., and Letaief, K. (2009). "Optimization of Cooperative Spectrum Sensing with Energy Detection in Cognitive Radio Networks," *IEEE Trans. Wireless Commun.* 8 (12), 5761–5766.

[42] Viswanathan, R. (2011), "Cooperative Spectrum Sensing for Primary User Detection in Cognitive Radio," Fifth International Conference on Sensing Technology, pp. 79–84.

[43] Cabric, D., Mishra, S., and Brodersen, R. W. (2004). "Implementation Issues in Spectrum Sensing for Cognitive Radios," Asilomar Conference on Signals, Systems and Computers, Vol. 1, pp. 772–776.

[44] Mishra, S. M., Sahai, A., and Brodersen R. W. (2006). "Cooperative Sensing among Cognitive Radios," IEEE International Conference on Communications, Vol. 4, pp. 1658–1663.

[45] Ma, J., and Li, Y. (2007). "Soft Combination and Detection for Cooperative Spectrum Sensing in Cognitive Radio Networks," IEEE Global Telecommunications Conference GLOBECOM, pp. 3139–3143.

[46] Guo, G., Peng, T., Xu, S., Wang, H., and Wang, W. (2009). "Cooperative Spectrum Sensing with Cluster-Based Architecture in Cognitive Radio Networks," IEEE 69th Vehicular Technology Conference, pp. 1–5.

[47] Zhang, W., Mallik, R. K., and Letaief, K. (2008). "Cooperative Spectrum Sensing Optimization in Cognitive Radio Networks," IEEE International Conference on Communications, pp. 3411–3415.

[48] Chaudhari, S., Lunden, J., Koivunen, V., and Poor, H. V. (2012). "Cooperative Sensing with Imperfect Reporting Channels: Hard Decisions or Soft Decisions?," *IEEE Trans. Signal Process.* 60 (1), 18–28.

[49] Lee, D. J. (2012). "Adaptive Cooperative Spectrum Sensing Using Random Access in Cognitive Radio Networks," IEEE 24th International Symposium on Personal Indoor and Mobile Radio Communications (PIMRC), pp. 1835–1839.

[50] Tian, Z., and Giannakis, G. (2007). "Compressive Sensing for Wideband Cognitive Radios," in Proc. IEEE International Conference on Acoustics, Speech, and Signal Processing, Honolulu, HI, USA, April 2007, pp. 1357–1360.

[51] Tian, Z., Tafesse, Y., and Sadler, B. M. (2012). "Cyclic Feature Detection with Sub-Nyquist Sampling for Wideband Spectrum Sensing", *IEEE J. Sel. Topics Sig. Proc.* 6 (1), Feb. 58–69.

[52] Zeng, F., Li, C., and Tian, Z. (2011). "Distributed Compressive Spectrum Sensing in Cooperative Multihop Cognitive Networks", *IEEE J. Sel. Topics in Signal Process.* 5 (1), Feb., 37–48.

[53] Tropp, J. A., Laska, J. N., Duarte, M. F., Romberg, J. K., and Baraniuk, R. G. (2010). "Beyond Nyquist: Efficient Sampling of Sparse Bandlimited Signals", *IEEE Trans. Information Theory*, 56 (1), Jan., 520–544.

[54] Mishali, M., and Eldar, Y. C. (2009). "Blind Multiband Signal Reconstruction: Compressive Sensing for Analog Signals", *IEEE Trans. Signal Process.* 57 (3), March, 993–1009.

[55] Venkataramani, R., and Bresler, Y. (2000). "Perfect Reconstruction Formulas and Bounds on Aliasing Error in sub-Nyquist Nonuniform Sampling of Multiband Signals", *IEEE Trans. Information Theory*, 46 (6), Sep. 2173–2183.

[56] Sun, H., Nallanathan, A., Wang, C. -X., Chen, Y. (2013). "Wideband Spectrum Sensing for Cognitive Radio Networks: A Survey", *Wireless Commun. IEEE*, 20 (2), 74–81, April.

[57] Chepuriy, S. P., Francisco, R., and Leusy, G. (2012). "Low-power Architecture for Wideband Spectrum Sensing", 3rd International Workshop on Cognitive Information Processing (CIP).

9

Compressive Sensing
for Wireless Sensor Networks

9.1 Introduction

Wireless sensor networks (WSNs) are spatially distributed autonomous sensors for monitoring physical or environmental conditions, such as temperature, sound, and pressure and for cooperatively passing their data through the network to a main location. The more modern networks are bi-directional, also enabling control of sensor activity. Though they are resource constrained based on the memory, bandwidth, and energy consumption, there is a wide variety of thrust applications which needs uninterrupted continuous monitoring of the remote areas, industries, battery field, home, machine health monitoring, etc. Hence, this chapter provides a brief overview of the WSN, its architecture, the associated wireless transmission technology, and protocols utilized along with the operating system being used for WSN. It also provides details about the application of CS to WSN and the significance related to it.

9.2 Sensor Networks Architecture

Deployment of WSN requires constituting nodes to be developed and available. These nodes have to meet the requirements coming from the specific requirements of a given application. They should be small, inexpensive, or energy efficient, equipped with the right sensors, the necessary computation, and memory resources. They also need adequate communication facilities. The elements required are discussed in this section.

9.2.1 Single-Node Architecture—Hardware Components

A wireless sensor node is characterized by its small size, ability to sense environmental phenomena through a set of transducers and a radio transceiver with autonomous power supply. The main components of a sensor node are a microcontroller, transceiver, external memory, power source, and one or more sensors which formalize the sensor-node architecture (Figure 9.1).

a) Controller

The controller performs tasks, processes data, and controls the functionality of other components in the sensor node, while the most common controller is a microcontroller, which is often used in many embedded systems such as sensor nodes due to its low cost, flexibility to connect to other devices, ease of programming, and low power consumption. A general-purpose microprocessor generally has power consumption higher than a microcontroller. Therefore, it is often not considered a suitable choice for a sensor node. Modern microcontrollers integrate flash storage, RAM, analog-to-digital converters, and digital I/O onto a single integrated circuit. Figure 9.2 shows the possible controllers available in the market. When selecting a microcontroller family, the key requirements to be considered are energy consumption, voltage requirements, cost, support for peripherals, and the number of external components required [1].

b) Transceiver

Transceiver circuitry enables the sensor mote to communicate with nearby units. Despite early projects being considered as using optical transmissions, current sensor hardware relies on RF communication. Optical communication is cheaper, is easier to construct, and consumes less power than RF, but requires visibility and directionality, which are extremely hard to provide in a sensor network. RF communication suffers from a high path loss and requires complex hardware, but is a more flexible and understood technology. WSNs tend to use license-free communication frequencies: 173, 433, 868, and 915 MHz; and 2.4 GHz. Figure 9.3 depicts the capabilities of current radio transceivers suitable for WSNs, their features, and power profile.

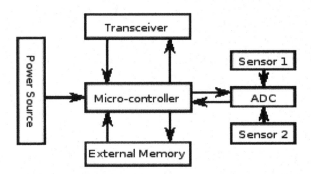

Figure 9.1   Basic architecture of sensor node.

Manufacturer	Device	RAM (kB)	Flash (kB)	Active (mA)	Sleep (µA)	Release
Atmel	AT90Ls8535	0.5	8	5	15	1998
	Mega 128	4	128	8	20	2001
	Mega165/325/645	4	64	2.5	2	2004
General Instruments	PIC	0.025	0.5	19	1	1975
Microchip	PIC Modern	4	128	2.2	1	2002
Intel	4004 4-bit	0.625	4	30	N/A	1971
	8051 8-bit Classic	0.5	32	30	5	1995
	8051 16-bit	1	16	45	10	1996
Philips	80 C51 16-bit	2	60	15	3	2000
Motorola	HC05	0.5	32	6.6	90	1988
	HC08	2	32	8	100	1993
	HCS08	4	60	6.5	1	2003
Texas Instruments	TSS400 4-bit	0.03	1	15	12	1974
	MSP430fl4x 16-bit	2	60	1.5	1	2000
	MSP430fl6x 16-bit	10	48	2	1	2004
Atmel	AT91 ARM Thumb	256	1024	38	160	2004
Intel	XScale PXA27X	256	N/A	39	574	2004

Figure 9.2   Features of different microcontrollers.

Currently available sensor nodes employ one of two types of radios. The simplest and cheaper alternative offers a basic carrier sense multiple access (CSMA) medium access control (MAC) protocol, operates in a license free band (315/433/868/916 MHz), and has a bandwidth in the range 20–50 kbps. Such radios usually offer a simple byte-oriented interface that permits software implementations of arbitrary (energy efficient) MAC protocols. Newer models support an 802.15.4 radio operating in the 2.4-GHz band and offering a 250-kbps bandwidth. The latter offers the possibility of using an internal (i.e., onboard) antenna which makes sensors more manageable and self-contained with respect to an external whip antenna. The radio range varies with a maximum of about 300 m (outdoor) for the first radio type and 125 m for the 802.15.4 radios.

The functionalities of both transmitter and receiver are combined into a single device known as a transceiver. Transceivers often lack unique identifiers. The operational states are transmit, receive, idle, and sleep. Most of the transceivers operating in idle mode have a power consumption almost equal to the power consumed in the receive mode. Thus, it is better to completely shut down the transceiver rather than leave it in the idle mode when it is not transmitting or receiving. A significant volume of power is used up when switching from sleep mode to transmit mode in order to transmit a packet.

Type	Narrowband				Wideband		
Vendor	RFM	Chipcon	Chipcon	Nordic	Chipcon	Motorola	Zeevo
Part no.	TR1000	CC1000	CC2400	nRF2401	CC2420	Mc13191_2	ZV4002
MaxData rate (kbps)	115.2	76.8	1000	1000	250	250	723.2
RX power(mA)	3.8	9.6	24	18(25)	19.7	37(42)	65
Tx power(mA/dBm)	12 / 1.5	16.5/10	19 / 0	13 / 0	17.4 / 0	34(30)/0	65 / 0
Powerdown power(μA)	1	1	1.5	0.4	0.58	1	140
Turn on time (ms)	0.02	2	1.13	3		20	*
Modulation	OOK/ASK	FSK	FSK,GFSK	GFSK	DSSS-O-QPSK	DSSS-O-QPSK	FHSS-GFSK
Packet detection	No	No	Programmable	Yes	Yes	Yes	Yes
Address decoding	No	No	No	Yes	Yes	Yes	Yes
Encryption support	No	No	No	No	128-bit AES	No	128-bit SC
Error detection	No	No	Yes	Yes	Yes	yes	Yes
Error correction	No	No	No	No	Yes	Yes	Yes
Acknowledgments	No	No	No	No	Yes	Yes	Yes
Interface	bit	Byte	Packet/byte	Packet /byte	Packet/byte	Packet/byte	Packet
Buffering (bytes)	No	1	32	16	128	133	Yes *
Time-sync	Bit	SFD/byte	SFD/packet	Packet	SFD	SFD	Bluetooth
Localization	RSSI	RSSI	RSSI	No	RSSI/LQI	RSSI/LQI	RSSI

*Manufacturer's documentation does not include additional information

Figure 9.3   Capabilities of current radio transceivers.

Hence, the radio subsystem is the most important system on a wireless sensor node since it is the primary energy consumer. Modern low-power and short-range transceivers consume between 15 and 300 mW of power when sending and receiving. A key hardware observation is that low-power radios consume approximately the same amount of energy when in receive or transmit mode. The factors that influence the choice of radio modules are transmission range, modulation type, power consumption, bit rate, and turn-on time.

c) External memory

From an energy perspective, the most relevant kind of memory is the on-chip memory of a microcontroller and Flash memory off-chip RAM is rarely, if ever, used. Flash memories are used due to their cost and storage capacity. Memory requirements are very much application dependent. Two categories of memory based on the purpose of storage are user memory used for storing application related or personal data, and program memory used for programming the device. Program memory also contains identification data of the device if present.

d) Power source

A wireless sensor node is a popular solution when it is difficult or impossible to run a mains supply to the sensor node. However, since the wireless sensor node is often placed in a hard-to-reach location, changing the battery regularly can be costly and inconvenient. An important aspect in the development of a wireless sensor node is ensuring that there is always adequate energy available to power the system. The sensor node consumes power for sensing, communicating, and data processing. More energy is required for data communication than any other process. Power is stored in either batteries or capacitors. Batteries, both rechargeable and non-rechargeable, are the main source of power supply for sensor nodes. They are also classified according to electrochemical material used for the electrodes such as NiCd (nickel–cadmium), NiZn (nickel–zinc), NiMII (nickel–metal hydride), and lithium–ion. Current sensors are able to renew their energy from solar sources, temperature differences, or vibration. Two power-saving policies used are dynamic power management (DPM) and dynamic voltage scaling (DVS). DPM conserves power by shutting down parts of the sensor node which are not currently used or active. A DVS scheme varies the power levels within the sensor node depending on the non-deterministic workload. By varying the voltage along with the frequency, it is possible to obtain quadratic reduction in power consumption.

e) Sensors

Sensors are hardware devices that produce a measurable response to a change in a physical condition like temperature or pressure. Sensors measure physical data of the parameter to be monitored. The continual analog signal produced by the sensors is digitized by an analog-to-digital converter and sent to controllers for further processing. As wireless sensor nodes are typically very small electronic devices, they can only be equipped with a limited power source of less than 0.5–2 Ah and 1.2–3.7 v. The last decade has seen an explosion in sensor technology. There are currently thousands of potential sensors ready to be attached to a wireless sensing platform. Additionally, advances in MEMS and carbon nanotubes technology are promising to create a wide array of new sensors. They range from simple light and temperature monitoring sensors to complex digital noses [2]. Figure 9.4 outlines a collection of common microsensors and their key characteristics.

Sensors are classified into three categories: passive, omnidirectional sensors; passive, narrow-beam sensors; and active sensors. Passive sensors sense the data without actually manipulating the environment by active probing. They are self-powered; that is, energy is needed only to amplify their analog signal. Active sensors actively probe the environment. A typical example is a sonar or radar sensor; they require continuous energy from a power source. Narrow-beam sensors have a well-defined notion of direction of measurement, similar to a camera. Omnidirectional sensors have no notion of direction involved in their measurements. The overall theoretical work on WSNs works with passive, omnidirectional sensors. Each sensor node has a certain area of coverage for which it can reliably and accurately report the particular quantity that it is observing. Several sources of power consumption in sensors are signal sampling and conversion of physical signals to electrical ones, signal conditioning, and analog-to-digital conversion. Spatial density

	Current	Discrete sample time	Voltage requirement	Manufacturer
Photo	1.9 mA	330 uS	2.7–5.5V	Taos
Temperature	1 mA	400 mS	2.5–5.5V	Dallas Semiconductor
Humidity	550 uA	300 mS	2.4–5.5V	Sensirion
Pressure	1 mA	35 mS	2.2V–3.6V	Intersema
Magnetic fields	4 mA	30 uS	Any	Honeywell
Acceleration	2 mA	10 mS	2.5–3.3V	Analog Devices
Acoustic	5 mA	1 mS	2–10V	Panasonic
Smoke	5 mA	--	6–12V	Motorola
Passive IR (motion)	0 mA	1 mS	Any	Melixis
Photosynthetic light	0 mA	1 mS	Any	Li-Cor
Soil moisture	2 mA	10 mS	2–5V	Ech2o

Figure 9.4   Characteristics of different microsensors.

of sensor nodes in the field may be as high as 20 nodes per cubic meter. The factors that characterize the sensor selection are interfaces, sample rate, turn-on time, and voltage requirements.

9.2.2 Network Architecture—Sensor Network Scenarios

The network has to be deployed to enable satisfaction of two main objectives: coverage and connectivity. Coverage pertains to the application-specific quality of information obtained from the environment by the networked sensor devices. Connectivity pertains to the network topology over which information routing can take place. Other issues, such as equipment costs, energy limitations, and the need for robustness, should also be taken into account. A number of basic questions must be considered when deploying a wireless sensor network [2]:

1. **Structured versus randomized deployment**: Does the network involve (a) structured placement, either by hand or via autonomous robotic nodes, or (b) randomly scattered deployment?
2. **Over-deployment versus incremental deployment**: For robustness against node failures and energy depletion, should the network be deployed a priori with redundant nodes, or can nodes be added or replaced incrementally when the need arises? In the former case, sleep scheduling is desirable to extend network lifetime.
3. **Network topology**: Is the network topology going to be a simple star topology, or a grid, or an arbitrary multi-hop mesh, or a two-level cluster hierarchy? What kind of robust connectivity guarantees is desired?
4. **Homogeneous versus heterogeneous deployment**: Are all sensor nodes of the same type or is there a mix of high- and low-capability devices? In the case of heterogeneous deployments, there may be multiple gateway/sink devices (nodes to which sensor nodes report their data and through which an external user can access the sensor network).
5. **Coverage metrics**: What is the kind of sensor information desired from the environment and how is the coverage measured? This could be on the basis of detection and false alarm probabilities or whether every event can be sensed by K distinct nodes, etc.

According to the Zigbee specifications, three different kinds of nodes can be used in a wireless network, namely (i) a router, (ii) a coordinator, and (iii) an end device. The coordinator can create the network, exchange the parameters used by the other nodes to communicate (e.g., network ID, beginning of a transmitted frame), relay packets received from remote nodes toward the correct destination, and collect data from the sensors. Only a single coordinator can be used in a network. A router, instead, relays the received packets and the control messages (in order to increase the network diameter), manages the routing tables, and, if required, can also collect data from a sensor [3]. The main difference between a coordinator and a router is that the former can create the network, while the latter cannot. Both these types of nodes are referred to as *full function devices* (FFDs): They can develop all the functions required by the Zigbee standard for setting up and managing the communication. On the other hand, end devices, also referred to as *reduced function devices* (RFDs), can act only as remote peripherals, which collect values from sensors and send them to the coordinator or other remote nodes. However, RFDs are not involved in network management and, therefore, cannot send or relay control messages.

The three different kinds of network topologies are shown in Figure 9.5: (i) *star*, (ii) *cluster-tree*, and (iii) *mesh*. (i) In a *star* network, there is a coordinator and one or many RFDs (end nodes) or FFDs (routers) which send messages directly to the coordinator (up to 65536 RFDs or FFDs). (ii) In a *cluster-tree* topology, instead, there are a coordinator which acts as a root and either RFDs or routers connected to it, for increasing the network dimension. The RFDs can only be the leaves of the tree, whereas the routers can also act as branches. In a cluster-tree topology, a beacon structure can be employed for obtaining improved battery conservation. (iii) In a *mesh* network, any source node can talk directly to any destination. The routers and the coordinator, in fact, are connected to each other, within their transmission ranges, in order to ease packet routing. The radio receivers at the coordinator and routers must be "on" all the time.

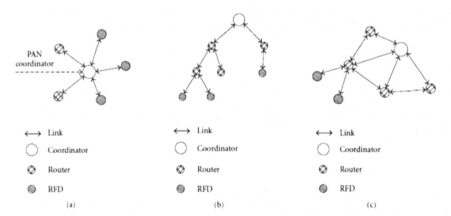

Figure 9.5 Possible network topologies (a) star, (b) cluster-tree, and (c) mesh.

9.2.3 Design Principles

Wireless sensor networks are interesting from an engineering perspective, as they present a number of serious challenges that cannot be adequately addressed by existing technologies:

Extended lifetime: WSN nodes generally face severe energy constraints due to the limitations of batteries. A typical alkaline battery, for example, provides about 50 W h of energy. This may translate to less than a month continuous operation for each node in a full active mode. Given the expense and potential infeasibility of monitoring and replacing batteries for a large network, much longer lifetimes are desired. Hardware improvements in battery design and energy harvesting techniques offer only limited solutions. This is the reason why most protocol designs in wireless sensor networks are designed explicitly with energy efficiency as the primary goal. Naturally, this goal must be balanced against a number of other concerns.

Responsiveness: A simple solution to extend network lifetime is to operate the nodes in a duty-cycled manner with periodic switching between sleep and wake-up modes. While synchronization of such sleep schedules is challenging by itself, a larger concern is that arbitrarily long sleep periods can reduce the responsiveness and effectiveness of the sensors. In applications where the detection and rapid development of certain events in the environment are critical, the latency induced by sleep schedules must be kept within strict bounds, even in the presence of network congestion.

Robustness: The vision of wireless sensor networks is to provide a large-scale, yet fine-grained, coverage. This motivates the use of large numbers of inexpensive devices which, however, can often be unreliable and prone to failures. Rates of device failure will also be high whenever the sensor devices are deployed in harsh or hostile environments. Protocol designs should, therefore, have built-in mechanisms to provide robustness. It is important to ensure that the global performance of the system is not sensitive to individual device failures. Further, it is often desirable that the performance of the system degrades as gracefully as possible with respect to component failures.

Synergy: Moore's law-type advances in technology have ensured rapid improvement in device capabilities in terms of processing power, memory, storage, radio transceiver performance, and even accuracy of sensing (given a fixed cost). However, when economic considerations dictate drastic reduction in cost per node, it is possible for the capabilities of individual nodes to remain constrained to some extent. The challenge is, therefore, to design synergistic protocols, which ensure higher capability of the system as a whole than the sum of the capabilities of its individual components.

Scalability: For many envisioned applications, a combination of fine granularity sensing and large coverage area implies the potential of having an extremely large-scale wireless sensor networks (tens of thousands, perhaps even millions of nodes in the long term). Protocols have to be inherently distributed, involving localized communication, while sensor networks must utilize hierarchical architectures for providing such scalability. However, visions of large numbers of nodes remain unrealized in practice until some fundamental problems, such as failure handling and *in-situ* reprogramming, are addressed even in small settings, involving tens to hundreds of nodes. There are also some basic limits on the throughput and capacity that impact the scalability of network performance.

Heterogeneity: There is heterogeneity of device capabilities (with respect to computation, communication, and sensing) in realistic settings. This heterogeneity can have a number of important design consequences. For instance, the presence of a small number of devices of higher computational capability along with a large number of low-capability devices can dictate a two-tier, cluster-based network architecture, while the presence of multiple sensing modalities requires pertinent sensor fusion techniques. A key challenge is often to determine the right combination of heterogeneous device capabilities for a given application.

Self-configuration: Wireless sensor networks are inherently unattended distributed systems due to their scale and the nature of their applications. Autonomous operation of the network is, therefore, a key design challenge. Nodes in a wireless sensor network should have the ability to configure their own network topology; localize, synchronize, and calibrate themselves; coordinate inter-node communication; and determine other important operating parameters.

Self-optimization and adaptation: Traditionally, most engineering systems are optimized a priori to operate efficiently in the face of expected or well-modeled operating conditions. In wireless sensor networks, there may often be significant uncertainty about operating conditions prior to deployment. Under such conditions, it is important that there must be in-built mechanisms for autonomous learning from sensor and network measurements collected over time and to use this learning for improvement of performance on a continuous basis. Moreover, WSN protocols should also be able to adapt to such environmental dynamics in an online manner.

Systematic design: Wireless sensor networks can often be perceptibly highly application specific. There is a challenging trade-off between (a) ad hoc, narrowly applicable approaches that exploit application-specific characteristics to offer performance gains and (b) more flexible, easy-to-generalize design methodologies that sacrifice some performance. While performance optimization is very important, given the severe resource constraints in wireless sensor networks, systematic design methodologies, allowing for reuse, modularity, and run-time adaptation are necessitated by practical considerations.

Privacy and security: The large-scale prevalence and sensitivity of the information collected by wireless sensor networks (as well as their potential deployment in hostile locations) give rise to the final key challenge of ensuring both privacy and security.

9.2.4 Development of Wireless Sensor Networks

Network has to be developed on the basis of the requirement of applications. If the aim is to minimize the duration and frequency of long-range communications, and limit such communications to the transmission of events (i.e., alarms, alerts, node health status), care must be taken to make the appropriate choice of the nodes in the network. The following methodology can be adopted:

1. Deploy a low-level prototype node in a controlled external environment to measure baseline operational parameters of the sensors and communications components. This has a simple self-diagnostic rule set and event-generating capability.
2. Develop a medium-level node that has full system functionality. This will have an adaptive self-diagnostic rule-set, full event-generating capability, and ability to receive updates from the management system. It will also incorporate two processors: a low-power microprocessor that performs low-level tasks and a higher power processor which is powered up intermittently to perform more CPU-intensive tasks.
3. Using data from the low-level and medium-level nodes, the hardware architecture and system parameters can be fully optimized for the high-level node, for maximizing the node lifetime.

9.2.5 Energy Consumption of Sensor Nodes

In case the of sensor networks, the task of providing a simple yet realistic energy model is relatively simple, as compared to the case of ad hoc networks. In fact, sensor networks are typically composed of homogeneous devices, which are usually very simple. Furthermore, since many sensor nodes have been designed in the research community, their features are very well known.

Table 9.1   Measured power consumption of a Rockwell's WINS sensor node

MCU Mode	Sensor Mode	Radio Mode	Total Power (mW)
On	On	Tx (power 36.3 mW)	1080.5
On	On	Tx (power 0.12 mW)	771.1
On	On	Rx	751.6
On	On	Idle	727.5
On	On	Sleep	416.3
On	On	Removed	383.3
Sleep	On	Removed	64.0

As a result, several sets of energy consumption measurements of wireless sensor nodes have been reported in literature [4]. Table 2.3 reports the power dissipation of a Rockwell's WINS sensor node [5]. The node is composed of three main components: the microcontroller unit (MCU), the sensing apparatus (sensor), and the wireless radio. When the power consumption of the wireless radio alone is considered, the following ratio is obtained (sleep: idle: receive: transmit) 0.09:1:1.07:2.02. These ratios are quite similar to the case of 802.11 wireless cards, except for higher power consumption, while the radio is transmitting at maximum power. When the transmit power is minimum (0.12 mW), the idle: transmit ratio in the WINS sensor is 1.12. So, there is an almost twofold increase in power consumption when varying the transmit power from the minimum to the maximum value. Hence, it is evident that varying the transmit power level has a considerable effect on the node's energy consumption.

9.3 Wireless Transmission Technology and Systems

9.3.1 Physical Layer and Transceiver Design Consideration

The physical layer is mostly concerned with modulation and demodulation of digital data. This task is carried out by the so-called **transceivers**. In sensor networks, the challenge is to find modulation schemes and transceiver architectures that are simple, of low cost, but still robust enough to provide the desired service. Some of the most crucial points influencing the PHY design in wireless sensor networks are as follows:

- Low power consumption.
- As a consequence: small transmit power and thus a small transmission range.
- Low duty cycle: Most hardware should be switched off or operated in a low-power standby mode for most of the time.

- Comparably low data rates, on the order of tens to hundreds kilobits per second, required.
- Lesser implementation complexity and cost.
- Low degree of mobility.
- A small form factor for the overall node.

In general, the challenge in sensor networks lies in finding modulation schemes and transceiver architectures that are simple, low-cost but still robust enough to provide the desired service [1].

9.3.2 Energy Usage Profile

The choice of a small transmit power leads to energy consumption profile different from the other wireless devices such as cell phones. First, the radiated energy is small, typically of the order of 0 dBm (corresponding to 1 mW). On the other hand, the overall transceiver (RF front end and baseband part) consumes much more energy than is actually radiated; Wang et al. [6] estimate that a transceiver working at frequencies beyond 1 GHz takes 10–100 mW of power to radiate 1 mW. The MICA motes consume 21 mW in the transmit mode and 15 mW in the receive mode [7].

A second key observation is the consumption of more or less the same power for the transmit and receive modes for small transmit powers. It is even possible that reception requires more power than transmission [8, 9]. Depending on the transceiver architecture, power consumption of the idle mode can be less or in the same range as the receive power [8]. Therefore, it is important to put the transceiver into a sleep state instead of just idling. However, there is the problem of the **startup energy/startup time**, which a transceiver has to spend following waking up from a sleep mode. No transmission or reception of data is possible during this startup time [9]. Therefore, going into sleep mode is unfavorable when the next wakeup comes fast.

A third key observation is the relative cost of communication versus computation in a sensor node. Clearly, a comparison of these costs depends for the communication part on the BER requirements, range, transceiver type, and so forth, and for the computation part on the processor type, the instruction mix, and so on.

9.3.3 Choice of Modulation

A crucial point is the choice of a modulation scheme. Several factors have to be balanced here: the required and desirable data rate and symbol rate, the implementation complexity, the relationship between radiated power and target BER, and the expected channel characteristics. Maximizing the time a receiver can spend in the sleep mode requires minimizing the transmission time. The higher the data rate offered by a transceiver/modulation, the smaller the time needed to transmit a given volume of data and, consequently, the smaller the energy consumption.

Power consumption of a modulation scheme depends much more on the symbol rate than on the data rate [10]. Obviously, the desire for "high" data rates at "low" symbol rates calls for *m*-ary modulation schemes. However, there are trade-offs:

- *m*-ary modulation requires more complex digital and analog circuitry than 2-ary modulation [9], for example, to parallelize user bits into *m*-ary symbols.
- Many *m*-ary modulation schemes require an increased E_b/N_0 ratio for increasing *m* and consequently an increased radiated power for achieving the same target BER, while others become less and less bandwidth efficient. This is shown as example for coherently detected *m*-ary FSK and PSK in Table 4.3, where, for different values of *m*, the achieved bandwidth efficiencies and the E_b/N_0 required to achieve a target BER of 10^{-6} are displayed. However, in wireless sensor network applications with only low-to-moderate bandwidth requirements, a loss in bandwidth efficiency can be more tolerable than an increased radiated power to compensate E_b/N_0 losses.
- In many wireless sensor network applications, most packets are expected to be short, in the order of tens to hundreds of bits. For such packets, the startup time easily dominates overall energy consumption, causing irrelevance in any efforts in reducing the transmission time by choosing *m*-ary modulation schemes.

Table 9.2 Bandwidth efficiency ηBW and Eb/N_0 [dB] required at the receiver to reach a BER of 10^{-6} over an AWGN channel for *m*-ary orthogonal FSK and PSK (adapted from reference [11])

M	2	4	8	16	32	64
m-ary PSK: η_{BW}	0.5	1.0	1.5	2.0	2.5	3.0
m-ary PSK: E_b/N_0	10.5	10.5	14.0	18.5	23.4	28.5
m-ary FSK: η_{BW}	0.40	0.57	0.55	0.42	0.29	0.18
m-ary FSK: E_b/N_0	13.5	10.8	9.3	8.2	7.5	6.9

9.4 Protocols for WSNs

A common challenge in any wireless network is the collision that occurs when any two nodes transmit data simultaneously over the transmission medium. The key concern of protocol design in wireless sensor networks is energy consumption, as sensor nodes are battery powered.

9.4.1 Medium Access Control Protocol

In general, wireless communication has a variety of MAC protocols, which can be classified into distinct groups on the basis of different criteria. Based on whether a central controller is involved in coordination, WSNs' MAC protocols can be categorized as centralized, distributed (decentralized), and hybrid. Basically, hybrid protocols attempt to combine the advantages of centralized and distributed schemes, but the relevant algorithm is more complex. Figure 9.6 shows the classification of MAC protocol for WSN.

9.4.1.1 Centralized MAC protocols

Centralized MAC protocols include polling algorithms and controlled multiplexing algorithms. A centralized controller is required for coordinating the channel access among different nodes and hence collision-free operation can be achieved. Thus, energy wasted due to collisions can be eliminated. However, due to high overhead and long delay, pure polling mechanisms are not suitable for large-scale WSNs. A controlled multiplexing mechanism such as frequency-division multiplexing access (FDMA), code-division

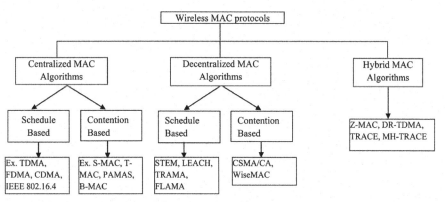

Figure 9.6 Classification of WSN MAC protocol.

multiplexing (CDMA), or time-division multiplexing access (TDMA) can be used depending on the manner in which bandwidth is assigned. This class of protocols is preferable in WSNs, since it is collision free and nodes can be turned off in unassigned slots, thus reducing energy expenditure arising out of idle sensing and overhearing. However, drawbacks do exist in channel partitioning schemes. When TDMA is used, the central controller consumes more energy than other nodes, and scheduling tends to be dynamic, leading to a more complex mechanism. Moreover, it requires clock synchronization among all nodes, which can dissipate some extra energy. On the other hand, when FDMA is used, it is not realistic to assign a unique frequency for each individual node due to the limited bandwidth in the system. Furthermore, bandwidth wastage occurs due to the low duty cycle. Similarly, when CDMA is used, all nodes can transmit at will, but some overhead results for each node encodes its data bits with its uniquely assigned code.

This scheme, therefore, lacks flexibility and scalability to adapt to the variation of WSN applications. Some efforts have been made to improve the performance in terms of energy efficiency. One way is to combine TDMA with other controlled multiplexing, such as self-organizing medium access control for sensor networks (SMACS), which is a combination of TDMA/FDMA MAC protocol. Low-energy adaptive clustering hierarchical (LEACH), on the other hand, combines TDMA and CDMA protocols; i.e., it uses TDMA protocol to prevent intra-cluster collisions and CDMA to avoid inter-cluster collisions.

Adaptive periodic threshold-sensitive energy-efficient sensor network protocol (APTEEN) also uses TDMA with CDMA. However, it adopts a modified TDMA in which the duration of slots assigned to idle nodes and sleeping nodes is different, and all the idle node slots are ordered to precede sleeping nodes.

9.4.1.2 Distributed MAC protocols

Distributed MAC protocols usually provide random multiple access to a wireless medium. Most prevailing MAC protocols in this category adapt carrier sensing and collision avoidance; i.e., it is based on CSMA/CA. A significant number of collisions can be eliminated through carrier sensing by deferring transmission when the channel is detected busy. Some collision avoidance measures can be taken up to enable further decrease in the probability of collisions such as a random back-off procedure as specified in IEEE 802.11 distributed coordination function (DCF). However, in some cases, this results in "hidden" and "exposed" terminal problems, which have a great impact on efficiency. Two types of CSMA/CA-based schemes have been proposed for solving these problems.

DCF, the exchange of "request to send and clear to send" (RTS-CTS) control messages, reserves the transmission space for subsequent data exchange, thereby eliminating the hidden terminal problem. Another scheme called dual busy tone multiple access (DBTMA) method that separates control and data channels to relieve the problems raised by hidden and exposed terminals by indicating the transmission or receiving status explicitly.

Using distributed MAC protocols, nodes operate in a decentralized manner. It is easy to implement and perform more flexible and scalable control mechanisms, which may suit the requirements of WSNs. However, the protocols which belong to this category are not collision-free. The listen-before-talk scheme calls for all nodes to keep sensing the channel. This results in high energy wastage due to collisions, idle listening, overhearing, and control message overhead.

9.4.1.3 Hybrid MAC protocols

Conventional centralized and distributed MAC layer protocols cannot provide optimal results in terms of energy efficiency in WSNs. Hybrid MAC protocols integrate the controllability of centralized protocols with the flexibility of distributed protocols.

9.4.2 Routing Protocols

Many routing protocols have been specifically designed for WSNs wherein energy awareness is an essential design issue. The focus, however, has been on the routing protocols which might differ depending on the application and network architecture. The design of routing protocols in WSNs is influenced by many challenging factors such as node deployment, energy consumption without losing accuracy, data reporting model, node/link heterogeneity, fault tolerance, scalability, network dynamics, connectivity, coverage, data aggregation, and quality of service. Generally, the routing techniques are classified into three categories based on the underlying network structure:

- Flat,
- Hierarchical, and
- Location-based routing.

Furthermore, these protocols can be classified into multi-path-based, query-based, negotiation-based, QoS-based, and coherent-based depending on the protocol operation.

9.4.2.1 Gossiping and agent-based unicast forwarding

These schemas are an attempt of working without routing tables in order for minimizing the overflow needed to build the tables. The simplest choice is flooding (forwarding each message received), but it is not very efficient. Gossiping avoids the problem of implosion by selecting a random node to which the packet has to be sent rather than blind broadcasting of the packet. However, this causes delays in propagation of data through the nodes [1].

9.4.2.2 Data-centric routing

Data-centric routing is one of the flat routing protocols, where the base station sends queries to certain regions and waits for data from the sensors located in the selected regions. As data are being requested through queries, attribute-based naming is necessary to specify the properties of data. Some of the protocols available under this category are discussed below.

Sensor protocols for information via negotiation (SPIN): Conventional protocols such as flooding- or gossiping-based routing protocols involve energy and bandwidth wastage when sending extra and unnecessary copies of data by sensors covering overlapping areas. The Classical flooding approach has deficiencies such as implosion, overlap, and resource blindness. SPIN addresses these problems and disseminates all the information at each node to every other node in the network assuming that all the nodes in the network are potential base stations. This enables a user to query any node and get the required information immediately. These protocols make use of the property of having similar data in nodes in close proximity, and hence, there is a need for distribution of the data that other nodes do not possess. The SPIN family of protocols uses data negotiation and resource-adaptive algorithms. Nodes running SPIN assign a high-level name for complete description of their collected data (called meta-data) and perform negotiations before transmission of any data. The semantics of the meta-data format is application-specific and is not specified in SPIN. Moreover, SPIN has access to the current energy level of the node and adapts the protocol it is running based on how much energy is remaining. The SPIN family of protocols is designed based on two basic ideas:

- Sensor nodes operate more efficiently and conserve energy by sending data that describe the sensor data instead of sending all the data.
- Nodes in a network must monitor and adapt to changes in their own energy resources to extend the operating lifetime of the system.

SPIN's metadata negotiation solves the classic problems of flooding, thereby achieving a lot of energy efficiency. SPIN nodes use three types of messages to communicate:

- ADV—new data advertisement. When a SPIN node has a data to share, it can advertise this fact by transmitting an ADV message containing metadata.
- REQ—request for data. A SPIN node sends a REQ message when it wishes to receive some data.
- DATA—data message. DATA messages contain actual sensor data with a metadata header.

In view of ADV and REQ messages containing only metadata, they are smaller, and cheaper to send and receive, than their corresponding DATA messages.

SPIN applications are resource-aware and resource-adaptive. They can poll their system resources to find out how much energy is available to them. They can also calculate the cost in terms of energy of performing computations and sending and receiving data over the network. The SPIN family of protocols includes many protocols. The two main protocols are called SPIN-1 and SPIN-2, which incorporate negotiation before transmitting data in order to ensure that only useful information will be transferred. Also, each node has its own resource manager which keeps track of resource consumption and is polled by the nodes before data transmission. The SPIN-1 protocol is a 3-stage protocol, as described below.

The SPIN-1 protocol is a simple handshake protocol for disseminating data through a lossless network. It works in three stages (ADV-REQ-DATA), with each stage corresponding to one of the messages described earlier. The protocol starts when a node obtains new data that it is willing to disseminate. It does this by sending an ADV message to its neighbors, naming the new data (ADV stage). Upon receiving an ADV, the neighboring node checks to see whether it has already received or requested the advertised data. If not, it responds by sending an REQ message for the missing data back to the sender (REQ stage). The protocol completes when the initiator of the protocol responds to the REQ with a DATA message, containing the missing data (DATA stage) [1].

An extension to SPIN-1 is SPIN-2, which incorporates threshold-based resource awareness mechanism in addition to negotiation. When energy in the nodes is abundant, SPIN-2 communicates using the 3-stage protocol of SPIN-1. However, when the energy in a node starts approaching a low-energy threshold, it reduces its participation in the protocol; i.e., it participates only when it believes that it can complete all the other stages of the protocol without going below the low-energy threshold. These protocols are well-suited for an environment where the sensors are mobile because they base their forwarding decisions on local neighborhood information.

One of the advantages of SPIN lies in the localization of the topological changes, since each node needs to know only its single-hop neighbors. SPIN provides much more energy savings than flooding, and metadata negotiation almost halves the redundant data. However, SPIN's data advertisement mechanism cannot guarantee delivery of the data.

Directed diffusion: Directed diffusion is a data-centric (DC) protocol in which data are named using attribute–value pairs. The main idea of the DC paradigm is to combine the data coming from different sources en route (in-network aggregation) by eliminating redundancy, minimizing the number of transmissions which in turn saves network energy and protocols its lifetime. Unlike traditional end-to-end routing, DC routing finds routes from multiple sources to a single destination that allows in-network consolidation of redundant data.

In directed diffusion, sensors measure events and create gradients of information in their respective neighborhoods. The base station requests data by broadcasting interests. Interest describes a task required to be done by the network. Interest diffuses through the network hop-by-hop and is broadcast by each node to its neighbors. As the interest is propagated throughout the network, gradients are set up to draw data satisfying the query toward the requesting node; i.e., a sink may query for data by disseminating interests and intermediate nodes propagate these interests. Each sensor that receives the interest sets up a gradient toward the sensor nodes from which it receives the interest.

This process continues until gradients are set up from the sources back to the sink. More generally, a gradient specifies an attribute value and a direction. The strength of the gradient may be different with different neighbors, resulting in different amounts of information flow. At this stage, loops are not checked, but are removed at a later stage. Figure 9.7 shows an example of the working of directed diffusion (a) interest propagation and gradient setup, (b) data propagation, and (c) data delivery along the reinforced path. When interests fit gradients, paths of information flow are formed from multiple paths and the best paths are then reinforced so as to prevent further flooding according to a local rule. Data are aggregated on the way for reducing communication costs. The goal is to find a good aggregation tree which gets the data from source nodes to the sink. The sink periodically refreshes and resends the interest when it starts to receive data from the source. This is necessary as interests are not reliably transmitted throughout the network.

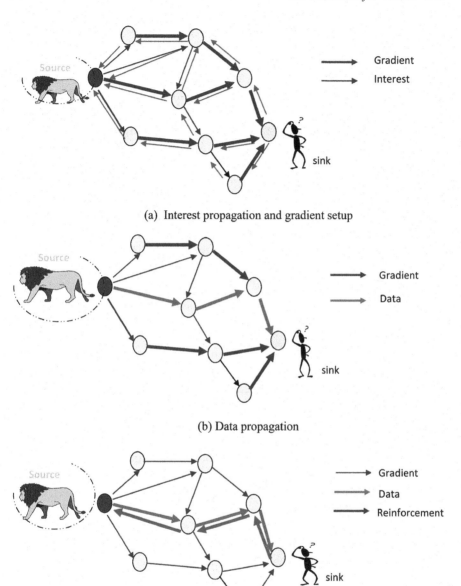

(a) Interest propagation and gradient setup

(b) Data propagation

(c) Data delivery along reinforced path

Figure 9.7 An example of interest diffusion in sensor network.

Directed diffusion differs from SPIN in two aspects. First, directed diffusion issues on demand data queries as the BS send queries to the sensor nodes by flooding some tasks. In SPIN, however, sensors advertise the availability of data allowing interested nodes to query that data. Second, all communication in directed diffusion is neighbor-to-neighbor with each node having the capability of performing data aggregation and caching. Unlike SPIN, there is no need to maintain global network topology in directed diffusion. However, directed diffusion may not be applied to applications that require continuous data delivery to the BS. This is because the query-driven on demand data model may not help in this regard. Moreover, matching data to queries may require some extra overhead at the sensor nodes.

Rumor routing: Rumor routing is a variation of directed diffusion and is mainly intended for applications where geographic routing is not feasible. The key idea is to route the queries to the nodes that have observed a particular event rather than flooding the entire network for retrieving information on the occurring events. In order to flood events through the network, the rumor routing algorithm employs long-lived packets, called agents. When a node detects an event, it adds such event to its local table, called event table, and generates an agent. Agents travel over the network for propagating information about the local events to distant nodes. When a node generates a query for an event, the nodes that know the route may respond to the query by inspecting its event table. Hence, there is no need to flood the entire network, which reduces communication cost. On the other hand, rumor routing maintains only one path between source and destination as opposed to directed diffusion where data can be routed through multiple paths at low rates. However, rumor routing performs well only when the number of events is small. Moreover, the overhead associated with rumor routing is controlled by different parameters used in the algorithm such as time-to-live (TTL) pertaining to queries and agents. Since the nodes become aware of events through the event agents, the heuristic for defining the route of an event agent highly affects the performance of next hop selection in rumor routing. When an agent finds a node whose route to an event is more expensive than its own, it will update the node's routing table to the more efficient path. So it is not necessary to produce more than a few agents for each event, since the trail will be picked up and propagated by other agents.

Minimum cost forwarding algorithm (MCFA): The MCFA exploits the fact that the direction of routing is always known, that is, toward the fixed external base station. Hence, a sensor node need not have a unique ID nor maintain a routing table. Instead, each node maintains the least cost estimate from itself to the base station. Each message to be forwarded by the sensor node is broadcast to its neighbors. When a node receives the message, it checks whether it is on the least cost path between the source sensor node and the base station. If this is the case, it re-broadcasts the message to its neighbors. This process is repeated until the base station is reached. In MCFA, each node should know the least cost path estimate from itself to the base station. This is obtained as follows [1].

COUGAR: Another data-centric protocol called COUGAR views the network as a huge distributed database system. The key idea is to use declarative queries in order to abstract, query processing from the network layer functions such as selection of relevant sensors and so on. COUGAR utilizes in-network data aggregation to obtain greater energy savings. The abstraction is supported through an additional query layer that lies between the network and application layers. COUGAR incorporates an architecture for the sensor database system where sensor nodes select a leader node for performing aggregation and transmit the data to the BS. The BS is responsible for generating a query plan, which specifies the necessary information on the data flow and in-network computation for the incoming query and send it to the relevant nodes. The query plan also describes how to select a leader for the query. However, COUGAR has some drawbacks. First, the addition of a query layer on each sensor node may add an extra overhead in terms of energy consumption and memory storage. Second, synchronization among nodes is required before sending the data to the leader node for obtaining successful in-network data computation. Third, the leader nodes should be dynamically maintained to prevent them from being hot spots (failure prone).

Active query forwarding in sensor networks (ACQUIRE): ACQUIRE views the network as a distributed database where complex queries can be further divided into several sub-queries. Operation of ACQUIRE can be described as follows. The BS node sends a query, which is then forwarded by each node receiving the query. Depending on the applications, there are likely to be different kinds of queries in these sensor networks. Figure 9.8 shows the types of queries which can be categorized in many ways, for example:

Continuous	One-shot
Aggregate	Non-aggregate
Complex	Simple
For replicated data	For unique data

Figure 9.8 Categorization of queries in sensor networks.

- Continuous queries: which result in extended data flows versus one-shot queries, which have a simple response.
- Aggregate queries: which require the aggregation of information from several sources versus non-aggregate queries which can be responded to by a single node.
- Complex queries: which consist of several sub-queries that are combined by conjunctions or disjunctions in an arbitrary manner versus simple queries, which have no sub-queries (e.g., "what is the value of the variable X?").
- Queries for replicated data: in which the response to a given query can be provided by many nodes and queries for unique data, in which the response to a given query can be provided only by one node.

During this process, each node tries to respond to the query partially by using its pre-cached information and then forwards it to another sensor node. If the pre-cached information is not up-to-date, the nodes gather information from their neighbors within a look ahead of d hops. Once the query is being resolved completely, it is sent back through either the reverse or shortest path to the BS. Hence, ACQUIRE can deal with complex queries by allowing many nodes to send responses. It can also provide efficient querying by adjusting the value of the look-ahead parameter.

9.4.2.3 Energy-aware routing

Energy-aware routing is a reactive routing protocol. It is a destination-initiated protocol where the consumer of data initiates the route request and maintains the route subsequently. Multiple paths are maintained from source to destination. These paths are maintained and chosen by means of a certain probability. The value of this probability depends on how low the energy consumption of each path can be achieved. Through paths chosen on different occasions, the energy of any single path will not deplete quickly. This can achieve longer network lifetime as energy is dissipated equally by all nodes. Network survivability is the main metric of this protocol. The protocol assumes that each node is addressable through a class-based addressing which includes the location and types of nodes.

The protocol operates in three phases:

- Setup phase or interest propagation—Localized flooding occurs to find all the routes from source to destination and their energy costs. This is when routing (interest) tables are built up.
- Data communication phase or data propagation—Data are sent from source to destination, using the information from the earlier phase. This

is when paths are chosen probabilistically according to the energy costs calculated earlier.

- Route maintenance—Route maintenance is minimal. Localized flooding is performed infrequently from destination to source to keep all the paths alive.

A. *Setup phase*

1. The destination node initiates the connection by flooding the network in the direction of the source node. It also sets the "cost" field to zero before sending the request.

$$\cos t(N_D) = 0$$

2. Every intermediate node forwards the request only to the neighbors that are closer to the source node than itself and farther away from the destination node. Thus, at a node N_i, the request is sent only to a neighbor N_j which satisfies:

$$d(N_i, N_S) \geq d(N_j, N_S)$$
$$d(N_i, N_D) \leq d(N_j, N_D)$$

where d (N_i, N_j) is the distance between N_i and N_j.

3. On receiving the request, the energy metric for the neighbor that sent the request is computed and is added to the total cost of the path. Thus, if the request is sent from node N_i to node N_j, N_i calculates the cost of the path as follows:

$$C_{N_j, N_i} = \text{Cost}(N_i) + \text{Metric}(N_j, N_i)$$

4. Paths that have a very high cost are discarded and not added to the forwarding table. Only the neighbors N_i with low cost paths are added to the forwarding table FT_j of N_j.

$$FT_j = \left\{ i \mid C_{N_j}, N_i \leq \alpha.\left(\frac{mn}{k} C_{N_j}, N_k\right) \right\}$$

5. Node N_j assigns a probability to each of the neighbors N_i in the forwarding table FT_j, with the probability inversely proportional to the cost.

$$P_{N_j, N_i} = \frac{1/C_{N_j, N_i}}{\sum_{k \in FT_j} 1/C_{N_j, N_k}}$$

6. Thus, each node N_j has a number of neighbors through which it can route packets to the destination. N_j then calculates the average cost of reaching the destination using the neighbors in the forwarding table.

$$\text{cost}(N_j) = \sum_{i \in FT_i} P_{.N_j}, \, N_i C_{N_j}, \, N_i$$

7. This average cost, $\text{cost}(N_j)$, is set in the ``cost'' field of the request packet and forwarded toward the source node as in Step 2.

B. Data communication phase

1. The source node sends the data packet to any of the neighbors in the forwarding table, with the probability of the neighbor being chosen equal to the probability in the forwarding table.
2. Each of the intermediate nodes forwards the data packet to a randomly chosen neighbor in its forwarding table, with the probability of the neighbor being chosen equal to the probability in the forwarding table.
3. This is continued till the data packet reaches the destination node.

Energy metric—The energy metric that is used to evaluate routes is provided

$$C_{ij} = e_{ij}^{\alpha} \, R_i^{\beta}$$

Here, C_{ij} is the cost metric between nodes i and j, e_{ij} is the energy used for transmitting and receiving on the link, while R_i is the residual energy at node i normalized to the initial energy of the node. The weighting factors α and β can be chosen for finding the minimum energy path or the path with nodes having the most energy or a combination of the above. Further study needs to be done for the best metric as it has a deep impact on the protocol performance. When compared to directed diffusion, this protocol provides an overall improvement of 21.5% energy saving and a 44% increase in network lifetime. However, it requires the gathering of location information and setting up the addressing mechanism for the nodes, which complicates route setup.

9.4.2.4 Gradient-based routing

Gradient-based routing (GBR) protocol memorizes the number of hops, while the interest is diffused through the entire network. Each node can calculate a parameter called the height of the node, which is the minimum number of hops to reach the BS. The difference between a node's height and that of its neighbor is considered as the gradient on that link. A packet is forwarded on a link with the largest gradient. GBR uses some auxiliary techniques such as data aggregation and traffic spreading in order to uniformly divide the traffic over the network. When multiple paths pass through a node, which acts as a relay node, that relay node may combine data according to a certain function. In GBR, three different data dissemination techniques have been discussed:

1. Stochastic scheme, where a node picks up one gradient at random when there are two or more next hops that have the same gradient,
2. Energy-based scheme, where a node increases its height when its energy drops below a certain threshold, so that other sensors are discouraged from sending data to that node.
3. Stream-based scheme, where new streams are not routed through nodes that are currently part of the path of other streams.

The main objective of these schemes is to obtain a balanced distribution of the traffic in the network, thus increasing the network lifetime. Simulation results of GBR show GBR outperforming directed diffusion in terms of total communication energy.

9.4.2.5 Hierarchical routing

Hierarchical or cluster-based routing, originally proposed in wire line networks, is well-known technique with special advantages related to scalability and efficient communication. In a hierarchical architecture, higher energy nodes can be used for processing and sending the information, while low-energy nodes can be used for performing the sensing in the proximity of the target. This means that creation of clusters and assigning special tasks to cluster heads can greatly contribute to overall system scalability, lifetime, and energy efficiency. Hierarchical routing is mainly two-layer routing where one layer is used to select cluster heads and the other layer is used for routing [1].

Low-energy-adaptive clustering hierarchy (LEACH) protocol: LEACH is a cluster-based protocol, which includes distributed cluster formation. LEACH randomly selects a few sensor nodes as cluster heads (CHs) and rotates this role for even distribution of the energy load among the sensors in the network. In LEACH, the cluster head (CH) nodes compress data arriving from nodes that belong to the respective cluster and send an aggregated packet to the base station in order to reduce the amount of information that must be transmitted to the base station. LEACH uses a TDMA/CDMA MAC to reduce inter-cluster and intra-cluster collisions. However, data collection is centralized and is performed periodically. After a given interval of time, a randomized rotation of the role of the CH is conducted so that uniform energy dissipation in the sensor network is obtained. Only 5% of the nodes need to act as cluster heads.

The operation of LEACH has two phases viz., the setup phase and the steady-state phase. In the setup phase, the clusters are organized and CHs are selected. In the steady-state phase, the actual data transfer to the base station takes place. The duration of the steady-state phase is longer than the duration of the setup phase for minimizing overhead. During the setup phase, predetermined fraction of nodes, p, elect themselves as CHs. Each elected CH broadcasts an advertisement message to the rest of the nodes in the network. All the non-cluster-head nodes, after receiving this advertisement, decide on the cluster to which they want to belong. This decision is based on the signal strength of the advertisement. The non-cluster-head nodes inform the appropriate cluster heads that they will be a member of its cluster. After receiving all the messages from the nodes that would like to be included in the cluster and based on the number of nodes in the cluster, the cluster head node creates a TDMA schedule and assigns each node a time slot for transmission. This schedule is broadcast to all the nodes in the cluster.

During the steady-state phase, the sensor nodes can begin sensing and transmitting data to the cluster heads. The cluster head node, after receiving all the data, aggregates it before sending it to the base station. After a certain duration, which is determined a priori, the network goes back into the setup phase again and enters another round of selecting new CH.

Despite the ability of LEACH to increase the network lifetime, there are still a number of issues about the assumptions used in this protocol. LEACH assumes that all nodes can transmit with enough power to reach the BS if needed and that each node has computational power to support different MAC protocols. Therefore, it is not applicable to networks deployed in large regions. It also assumes that nodes always have data to send, and nodes located close to each other have correlated data. Therefore, there is the possibility that the elected CHs will be concentrated in one part of the network. Hence, some nodes do not have any CHs in their vicinity. Furthermore, the idea of dynamic clustering brings extra overhead. Finally, the protocol assumes that all nodes begin with the same amount of energy capacity in each election round, assuming that being a CH consumes approximately the same amount of energy for each node.

Power-efficient gathering in sensor information systems (PEGASIS): It is an enhancement over LEACH protocol which is a near optimal chain-based protocol. The basic idea of the protocol is that in order to extend network lifetime, nodes need only communicate with their closest neighbors and they take turns in communicating with the base station. This approach will distribute the energy load evenly among the sensor nodes in the network. When the round of all nodes communicating with the base station ends, a new round will start and so on. This reduces the power required to transmit data per round as the power draining is spread uniformly over all nodes. Hence, PEGASIS has two main objectives: first, to increase the lifetime of each node by using collaborative techniques and, second, to allow only local coordination between nodes that are close together to ensure reduction in bandwidth consumed in communication.

PEGASIS performs data fusion at every node except at the end nodes in the chain. Each node fuses its neighbor's data with its own for generating a single packet of the same length and then transmits that to its other neighbor (if it has two neighbors). In Figure 9.9, node $C0$ passes its data to node $C1$. Node $C1$ fuses node $C0$'s data with its own and then transmits to the leader. After node $C2$ passes the token to node $C4$, node $C4$ transmits its data to node $C3$. Node $C3$ fuses data of node $C4$ with its own and then transmits it to the leader. Node $C2$ waits to receive data from both neighbors and then fuses its data with its neighbors' data. Finally, node $C2$ transmits one message to the BS.

For locating the closest neighbor node in PEGASIS, each node uses the signal strength to measure the distance to all neighboring nodes and then adjusts the signal strength so that only one node can be heard. The chain in PEGASIS consists of those nodes that are closest to each other and form a path to the base station. The study shows the capacity of PEGASIS to increase the lifetime of the network twice as much as the lifetime of the network under the LEACH protocol, by reducing the clustering overhead. However, it requires dynamic topology adjustment which introduces a significant overhead especially for highly utilized networks. Moreover, PEGASIS assumes that each sensor node is able to communicate with the BS directly. In practical cases, sensor nodes use multi-hop communication to reach the base station. Also, PEGASIS assumes that all nodes maintain a complete database about the location of all other nodes in the network. It introduces excessive delay for distant node on the chain.

Threshold-sensitive energy-efficient protocols (TEEN and adaptive periodic APTEEN): In TEEN, sensor nodes sense the medium continuously,

$$\frac{C0 \rightarrow \quad 1 \rightarrow \quad 2 \rightarrow \quad 3 \rightarrow \quad 4}{\downarrow}$$

BaseStation

Figure 9.9 Token passing approach.

but data transmission is done less frequently. A cluster head sensor sends its members a hard threshold, which is the threshold value of the sensed attribute and a soft threshold, which is one with a small change in the value of the sensed attribute that triggers the node to switch on its transmitter and transmit. Thus, the hard threshold tries to reduce the number of transmissions by allowing the nodes to transmit only when the sensed attribute is in the range of interest. The soft threshold further reduces the number of transmissions that might have otherwise occurred when there is little or no change in the sensed attribute. A smaller value of the soft threshold gives a more accurate picture of the network, at the expense of increased energy consumption. Thus, the user can control the trade-off between energy efficiency and data accuracy. When cluster heads are to change (see Figure 9.10 (a)), new values for the above parameters are broadcast. The main drawback of this scheme is that the nodes never communicate when the thresholds are not received and the user does not get any data from the network at all.

The nodes sense their environment continuously. The first time a parameter from the attribute set reaches its hard threshold value, the node switches its transmitter on and sends the sensed data. The sensed value is stored in an internal variable, called sensed value (SV). The nodes transmit data in the current cluster period only when the following conditions are true:

1. The current value of the sensed attribute is greater than the hard threshold.
2. The current value of the sensed attribute differs from SV by an amount equal to or greater than the soft threshold.

Important features of TEEN include its suitability for time critical sensing applications. Also, since message transmission consumes more energy than data sensing, the energy consumption in this scheme is less than the proactive networks. The soft threshold can be varied.

APTEEN, on the other hand, is a hybrid protocol that changes the periodicity or threshold values used in the TEEN protocol according to the user needs and the type of the application. In APTEEN, the cluster heads

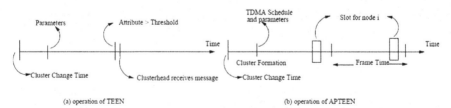

(a) operation of TEEN   (b) operation of APTEEN

Figure 9.10   Time line for the operation of (a) TEEN and (b) APTEEN.

broadcast the parameters such as attributes, thresholds, schedule, and count time. The main features of the APTEEN scheme include the following. It combines both proactive and reactive policies. It offers a lot of flexibility by allowing the user to set the count-time interval (CT), and the threshold values for the energy consumption can be controlled by changing the count time as well as the threshold values. The main drawback of the scheme is the additional complexity required for implementing the threshold functions and the count time. Simulation of TEEN and APTEEN has shown these two protocols outperforming LEACH. The main drawbacks of the two approaches are the overhead and complexity associated with forming clusters at multiple levels, the method of implementing threshold-based functions, and the manner of dealing with attribute-based naming of queries.

Self-organizing protocol (SOP): Self-organizing protocol is used to build an architecture that supports heterogeneous sensors, which can be mobile or stationary. Some sensors probe the environment and forward the data to a designated set of nodes that act as routers. Router nodes are stationary and form the backbone for communication. Collected data are forwarded through the routers to the more powerful BS nodes. Each sensing node should be able to reach a router in order to be part of the network. Sensing nodes are identifiable through the address of the router node they are connected to. The routing architecture is hierarchical where groups of nodes are formed and merge when needed.

Local Markov loops (LML) algorithm, which performs a random walk on the spanning trees of a graph, has been used for supporting fault tolerance and also as a means of broadcasting. Such approach is similar to the idea of virtual grid used in some location-based routing protocols. In this approach, sensor nodes can be addressed individually in the routing architecture, thereby making it suitable for applications where communication to a particular node is required. Furthermore, this algorithm incurs a small cost for maintaining routing tables and keeping a balanced routing hierarchy. The energy consumed for broadcasting a message is found to be less than that consumed in the SPIN protocol. This protocol, however, is not an on-demand protocol especially in the organization phase of algorithm. There may be many cuts in the network, during the formation of hierarchy, and hence, the probability of applying reorganization phase increases, which will be an expensive operation due to extra overhead.

Sensor aggregates routing: A set of algorithms for constructing and maintaining sensor aggregates are used. They collectively monitor the target activity in a certain environment (target tracking applications). A sensor aggregate comprises those nodes in a network that satisfy a grouping predicate for a collaborative processing task. The parameters of the predicate depend on the task and its resource requirements. The formation of appropriate sensor aggregates is dependent on the allocation of resources pertaining to sensing and communication tasks. Sensors in a sensor field are divided into clusters according to their sensed signal strength, so that there is only one peak per cluster. Local cluster leaders are elected later. One peak may represent one target, multiple targets, or even no target in case the peak is generated by noise sources. Information exchanges between neighboring sensors are necessary for electing a leader. If a sensor, after exchanging packets with all its one-hop neighbors, finds itself higher than all its one-hop neighbors on the signal field landscape, it declares itself a leader. This leader-based tracking algorithm assumes the unique leader knows the geographical region of the collaboration [1].

Distributed aggregate management (DAM) protocol for forming sensor aggregates for a target monitoring task is utilized. This protocol comprises a decision predicate P for each node to decide the requirement to participate in an aggregate and a message exchange scheme M about how the grouping predicate is applied to nodes. A node determines whether it belongs to an aggregate based on the result of applying the predicate to the data of the node as well as information from other nodes. Aggregates are formed when the process eventually converges.

Energy-based activity monitoring (EBAM) algorithm estimates the energy level at each node by computing the signal impact area, combining a weighted form of the detected target energy at each impacted sensor assuming that each target sensor has equal or constant energy level. The third algorithm, expectation–maximization like activity monitoring (EMLAM), removes the constant and equal target energy-level assumption. EMLAM estimates the target positions and signal energy using received signals, and uses the resulting estimates to predict how signals from the targets may be mixed at each sensor. This process is iterated, until the estimate is sufficiently good.

Virtual grid architecture (VGA) routing: An energy-efficient routing paradigm utilizes data aggregation and in-network processing for maximizing network lifetime. A reasonable approach is needed to arrange nodes in a fixed topology considering the stationary and extremely low mobility in many applications in WSNs. A GPS-free approach is used for building clusters that are fixed, equal, adjacent, and non-overlapping with symmetric shapes. Square clusters are used for obtaining a fixed rectilinear virtual topology as shown in Figure 9.11. Inside each zone, a node is optimally selected to act as cluster head.

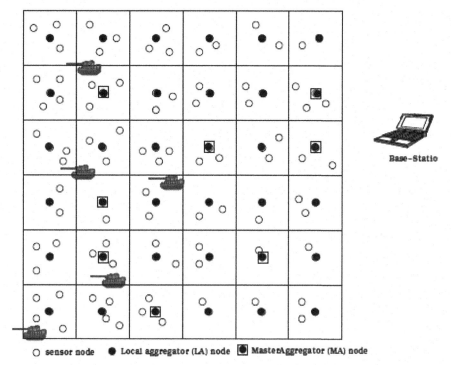

O sensor node   ● Local aggregator (LA) node   ▣ MasterAggregator (MA) node

Figure 9.11   Regular shape tessellation applied to the network area.

Data aggregation is performed at two levels: local and then global. The set of cluster heads, also called local aggregators (LAs), perform the local aggregation, while a subset of these LAs is used for performing global aggregation. However, the determination of an optimal selection of global aggregation points, called master aggregators (MAs), is NP-hard problem. Figure 9.11 illustrates an example of fixed zoning and the resulting virtual grid architecture (VGA) used for performing two-level data aggregation. The location of the base station can be at any arbitrary place.

In each zone, a cluster head is selected for local aggregation. A subset of those cluster heads, called master nodes, is optimally selected for doing global aggregation. Two solution strategies for the routing with data aggregation problem are used: (a) an exact algorithm using an integer linear program (ILP) formulation and (b) several near optimal, but simple and efficient, approximate algorithms, namely a genetic algorithm-based heuristic, a k-means heuristic, and a greedy-based heuristic. Another efficient heuristic, called clustering-based aggregation heuristic (CBAH), is also used for minimizing energy consumption in the network, and hence prolong the network lifetime. The objective of all algorithms is to select a number of MAs out of the LAs that maximize the network lifetime.

9.4.2.6 Location-based routing

In this kind of routing, sensor nodes are addressed by means of their locations. The distance between neighboring nodes can be estimated on the basis of incoming signal strengths. Relative coordinates of neighboring nodes can be obtained by exchanging such information between neighbors. Alternatively, the location of nodes may be available directly by communicating with a satellite, using GPS (global-positioning system), if nodes are equipped with a small low-power GPS receiver. Some location-based schemes demand nodes to go to sleep if there is no activity. This is done for saving energy. More energy savings can be obtained by having as many sleeping nodes in the network as possible. The problem of designing sleep period schedules for each node in a localized manner was addressed. In the rest of this section, most of the location- or geographic-based routing protocols are reviewed.

Geographic adaptive fidelity (GAF): GAF is an energy-aware location-based routing algorithm designed primarily for mobile ad hoc networks, but can be applicable to sensor networks as well. The network area is first divided into fixed zones and forms a virtual grid as shown in Figure 9.12. Inside each zone, nodes collaborate with each other to play different roles. For example, nodes elect one sensor node to stay awake for a certain duration and then go to sleep. This node is responsible for monitoring and reporting data to the BS on behalf of the nodes in the zone. Hence, GAF conserves energy by turning off superfluous nodes in the network without affecting the level of routing fidelity. Each node uses its GPS-indicated location for associating itself with a point in the virtual grid. Nodes associated with the same point on the grid are considered equivalent in terms of the cost of packet routing. Such equivalence is exploited in keeping some nodes located in a particular grid area in sleeping state in order to save energy. There are three states defined in GAF. These states are discovery, for determining the neighbors in the grid, active reflecting participation in routing, and sleep when the radio is turned off. In order to handle the mobility, each node in the grid estimates its leaving time of grid and sends this to its neighbors. The sleeping neighbors adjust their sleeping time accordingly in order to keep the routing fidelity. Before the leaving time of the active node expires, sleeping nodes wake up and one of them becomes active.

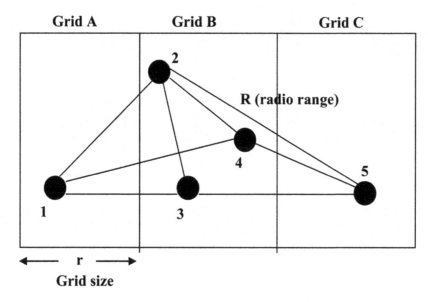

Figure 9.12 Virtual grid structure in the GAF protocol.

Advantages: The performance is similar to that of a normal ad hoc routing protocol in terms of latency and packet loss and increases the lifetime of the network by saving energy. Although GAF is a location-based protocol, it may also be considered as a hierarchical protocol, where the clusters are based on geographic location. For each particular grid area, a representative node acts as the leader to transmit the data to other nodes.

Disadvantages: The leader node, however, does not do any aggregation or fusion as in the case of other hierarchical protocols discussed earlier in this article.

Geographic and energy-aware routing (GEAR): The use of geographic information while disseminating queries to appropriate regions yields fruitful result as the data queries often include geographic attributes. The protocol, called geographic and energy-aware routing (GEAR), uses energy-aware and geographically informed neighbor selection heuristics to route a packet toward the destination region. The key idea is to restrict the number of interests in directed diffusion by considering only a certain region rather than sending the interests to the whole network. By doing this, GEAR can conserve more energy than directed diffusion.

Each node in GEAR keeps an estimated cost and a learned cost of reaching the destination through its neighbors. The estimated cost is a combination of residual energy and distance to destination. The learned cost is a refinement of the estimated cost that accounts for routing around holes in the network. A hole occurs when a node does not have any neighbor closer to the target region than itself. In the absence of holes, the estimated cost is equal to the learned cost. The learned cost is propagated one hop back every time a packet reaches the destination so that route setup for next packet will be adjusted. There are two phases in the algorithm: (1) forwarding packets toward the target region: Upon receiving a packet, a node checks its neighbors to see whether there is one neighbor, which is closer to the target region than itself. If there is more than one, the nearest neighbor to the target region is selected as the next hop. If they are all farther than the node itself, this means there is a hole. In this case, one of the neighbors is picked to forward the packet based on the learning cost function. This choice can then be updated according to the convergence of the learned cost during the delivery of packets. And (2) forwarding the packets within the region: If the packet has reached the region, it can be diffused in that region by either recursive geographic forwarding or restricted flooding [1]. Restricted flooding is good when the sensors are not densely deployed. In high-density networks, recursive geographic flooding is more energy efficient than restricted flooding. For an uneven traffic distribution, GEAR delivers 70–80% more packets than GPSR a non-energy-aware routing protocol. For uniform traffic pairs, GEAR delivers 25–35% more packets than GPSR.

MFR, DIR, and GEDIR: These protocols deal with basic distance, progress, and direction based methods. The key issues are forward direction and backward direction. A source node or any intermediate node selects one of its neighbors according to a certain criterion. The routing methods, which belong to this category, are MFR (most forward within radius), GEDIR (the geographic distance routing) that is a variant of greedy algorithms, 2-hop greedy method, alternate greedy method, and DIR (compass routing method). GEDIR algorithm is a greedy algorithm that always moves the packet to the neighbor of the current vertex whose distance to the destination is minimized. The algorithm fails when the packet crosses the same edge twice in succession. In most cases, the MFR and greedy methods have the same path to destination. In the DIR method, the best neighbor has the closest direction (i.e., angle) toward the destination. It means that the neighbor with the minimum angular distance from the imaginary line joining the current node and the destination is selected. In the MFR method, the best neighbor A minimizes the dot product DA: DS, where S;D are the source and destination nodes, respectively, and SD represents the Euclidean distance between the two nodes S;D. Alternatively, maximization of the dot product SD: SA can be done. Each method stops forwarding the message at a node for which the best choice is to return the message back to a previous node. GEDIR and MFR methods are loop-free, while the DIR method may create loops, unless past traffic is memorized or a time stamp is enforced.

9.4.3 Transport Control Protocols

Transport protocols are used for eliminating or mitigating congestion and reduce packet loss, to provide fairness in bandwidth allocation, and to guarantee end-to-end reliability. However, the traditional transport protocols that are designed for the Internet, i.e., UDP and TCP, cannot be directly applied to WSNs [12].

The transport protocol runs over the network layer protocol. It enables end-to-end message transmission, where messages are fragmented to chains of segments at senders and reassembled at receivers. The transport protocol usually provides the following functions: orderly transmission, flow control and congestion control, loss recovery, and possibly QoS guarantee such as timing requirement and fairness. In WSNs, many new factors such as the convergent nature of upstream traffic and limited wireless bandwidth can cause congestion and packet loss. In order to design an efficient transport protocol for WSNs, several factors must be taken into consideration including the topology, diversity of applications, traffic characteristics, and resource constraints. The two most significant constrains introduced by WSNs are the energy constrains and fairness among different geographically placed sensor nodes [13]. The transport protocol needs to provide high energy efficiency and flexible reliability and sometimes the traditional QoS in terms of throughput, packet loss rate, and end-to-end delay. Therefore, transport protocols for WSNs should have components including congestion control and loss recovery, since the two components have direct impact on energy efficiency, reliability, and application's QoS as explained above.

There are generally two approaches to perform this task. First, design separated protocols or algorithms, respectively, for congestion control and loss recovery. Most existing protocols use this process and address either congestion control or reliable transport. With this separated and modular design, applications that need reliability can invoke only a loss recovery algorithm, or invoke a congestion control algorithm when they need to control congestion otherwise. Their combined use could provide the entire gamut of functions

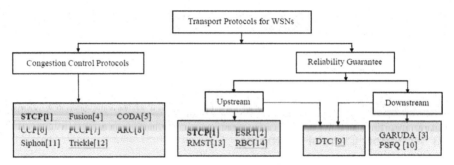

Figure 9.13   Classification of transport protocols.

required by a transport protocol for WSNs. Second, design, if possible, is a full-fledged transport protocol that provides congestion control and loss control in an integrated way. For example, STCP (sensor transmission control protocol) [1] implements both congestion control and flexible reliability in a single protocol. For different applications, STCP offers different control policies in a way to guarantee application requirements and also improve energy efficiency. Several transport protocols have been designed for WSNs (Figure 9.13). Some of which addressed congestion or reliability only; others examined both of them. They are categorized into three types: (1) congestion control protocols; (2) protocols for reliability; and (3) protocols considering both congestion control and reliability [14].

9.5 Applications of Wireless Sensor Networks

The several envisioned applications of WSN are still very much under active research and development, in both academia and industry. Brief details of a few applications from different domains are provided to give a sense of the wide-ranging scope of this field:

Ecological habitat monitoring: Scientific studies of ecological habitats (animals, plants, microorganisms) are traditionally conducted through hands-on field activities by the investigators. One serious concern in these studies is that the very presence and potentially intrusive activities of the field investigators may affect the behavior of the organisms in the monitored habitat and thus bias the observed results. Unattended wireless sensor networks promise a cleaner, remote-observer approach to habitat monitoring. Further, sensor networks, due to their potentially large scale and high spatiotemporal density, can provide experimental data of an unprecedented richness.

Military surveillance and target tracking: As with many other information technologies, wireless sensor networks originated primarily in military-related research. Unattended sensor networks are envisioned as the key ingredients in moving toward network-centric warfare systems. They can be deployed rapidly for surveillance and used to provide battlefield intelligence regarding the location, numbers, movement, and identity of troops and vehicles, and for detection of chemical, biological, and nuclear weapons.

Structural and seismic monitoring: Another class of applications for sensor networks pertains to monitoring the condition of civil structures. The structures could be buildings, bridges, and roads; or even aircrafts. The health of such structures is currently monitored primarily through manual and visual inspections or occasionally through expensive and time-consuming technologies, such as X-rays and ultrasound. Unattended networked sensing techniques can automate the process, providing rich and timely information about incipient cracks or about other structural damage. Researchers envision deploying these sensors densely on the structure—either literally embedded into the building material such as concrete, or on the surface. Such sensor networks have the potential for monitoring the long-term wear of structures as well as their condition after destructive events, such as earthquakes or explosions. A particularly compelling futuristic vision for the use of sensor networks involves the development of controllable structures, which contain actuators that react to real-time sensor information to perform "echo-cancellation" on seismic waves so that the structure is unaffected by any external disturbance.

Industrial and commercial networked sensing: In industrial manufacturing facilities, sensors and actuators are used for process monitoring and control. For example, in a multistage chemical processing plant, sensors may be placed at different points in the process for monitoring temperature, chemical concentration, pressure, etc. The information from such real-time monitoring may be used for variation of process controls, such as adjusting the amount of a particular ingredient or changing the heat settings. The key advantage of creating wireless networks of sensors in these environments is that they can significantly improve both the cost and the flexibility associated with installing, maintaining, and upgrading wired systems.

Ocean temperature monitoring for improved weather forecast: It is known that the evolution of weather conditions is strongly influenced by the temperature of large water masses such as the oceans. However, currently, the ability to perform a large-scale monitoring of the ocean temperature is very little. Sensor networks can be used for this purpose. By dropping a large number of tiny sensors into the sea, water temperature and ocean currents can be accurately monitored, helping the scientists in the task of providing more accurate weather forecast [5].

Intrusion detection: Camera-equipped sensors can be used for forming a network that monitors an area with restricted access. If the network is properly deployed, intruders can be detected and an alarm message quickly propagated to the external observer.

Avalanche prediction: Sensors equipped with location devices (such as GPS) can be used to monitor the movements of large snow masses, thus allowing a more accurate avalanche prediction.

9.6 Sensor Networks—Operating Systems

In general, traditional programming technologies rely on operating systems to provide abstraction for processing, I/O, networking, and user interaction hardware, as shown in Figure 9.14.

By applying such a model to sensor networks, the application programmers necessarily deal with message passing, event synchronization, interrupt handing, and sensor reading. As a result, an application will be typically implemented as a finite-state machine (FSM) which covers all extreme cases such as unreliable communication channels, long delays, irregular arrival of messages, simultaneous events, and so on.

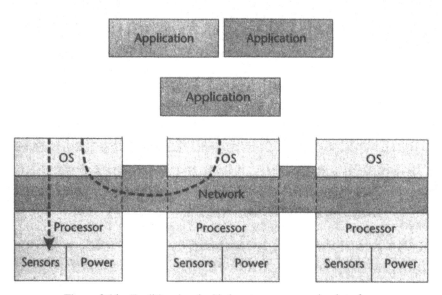

Figure 9.14   Traditional embedded system programming interface.

For resource-constrained embedded systems with real-time requirements, several mechanisms are used in embedded operating systems to reduce code size, improve response time, and reduce energy consumption. Although these techniques may work well for small, stand-alone embedded systems, they do not scale up for the programming of sensor networks for two reasons. Sensor networks are large-scale distributed systems, where global properties are derivable from program execution in a massive number of distributed nodes. Distributed algorithms by themselves are hard to implement, especially when infrastructure support is limited due to the ad hoc formation of the system and constrained power, memory, and bandwidth resources. Sensor nodes deeply embed into the physical world. So, a sensor network should be able to respond to multiple concurrent stimuli at the speed of changes of the physical phenomena of interest.

The design methodologies used for sensor network software are node-centric. A node-level platform can be a node-centric operating system, which provides hardware and networking abstractions of a sensor node to programmers. Alternatively, it can be a language platform, which provides a library of components to programmers.

A typical operating system abstracts the hardware platform by providing a set of services for applications, including file management, memory allocation, task scheduling, peripheral device drivers, and networking. Operating systems developed for embedded systems make different trade-offs while providing these services. This is due to their highly specialized applications and limited resources. A highly elaborate priority scheduling mechanism may be added when prioritization among tasks is critical. TinyOS and TinyGALS are two representative examples of node-level programming tools.

9.6.1 Operating System: TinyOS

TinyOS aims at supporting sensor network applications on resource-constrained hardware platforms, such as the Berkeley motes. It has following features:

- No file system
- Supports only static memory allocation
- Implements a simple task model
- Provides minimal device and networking abstractions

Furthermore, it takes a language-based application development approach, so that only the necessary parts of the operating system are compiled with the application. To a certain extent, each TinyOS application is built into the operating system. TinyOS organizes components into layers and has a unique component architecture providing a library as a set of system software components. A component specification is independent of component implementation. Although most components encapsulate software functionalities, some are just thin wrappers around hardware. An application, typically developed in the nesC language, *wires* these components together with other application-specific ones.

The design decisions in TinyOS are made to ensure it is extremely lightweight. Using a component architecture contains all variables inside the components and disallowing dynamic memory allocation, reduces the memory management overhead, and makes the data memory usage statically analyzable. The simple concurrency model allows high concurrency with low thread maintenance overhead. As a consequence, the entire field monitor system takes only 3 KB of space for code and 226 bytes for data. However, the advantage of being lightweight is not without cost. The hardware complexities of concurrency management are left for the application programmers to handle. Several tools have been developed for providing language-level support for improving programming productivity and code robustness. The special-purpose language called nesC is used for programming sensor network nodes.

9.6.1.1 Imperative language: nesC

nesC is a programming language designed to build applications for the TinyOS platform. nesC is built as an extension to the C_Programming_Language with components "wired" together to run applications on TinyOS. A key focus of nesC is a holistic system design. Mote applications are deeply tied to hardware, with each mote running a single application at a time. This approach has three important properties. First, all resources are known statically. Second, rather than employing a general-purpose OS, applications are built from a suite of reusable system components coupled with application-specific code. Third, the hardware/software boundary varies depending on the application and hardware platform; it is important to design for flexible decomposition.

There are a number of unique challenges that nesC should address:

- **Driven by interaction with environment**: Unlike traditional computers, motes are used for data collection and control of the local environment, rather than general-purpose computation. This focus leads to two observations. First, motes are fundamentally event driven, reacting to changes in the environment (message arrival, sensor acquisition) rather than driven by interactive or batch processing. Second, event arrival and data processing are concurrent activities, demanding an approach to concurrency management that addresses potential bugs such as race conditions.

- **Limited resources**: Motes have very limited physical resources, due to the goals of small size, low cost, and low power consumption.
- **Reliability**: Despite the failure of the individual motes due to hardware issues, they should be enabled for very long-lived applications. For example, environmental monitoring applications should collect data without human interaction for months at a time. An important goal is to reduce run-time errors, since there is no real recovery mechanism in the field except for automatic reboot.
- **Soft real-time requirements**: Despite the presence of few time critical tasks such as radio management or sensor polling, the timing constraints are easily met by having a complete control over the application and OS, and limiting utilization. One of the few timing-critical aspects in sensor networks is radio communication. However, given the fundamental unreliability of the radio link, there is no need to meet hard deadlines in this domain.

Although nesC is a synthesis of many existing language concepts targeted at the above problems, it provides three broad contributions [1]. The basic concepts behind nesC are as follows:

- Separation of construction and composition.
- Specification of component behavior in terms of set of interfaces.
- Interfaces are bidirectional.
- Components are statically linked to each other via their interfaces.
- nesC simplifies application development, reduces code size, and eliminates many sources of potential bugs.

9.6.2 Contiki OS

Contiki is an operating system developed for such constrained environments. It is an open-source, highly portable, multi-tasking operating system for memory-efficient networked embedded systems and wireless sensor networks, providing dynamic loading and unloading of individual programs and services. Contiki has been used in a variety of projects from road tunnel fire monitoring, intrusion detection, water monitoring, to surveillance networks. The kernel is event-driven, but the system supports preemptive multi-threading that can be applied on a per-process basis. Preemptive multi-threading is implemented as a library that is linked only to programs that explicitly require multi-threading.

Contiki is implemented in the C language and has been ported to a number of microcontroller architectures, including the Texas Instruments MSP430 and the Atmel AVR. Currently, it is running on the ESB platform which uses the MSP430 microcontroller with 2 KB of RAM and 60 KB of ROM running at 1 MHz. The microcontroller has the ability to selectively reprogram parts of the on-chip flash memory. The basic features of Contiki are as follows:

- TCP/IP communication with uIP stack.
- Loadable modules.
- Event-driven kernel.
- Protothreads.
- Protocol-independent radio network with the Rime stack.
- Cross-layer network simulation with Cooja.
- Networked shell.
- Memory-efficient flash-based Coffee file system.
- Software-based power profiling.

9.6.2.1 System overview

A running Contiki system consists of the kernel, libraries, the program loader, and a set of processes. A process may be either an application program or a service. A service implements functionality used by more than one application process. All processes, both application programs and services, can be dynamically replaced at run time. Communication between processes always goes through the kernel, which does not provide a hardware abstraction layer, but lets device drivers and applications communicate directly with the hardware.

A process is defined as an event handler function and an optional poll handler function. The process state is held in the private memory of the process. The kernel only keeps a pointer to the process state. On the ESB platform, the process state consists of 23 bytes. All processes share the same address space and do not run in different protection domains. Inter-processor communication is done by posting events.

A Contiki system is partitioned into two parts: the core and the loaded programs as shown in Figure 9.15. The partitioning is made at compile time and is specific to the deployment in which Contiki is used. Typically, the core consists of the Contiki kernel, the program loader, the most commonly used parts of the language run time and support libraries, and a communication stack with device drivers for the communication hardware. The core is compiled into a single binary image that is stored in the devices prior to deployment.

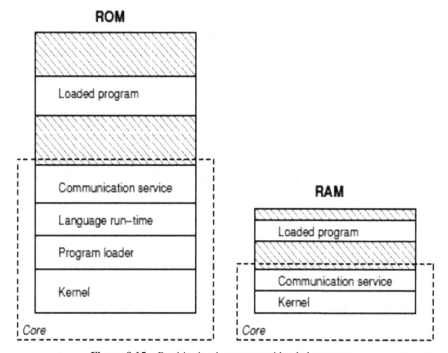

Figure 9.15 Partitioning into core and loaded programs.

The core is generally not modified after deployment, despite its possibility to use a special boot loader to overwrite or patch the core.

Programs are loaded into the system by the program loader. The program loader may obtain the program binaries either by using the communication stack or by using directly attached storage such as EEPROM. Typically, programs to be loaded into the system are first stored in EEPROM before they are programmed into the code memory.

9.7 Compressive Wireless Sensing

WSNs are critically resource constrained by limited power supply, memory, processing performance, and communication bandwidth [15]. Due to limited power supply, energy consumption is a key issue in the design of protocols and algorithms for WSNs. Energy efficiency is necessary in every level of WSN operations (e.g., sensing, computing, switching, transmission). In the conventional view, energy consumption in WSNs is dominated by radio communications [16–18]. In most cases, computational energy cost is insignificant compared to communication cost. Therefore, using compression to reduce the number of bits to be transmitted has the potential of drastic reduction in communication energy costs and increase network lifetime. Thus, researchers have investigated optimal algorithms for the compression of sensed data, communication, and sensing in WSNs [18, 19]. Most existing data-driven energy management and conservation approaches for WSNs target reduction in communication energy at the cost of increased computational energy. In principle, most compression techniques work on reducing the number of bits needed to represent the sensed data, not on the reducing the amount of sensed data. Hence, they are unable to utilize sensing energy costs efficiently in WSNs. Importantly, in most cases, these approaches assume that sensing operations consume significantly less energy than radio transmission and reception [19, 20]. In fact, the energy cost of sensing is also not always insignificant.

However, CS and DCS (distributed compressed sensing) exploit the information rate within a particular signal. Unlike other compression algorithms, they remove redundancy in the signal during the sampling process, leading to a lower effective sampling rate. The asymmetric computational nature of CS and DCS makes them even more attractive for compression in WSNs. Most computation in CS and DCS takes place at the decoder (sink), rather than at the encoder (sensor nodes); thus, sensor nodes with minimal computational performance can efficiently encode data. In addition, CS has two additional advantages: graceful degradation in the event of abnormal sensor readings and low sensitivity to packet loss. Hence, CS and DCS are promising approaches [21, 22] for removing redundancy during sensing operations in WSNs, and, hence, for energy-efficient sensing. CS for WSNs exploits only temporal (intra-signal) structures within multiple sensor readings at a single sensor and does not exploit spatial (inter-signal) correlations among nearby sensors [23]. DCS works on multisensor scenarios considering only standard CS for the joint measurements at single-time instances [24]. These schemes ignore the intra-signal or temporal correlations. On the other hand, some DCS approaches (spatiotemporal) [25, 26] exploit the spatial correlation structures between nearby sensors and the temporal correlation of each sensor's time-variant readings.

9.7.1 Spatial Compression in WSNs

The efficacy of data aggregation in sensor networks is a function of the degree of spatial correlation in the sensed phenomenon. The spatial correlation is due to the densely deployed motes to provide a wider coverage area.

The temporal compression can be carried out to eliminate the redundancy. Hence, the spatiotemporal compression is used to improve the efficiency, to maximize the utilization of the network, and to reduce energy consumption. The various techniques are used for compression such as buffering method, suppression method, ordering method, and clustering method [27]. Clustering is meant to satisfy scalability technique by grouping of sensor nodes into clusters, in which each cluster has a leader referred as cluster head, and it performs the special function such as fusion and aggregation. Clustering reduces the number of nodes taking part in the transmission. Aggregation processes the query to obtain aggregate results from data collected by sensors. The aggregate queries execute functions such as min, max, sum, count, avg, median, and the histogram [28]. Ordering is mainly used for encoding the specific values, to locate the data history table, and to optimize the data history table [29].

Matrix completion and compressive sensing-based data aggregation:
In wireless sensor networks, usually there are intra-temporal and inter-spatial correlations in the sensed data. Low-rank matrix completion theory can be used for exploring the inter-spatial correlation. The compressive sensing theory can be used for availing intra-temporal correlation. Dubbed matrix completion and compressive sensing (MCCS) method can significantly reduce the amount of data that each sensor must send through network and to the sink, thus prolong the lifetime of the whole networks. Matrix completion (MC) is the theory to recover of a full data matrix from part of its entries. Recently, candès et al. prove that if the data matrix is a low-rank or approximately low-rank matrix, missing entries can be recovered from an incomplete set of entries [30].

A. Encoding algorithm at each sensor node

The details of the encoding algorithm utilized at each sensor are provided below [31]. The algorithm tries to reduce the energy cost in all possible aspects including sample energy cost, communication cost, and computation complex.

Step 1: Each sensor node randomly generates a binary sampling position vector P_N with only
$q(q < N)$ q nonzero entries. Here, $\lambda = [N/q]$ is the MC (matrix completion) sample ratio.

Step 2: Each sensor node then scans the binary sampling position vector and only samples when the corresponding entry is nonzero. At the end, those sampled readings form a vector $x_q \in R^q$. At this step, MC-based compression is applied. Each node only samples q readings instead of N which results to $\lambda = [N/q]$ compression ratio. Thus, the energy cost of sampling is reduced.

Step 3: At the first time, each sensor node generates and stores the same sparse binary matrix $B_{p \times q}(P < q)$ using the seed K pre-shared among all sensors and sinks. There are only small numbers $d(p > d \geq 1)$: nonzero elements random located in each column of $B_{p \times q}$. Also, $\eta = [p/q]$ is called as the CS compression ratio.

Step 4: Each sensor node gets CS measurements y_p from x_q according below operation $y_p = B_{p \times q} \times x_q$. After sending out y_p and the sampling position, vector P_N (using bit mode) is sent out, and then goes back to step 1. At this step, since our measurement matrix is a sparse binary matrix, the energy expenditure for this CS compression involves only a simple addition operation. Although x_q is randomly sampled from N continuous readings, due to the temporal correlation, x_q is still found to be sparse under certain transform basis. That is why CS is used for compressing the readings after matrix completion-based compression. At the end, each sensor only needs to send out p readings instead of N, which results to $p/N = \lambda \times \eta$ total compression ratio.

B. Recovering algorithm at sink node

At the beginning, the sink node generates and stores the same sparse binary matrix $B_{p \times q}$ using shared seed K. After receiving CS measurements y_p from each node, sink node is able to reconstruct the partial readings x_q through solving a 1 l-minimization problem:

$$\min \|\theta_q\|_{l_1} \quad s.t. \quad y_p = B_{p \times q} x_q, \quad x_q = \Psi_{q \times q} \theta_q \qquad (9.1)$$

Here, $\Psi_{N \times N}$ is a transform basis that can make x_q sparsely represented as θ_q. Suppose $\tilde{\theta}_q$ is the solution to the convex optimization problem, then the original partial readings are $\tilde{x}_q = \Psi_{q \times q} \tilde{\theta}_q$. With proper values of p and q, the error between $\tilde{x}_q$ and x_q can be very small. After recovering from CS compression, the sink node uses $\tilde{x}_q$ and the binary sampling position vector P_N from each sensor node forming an incomplete readings matrix $\tilde{x}_{J \times N}$ where J is the number of sensor nodes. According to the spatial correlation, $\tilde{x}_{J \times N}$ is an approximate low-rank matrix. Hence, the full reading matrix can be recovered from convex optimization problem [11]:

$$\text{minimize } \|X\|^* \text{ subject to } X_{ij} = \tilde{X}_{ij}(i,j) \in \Omega \qquad (9.2)$$

Here, $\|.\|^*$ is the nuclear norm, and the set W is the locations corresponding to the partially sampled sets of readings.

Spatial correlation-based distributed compressed sensing (SCDCS) model:

Spatial correlation and joint sparse models between the sensor nodes can be exploited for compression and reconstruction of sensor observations in an energy-efficient manner. Data fusion techniques can be used for collecting the observations of sensor nodes before they are sent to the sink. CS usage in compress includes the following advantages: First, the encoders are very simple: Each sensor obtains compressive measurements by projecting its signal onto another random basis. Second, it provides effective data compression, which can reconstruct k-sparse signal from a small number of measurements. When the sensor observations are encoded by projecting them onto a random basis, just a few of the measurements need transmission to the sink node, and a decoder at sink can reconstruct the sensor observations without collaboration among the sensors. This process thereby assists significant saving in scarce resources such as energy and bandwidth in WSN.

The sink in WSN performs the sensed data of corresponding sensors gathering in a specified event area for estimating the event, S, within a reliability measure required by the application. The model for the information collected by N sensors in an event area [32] is shown in Figure 9.16.

The sink is interested in estimating the event in the sensor field, on the basis of event information of the sensor nodes, in the event area, which is spatially correlated to the event source S, due to spatially dense sensor deployment. The event source S is assumed to be located at (0,0). Information variogram is calculated on the basis of the location of the event. Higher value of the

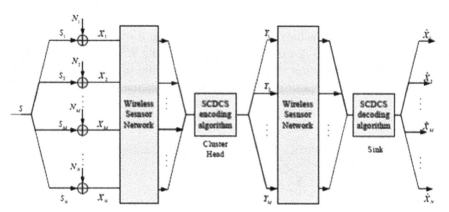

Figure 9.16 Spatial correlation-based distributed compressed sensing model and architecture.

variogram implies a lower value of the covariance and correlation. It is often used to characterize spatial correlation in the sensor field of event area. The size of event area can be obtained according to the error threshold given by the sensor application.

Sensor nodes of event area form a cluster, and a cluster head is selected for collecting the sensed data from the sensors in the event area. The sensed data have a k-sparse representation in wavelet basis by exploiting spatial correlations. SCDCS encoding algorithm is performed when the cluster head obtains the sensed data and the measurements of the k-sparse vector are obtained. When the cluster head calculates the measurements, it tries to forward it to the sink along a wireless multi-hop transmission route. The sensed data of N sensors are sparsely represented in the wavelet basis, and the random measurement matrix is incoherent with the basis matrix, and SCDCS decoding algorithm (l1 minimization) is performed at the sink node.

9.7.2 Projections in WSNs

A distinctive feature of compressive sensing is that it uses projections to collect information. For a snapshot of the noisy data field $\{yi\}$, the projection of the vector y on a projection vector $p = [p1, p2, \ldots, pn]^T$ is defined by the inner product $p^T y = \sum_{i=1}^{n} p_i y_i$. Let us illustrate the concept of projection vectors and how projections can be calculated in a WSN with a few examples [33]. Consider the network shown in Figure 9.17 with 4 sensor nodes $\{1, 2, 3, 4\}$ and sink node s.

Example 1: If the projection vector p is $[0.2, 0.3, 0.4, 0.1]$, then the projected value is $p^T y = 0.1y1 + 0.3y2 + 0.4y3 + 0.1y4$. The sink can obtain this projected value without the sensors sending their sensor readings to the sink. This can be achieved by the sink passing a message along the tour $S - 1 - 2 - 3 - 4 - S$ using source routing in the WSN. The message contains the entire projection vector p as well as a field in the message for storing the intermediate result of the projection calculation. As the message

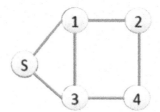

Figure 9.17   Sample WSN scenario.

travels through the tour, each sensor computes its contribution to the projected value and adds it to the intermediate result. Following this, the sensor writes the new intermediate result to the message and forwards the message to the next hop. For example, sensor node 2 receives a message with $0.2y1$ as the intermediate result from sensor node 1; sensor 2 computes $0.3y2$ and adds this to $0.2y1$; then, it writes the sum to the message and then passes it on to the next hop. Note that the computation of this projection requires 5 wireless transmissions.

Example 2: When the projection vector p is $[0.1, 0.2, 0, 0]^T$, the projected value is $p^T y = 0.1y1 + 0.2y2$. This can be computed by passing a message along the tour $S - 1 - 2 - 1 - S$ since sensor readings from sensors 3 and 4 are not needed for computing this projection. Therefore, the calculation of this projection requires only 4 wireless transmissions. In general, the calculation of a projection requires only a message to be passed along those sensor nodes with a nonzero projection vector coefficient. From an energy efficiency point of view, one would therefore aim to find the shortest tour (or aggregation tree) which passes through all sensor node required.

Example 3: If the projection vector p is $[0, 0, 1, 0]^T$, then the projected value $p^T y = y3$. Therefore, this projection vector corresponds to collecting the noisy sensor reading from sensor 3. This example aims to show that a projection is a general method of collecting data and collecting a sensor reading from a sensor is in fact a special case of performing a projection.

By choosing the coefficients in the projection vector for maximizing the information content, it is possible to collect more information (or achieve a smaller relative error in estimating x) using a smaller number of projections. The generic adaptive compressive sensing algorithm analyzed is outlined as follows.

Algorithm 1: Generic adaptive compressive sensing

1. Each sensor randomly decides whether to send its reading to the sink, and if so, it sends its reading to the sink. Note that this can be realized by a node having a probability to send data to the sink and this probability can be determined by the sink and change over time.
2. while
   The sink is not satisfied with the accuracy of the estimated data field do
3. The sink determines a projection vector and the corresponding tour to use.

4. The sink sends a message along the tour and waits for the projected value to return.
5. The sink updates the estimate of the unknown data field and determines its accuracy.
6. end while

The general idea of the generic adaptive compressive sensing algorithm is as follows. In line 1, a small number of sensors randomly decide to return their sensor readings to the sink. These random samples allow the sink to estimate the data field and decide on the need to use the iteration in lines 2–6 to collect further information from the sensor field. This iteration continues until the sink is satisfied with the accuracy of the estimated data field. The rationale behind the above adaptive compressive sensing is the same as that of the optimal experiment design in statistics [34] and active learning in machine learning [35]. A key step in the above algorithm is line 3 where a good projection vector has to be determined. This step is also studied in the adaptive compressive sensing algorithms in [36, 37] where a projection vector which maximizes the amount of information gain is determined. However, these earlier works are not suitable for WSNs as they do not take into consideration the energy required to acquire a projection. The choice of the coefficient of a projection vector can affect both the information content and the energy expenses (via the choice of route to obtain the projection). This means that a projection vector should be chosen for collecting all possible information on the unknown data field with as little energy as possible, which is a design problem requiring information from both the application and routing layers. The energy consumption in WSNs is measured by counting the number of packet transmissions; e.g., a node which is k hops away from the sink will require k packet transmissions to reach the sink. In line 1 of Algorithm 1, energy is consumed for sending packets to the sink, while in lines 2–6, energy is consumed for sending packets along the tour. It can be seen that when a good probability can be found in line 1, it minimizes the energy consumption in lines 2–6. A possibility is to determine this probability from the past history of the data. It is assumed to have a reasonable estimate of this probability and focus on the design of a projection vector which balances information gain and energy consumption. It is also possible to modify line 1 to have a random number of sensors initiating a number of random projections toward the sink; this improves the information content at the sink. It is assumed that the sink (or cluster head) knows the topology of the network (or cluster).

Minimum-spanning tree projection (MSTP) method:

MSTP creates a number of minimum-spanning trees (MSTs), each rooted at a randomly selected projection node, which in turn aggregates sensed data from sensors using compressive sensing. MSTP unlike [38] uses CDG [39] for each projection node for collecting one weighted sum by constructing independent forwarding tree which ensures fewer transmissions. MSTP gathers a weighted sum from nodes step by step starting from leaf nodes following the routing paths toward the projection node. Finally, each projection node sends the gathered data in one packet through a shortest path to the sink. The novelty of our approach lies in utilizing independent forwarding trees, where each forwarding tree carries one weighted sum (compressed data rather than native) to the sink. These forwarding trees are constructed to ensure fewer numbers of transmissions as well as evenly distribute the transmission load across the network.

Collection of data with sparse random projection was first introduced in [38], where m nodes are selected at random to gather m weighted sums in the network. Each projection node gathers one weighted sum for the sink and each row of the basis matrix U is assigned to one projection node. First, a projection node i asks the nodes whose coefficients ϕ_{ij} are nonzero to send their data readings by one packet each through shortest path to it, and after receiving all the packets, the projection node gathers all the data with its own data reading and sends the result through shortest path to the sink by a single packet. Similarly, all the other projection nodes gather and send the weighted sums to the sink. Illustration of the data gathering for one random projection node is shown in Figure 9.18. In this figure, node 5 initializes the projection by sending requests to nodes 11, 15, and 20, where their $\phi_{ij} \neq 0$. These nodes

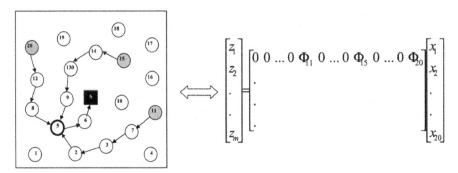

Figure 9.18 Illustration of the data gathering for one random projection node.

reply to the request by sending their data readings x_j to node 5 (marked by arrows). Then, node 5 computes $\sum_{j=1}^{n} \phi_{ij} \chi_j$ and transmits it to the sink.

Consider a WSN of size n with correlated reading values $x_j, j = 1; 2 \ldots; n$. According to the theory of CS, the sink needs to receive only m weighted sums (sample measurements) to recover the readings from all the nodes; i.e., $Z_i = \sum_{j=1}^{n} \varphi_{ij} \chi_j, i = 1, \ldots, m$). Since there are m projection nodes in the network, each projection node i gathers one sample measurement from all the nodes in the network. Each row vector of the matrix U is assigned to one projection node for performing this. The size of each row vector of U is n (related to n nodes in the network), and the nodes whose coefficient $\phi_{ij} \neq 0$ represents the interest nodes for that projection node i. When projection node i retrieves its interest nodes from the matrix U, it uses the minimum-spanning tree (MST) algorithm (plus all-pairs-shortest-paths (APSP) algorithm if needed) to construct a tree i, which connects all interest nodes to the projection node. Projection node i first considers itself as a one-node tree for constructing the tree. Next, it expands the tree, using the MST algorithm, adding all interest nodes that can be reached directly without any multi-hop. Then, if there are more interest nodes that have not been added to the tree, using the APSP algorithm, the nearest interest node through the shortest path is added to the tree. The APSP algorithm uses breath-first search (BFS) algorithm for finding the shortest path. Next, if still more interest nodes remain, the algorithm continues using the same strategy (using MST and APSP algorithm) until it connects all the interest nodes to the tree. The projection node i represents the root for the current tree. Then, according to the routing tree, each node knows its parent and child nodes and, similar to CDG, it multiplies its reading x_j with its coefficient φ_{ij} and gathers its data $\varphi_{ij} \chi_j$ with its descendants and sends the weighted sum to the parent node. When the root (projection node i) receives the weighted sum from its children, it transmits this sum $\sum_{j=1}^{n} \phi_{ij} \chi_j$ in one packet through the shortest path to the sink. There are m such trees in the network in all. Each tree represents one projection (weighted sum or sample measurement z_i) [40].

9.8 Summary

WSN is a promising solution for several real-life problems. Moreover, in recent years, it finds profound usage in several on-field applications. Due to the advent of the CMOS technology and the availability of low-cost imaging sensors, it is used in imaging applications also. As these applications are resource constrained, they pave the way to lifetime issues which can be resolved with the help of compressed sensing. CS can be used in data acquisition, data aggregation, routing, and recovery in WSN to enhance the performance and operational lifetime of the network.

References

[1] Karl, H., Willig, A. (2005). "Protocols and Architectures for Wireless Sensor Networks". (New York: John Wiley and Sons, Ltd). ISBN: 0-470-09510-5.

[2] Krishnamachari, B. (2005). "Networking Wireless Sensors", Cambridge University Press.

[3] Ferrari, G., Medagliani, P., Di Piazza, S., and Martal, M. (2007). "Wireless Sensor Networks: Performance Analysis in Indoor Scenarios", *Wireless EURASIP J Wireless Commun. Network.*, 81864, 14.

[4] Raghunathan, V., Schurgers, C., Park, S., and Srivastava, M. (2002). "Energy-Aware Wireless Microsensor Networks", *IEEE Signal Process. Mag.* 19, 2, 40–50.

[5] Santi, P. (2005). "Topology Control in Wireless Ad hoc and Sensor Networks", (New York: John Wiley & Sons, Ltd.).

[6] Wang, A., Cho, S.-H., Sodini, C. G., and Chandrakasan, A. P. (2001). "Energy-Efficient Modulation and MAC for Asymmetric Microsensor Systems", In *Proceedings of ISLPED 2001*, Huntington Beach, CA, August.

[7] Hill, J., and Culler, D. (2002). "MICA: A Wireless Platform for Deeply Embedded Networks", *IEEE Micro*, 22 (6), 12–24.

[8] Raghunathan, V., Schurgers, C., Park, S., and Srivastava, M. B. (2002). "Energy—Aware Wireless Microsensor Networks", *IEEE Signal Process. Mag.* 19, 40–50.

[9] Shih, E., Cho, S.-H., Ickes, N., Min, R., Sinha, A., Wang, A., and Chandrakasan, A. (2001) "Physical Layer Driven Protocol and Algorithm Design for Energy-Efficient Wireless Sensor Networks", In *Proceedings of the Seventh Annual International Conference on Mobile Computing and Networking 2001* (MobiCom), 272–286, Rome, Italy, July.

[10] Callaway, E. H. (2003) "Wireless Sensor Networks—Architectures and Protocols", Auerbach, Boca Raton, FL.

[11] Rappaport, T. S. (2002). "Wireless Communications—Principles and Practice", Prentice Hall, Upper Saddle River, NJ.

[12] Akyildiz, I. F., and Wang, X., (2005) "A Survey on Wireless Mesh Networks", IEEE Commun. Magazine, 43(9), S23–S30, September.

[13] Wang, C., Daneshmand, M., Li, B., and Sohraby, K. "A Survey of Transport Protocols for Wireless Sensor Networks".

[14] Wang, C., Sohraby, K., Hu, Y., Li, B., and Tang, W. (2005). "Issues of Transport Control Protocols for Wireless Sensor Networks", In Proc of 2005 International Conference on Communications, Circuits and Systems, 1, 422–426, May

[15] Pottie, G. J. and Kaiser, W. J. (2000) "Wireless Integrated Network Sensors", *Commun. ACM*, 43, 51–58.

[16] Barr, K. C., and Asanovi C. K. (2006) "Energy-Aware Lossless Data Compression", *ACM Trans. Comput. Syst.*, 24, 250–291.

[17] Anastasi, G., Conti, M., Di Francesco, M., and Passarella, A. (2009) "Energy Conservation in Wireless Sensor Networks: A Survey" *Ad Hoc Netw.*, 7, 537–568.

[18] Razzaque, M. A, Bleakley, C, and Dobson, S. (2013). "Compression in Wireless Sensor Networks: A Survey and Comparative Evaluation" *ACM Trans. Sens. Netw.* 10, 5:1–5:44.

[19] Alippi, C., Anastasi, G., Francesco, D., and Roveri, M. (2009). "Energy Management in Wireless Sensor Networks with Energy-Hungry Sensors", *IEEE Instrum. Meas. Mag.*, 12, 16–23.

[20] Cardei, M., Thai, M. T., Li, Y., and Wu, W. (2005) "Energy-Efficient Target Coverage in Wireless Sensor Networks", In *Proceedings of the IEEE Infocom*, Miami, FL, USA, 13–17 March, 1976–1984.

[21] Subramanian, R., and Fekri, F. (2006). "Sleep Scheduling and Lifetime Maximization in Sensor Networks: Fundamental Limits and Optimal Solutions", In *Proceedings of the 5th International Conference on Information Processing in Sensor Networks*, TN, USA, 19–21 April; 218–225.

[22] Wu, X., and Liu, M. (2012) "*In-situ* Soil Moisture Sensing: Measurement Scheduling and Estimation Using Compressive Sensing", In *Proceedings of the 11th International Conference on Information Processing in Sensor Networks*, Beijing, China, 16–19, 1–12.

[23] Bajwa, W. U, Haupt, J. D, Sayeed, A. M, and Nowak, R. D. (2007) "Joint Source-Channel Communication for Distributed Estimation in Sensor Networks", *IEEE Trans. Inform. Theory*, 53, 3629–3653.

[24] Vuran, M. C, Akan, O. B, and Akyildiz, I. F. (2004) "Spatio-Temporal Correlation: Theory and Applications for Wireless Sensor Networks", *Comput. Networking*, 45, 245–259.

[25] Baron, D., Wakin, M. B., Duarte, M. F., Sarvotham, S., and Baraniuk, R. G. (2005). "Distributed Compressed Sensing", Technical Report, Rice University: Houston, TX, USA.

[26] Abdur Razzaque, M., and Dobson, S. (2014). "Energy-Efficient Sensing in Wireless Sensor Networks Using Compressed Sensing", Sensors, 14, 2822–2859.

[27] Boopal, N., Gunasekaran, S., and Alamelu Mangai, V. (2015). "A Survey of Spatiotemporal Data Compression in Wireless Sensor Networks", *Int. J. Adv. Res. Comp. Eng. Tech.* (IJARCET), 4 (4), April.

[28] Kimura, L. S. (2005). "A Survey on Data Compression in Wireless Sensor Networks", *Proc. Int. Conf. Inform. Tech.: Coding Comput.* (ITCC'05), 1–6.

[29] Yang, C., Yang, Z., Ren, K., and Liu, C. (2011) "Transmission Reduction Based on Order Compression of Compound Aggregate Data Over Wireless Sensor Networks", *Int. Conference Pervasive Comput. App.* (ICPCA "11), 335–342.

[30] Candes, E., and Recht, B. (2009). "Exact Matrix Completion via Convex Optimization," Found. Comput. Math., 9, 717–772.

[31] Xiong, J., Member, IEEE, Zhao, J., and Chen, L. "Efficient Data Gathering in Wireless Sensor Networks based on Matrix Completion and Compressive Sensing", IEEE Communications Letters, Submitted Paper for Review.

[32] Hu, H., and Yang, Z. (2010). "Spatial Correlation-based Distributed Compressed Sensing in Wireless Sensor Networks", In *Proceedings of the 6th International Conference on Wireless Communications Networking and Mobile Computing (WiCOM)*, pp. 1–4, Sept.

[33] Tung Chou, C., Rana, R., and Hu, W. (2009). "Energy Efficient Information Collection in Wireless Sensor Networks Using Adaptive Compressive Sensing", In *Proceedings of IEEE 34th Conference on Local Computer Networks*, Zurich, Switzerland, pp. 443–450, October.

[34] Fedorov, V. V. (1972). "Theory of Optimal Experiments", Academic Press.

[35] MacKay, D. J. C. (1991). "Bayesian Interpolation", Neural Comp., 4.

[36] Ji, S., Xue, Y., and Carin, L. (2008). "Bayesian Compressive Sensing", *IEEE Trans. Signal Processing*.

[37] Seeger, M. W. (2008). "Bayesian Inference and Optimal Design for the Sparse Linear Model", *J. Machine Learn. Res*, 9, 759–813.

[38] Wang, W., Garofalakis, M., and Ramchandran, K. (2007). "Distributed Sparse Random Projections for Refinable Approximation", In: 6th International Symposium on Information Processing in Sensor Networks, 2007 (IPSN 2007), pp. 331–339.

[39] Luo, C., Sun, J., Wu, F., and Chen, C. W. (2009). "Compressive Data Gathering for Large-Scale Wireless Sensor Networks", In: Proc. ACM Mobicomn, pp. 145–156.

[40] Ebrahim, D., Ass, C. (2014). "Compressive Data Gathering Using Random Projection for Energy Efficient Wireless Sensor Networks", *Ad Hoc Networks*, 16, 105–119.

10

Efficient Sampling and Reconstruction Strategy for CS in WMSNS

10.1 Introduction

Tremendous growth in microelectronics has led to the development of low-cost and miniaturized cameras which paved way for the evolution of wireless multimedia sensor networks (WMSNs). However, it is still resource-constrained in terms of energy, memory, and bandwidth. Compressed sensing (CS) is a promising method that recovers the sparse/compressible signals from severely undersampled measurements. Its usage can considerably reduce the bandwidth and memory requirements. Hence, this chapter focuses on efficient strategies to improve the sampling and reconstruction process in CS.

10.2 Block CS and Non-Uniform Sampling CS

Block compressed sensing (BCS) is the methodology used to reduce storage requirement of the measurement matrix utilized in CS process and to enable parallel computing. Figure 10.1 depicts the BCS process [1].

The entire image is divided into small blocks, and CS measurements are taken for each block. At the receiver end, the elements of each block are recovered using recovery algorithms, and then, the inverse transform (IDCT/IDWT) is applied. However, in this chapter, DCT-based analysis is performed.

Cheng et al., have proposed NUS CS [2]. The image is divided into blocks, DCT transformed, and zigzag ordered, and then classified as important (coarse) and unimportant (fine) coefficients. Important coefficients (IC) are transmitted as such, and the CS measurements (Y) are taken over the unimportant coefficients. IC and Y are combined, and the combined measurements (CM) are transmitted to the receiving end. Figure 10.2 illustrates the non-uniform sampling CS process.

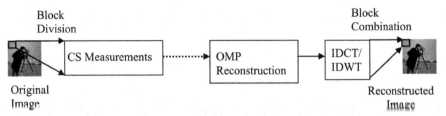

Figure 10.1 Block diagram of block compressed sensing.

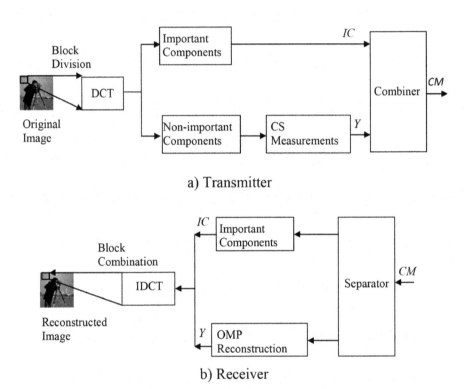

a) Transmitter

b) Receiver

Figure 10.2 Block diagram of non-uniform-sampling-based compressed sensing (a) transmitter (b) receiver.

Though it provides good quality with reduced measurements, it has increased complexity and measurement matrix changes from block to block, further increasing the burden. The number of measurements taken from each block remains the same, irrespective of the nature of the image.

10.3 Content-based CS

Block CS and NUS CS recover the image with reduced number of measurements. However, the size of the measurement matrix is the same for all blocks and this increases the storage requirement. Therefore, it is essential to use different measurement matrices for achieving an efficient recovery. To reduce the storage requirement and the number of measurements required for the recovery, the measurements taken from each block can be varied in proportion to the content of the image. Hence, content-based measurement matrix allocation using two-measurement matrix algorithm is proposed. The algorithm is utilized in both CS and NUS CS strategies.

When an image is divided into blocks of small size, the frequency component in each block will not be the same. A block with smooth variations in the gray level contains more large amplitude DCT coefficients. It is called the low-frequency component of the image. A block with sharp variations in the gray level contains only very few large amplitude DCT coefficients. It is called the high-frequency component of the image. This is shown in the following cameraman image in Figure 10.3. Block 1 is one of the high-frequency content blocks, and it has only one large amplitude DCT coefficient. Block 2 is one of the low-frequency content blocks, and it has approximately 28 large amplitude DCT coefficients. If the same-sized measurement matrix is used to measure both low- and high-frequency content blocks, a large number of measurements are needed. The proposed method uses two measurement matrices with different sizes.

The sizes are determined on the basis of the number of nonzero (N_{nz}) elements present in the quantized vector of a block. They are compared with a predetermined threshold (T) to fix the sparsity level that has to be retained in the reconstruction side. Low-frequency component of an image is assigned a higher sparsity level (S_h) and high-frequency component is assigned a lower sparsity level (S_l). The threshold is fixed after several trials with different set of images.

The measurement matrix with a large number of rows (larger K) is used to measure the low-frequency content blocks, and the measurement matrix with smaller number of rows (smaller K) is used to measure the high-frequency content blocks, thereby reducing the total number of measurements.

Based on the application (e.g., object detection), the number of measurements can be reduced further by sacrificing the quality of the image. The procedure for the two-measurement matrix allocation is summarized in Table 10.1. This variation leads to reduction in measurements since the nature of the image is exploited to take measurements.

Figure 10.3   Example of low-frequency and high-frequency blocks.

Table 10.1   Two-measurement matrix allocation algorithm

For each block in an image
Input: Quantization level n (≥ 1), T
Output: y

1. Convert the block of an image to a column vector X of size 64×1.
2. Quantize: $X_q = \frac{X}{2^n}$
3. Find the number of nonzero elements N_{nz}
4. Compare N_{nz} with a pre-fixed threshold, T, and set the sparsity level

$$K = \begin{cases} S_h & \text{if} \quad N_{nz} > T \\ S_l & \text{if} \quad N_{nz} \leq T \end{cases}$$

5. Generate a measurement matrix φ of size $K log(64) \times 64$. Take linear measurements, $y = \varphi x_q$

10.3.1 Binary DCT

The conventional floating-point method of DCT computation is very exhaustive as it uses multiplications and additions. The energy consumed for multiplications is higher than the energy consumed for addition or shifting operations. Tran has proposed DCT with lifting scheme called as binDCT [3]. It is derived from Chen's plane rotation-based factorizations of the DCT matrix. The forward and the inverse transforms in binDCT are implemented using only binary shift and addition operations. BinDCT is used in this thesis to attain energy efficiency. There are several families of such multiplier-less binDCT computations. The configuration employed in this work is that of C1-Chen's factorization [4].

10.3.2 Two-Measurement Matrix-based CS

The captured input image is divided into blocks of size $s \times s$ and spatially transformed using binDCT. DCT is chosen since it has high energy compaction. Block size is chosen as $s = 8$, to reduce the blocking artifacts and blurring effect in the reproduced image. Moreover, as the measurement matrix size is $M \times 64$ ($M \times s^2$), the increase in the block size will have a reasonable increase in the size and storage requirement of the measurement matrix.

The measurement matrix for an individual block is assigned by the two-measurement matrix allocation algorithm. The measurement matrix with a large number of rows is used to measure the low-frequency content blocks (φ_1), and the measurement matrix with less number of rows is used to measure the high-frequency content blocks (φ_2), thereby reducing the total number of measurements. Measurement matrix variation from block to block is passed as side information to the receiving end. Whenever (φ_1) is used, the side information is "1", and it is "0" if (φ_2) is used.

At the receiving end, the side information and measurement matrices are used to reconstruct the individual blocks. OMP is used as the reconstruction algorithm. The reconstructed blocks are inverse transformed and rearranged to get back the original image.

The system model used to reduce the total number of measurements on content basis is shown in Figure 10.4.

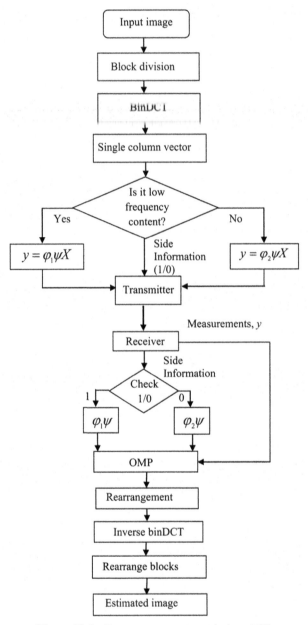

Figure 10.4 Two-measurement matrix-based CS.

10.3.3 Two-Measurement Matrix-based Non-Uniform Sampling CS

In NUS-based CS, the transformed coefficients in each block are separated into important and unimportant components [2]. Important components are the first few elements taken after zigzag ordering of the block. They are transmitted as such, whereas the remaining unimportant components undergo CS measurement. If the same measurement matrix is used to measure the unimportant components of all blocks of an image (i.e., the unimportant components of low-frequency content blocks and high-frequency content blocks are weighted equally), it results in large number of measurements. The proposed method uses two measurement matrices in non-uniform-sampling-based CS as shown in Figure 10.5.

The measurement matrix with a large number of rows is used to measure the unimportant component of low-frequency content blocks (φ_1) and the measurement matrix with less number of rows is used to measure the unimportant component of high-frequency content blocks (φ_2), thereby reducing the total number of measurements. The size of the important components of all blocks is kept constant. The important component is transmitted to the receiver directly along with CS measurements of unimportant components. The measurements are encoded using the conventional JPEG Huffman table [5], and the coded data are transmitted to the sink. The measurement matrices are assumed to be known to the transmitter and receiver. One extra bit (side information) per block is transmitted to inform the receiver that which measurement matrix is used for that block.

At the sink, the received data are decoded and the unimportant component is estimated using OMP algorithm. Then, the important component is concatenated with it.

The resultant single column vector is rezigzag ordered to form a matrix, and inverse transformation is applied to get the estimated block. These blocks are rearranged to form the estimated image. Here, the following two-level compression is performed to get better compression rate.

i. The required measurements to represent the image are reduced by CS.
ii. The redundancy in the measurements is removed by encoding it using the standard JPEG Huffman table.

The entire process is formulated to reduce the computational complexity, the amount of data to be processed and transmitted, and the storage requirement.

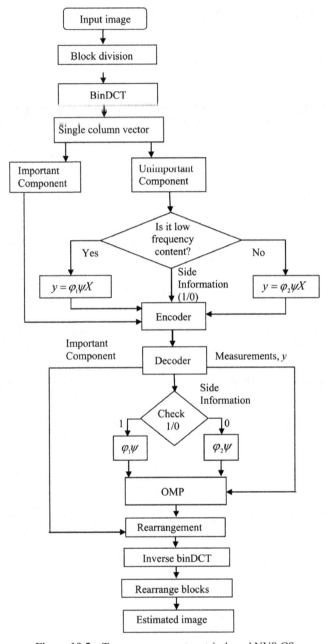

Figure 10.5 Two-measurement matrix-based NUS CS.

10.4 Performance Evaluations

Performance analysis for the proposed content-based model is evaluated based on the reduction in the number of measurements, image quality, and bit rate. The total number of measurements required for the reconstruction of an image is considered. PSNR is used to evaluate the image quality. The entire analysis is performed for both methods, namely two-measurement matrix with CS and two-measurement matrix with NUS CS in MATLAB.

10.4.1 Two-Measurement Matrix-based CS

The performance of the proposed method with CS measurements alone is analyzed in this section. Fixed sparsity level (K) is assumed for each block in single measurement matrix-based CS irrespective of the nature of the block in the image, whereas different sparsity levels are allotted in the two-measurement matrix-based CS. Simulations are performed and the performance of the proposed method is evaluated.

The results are shown for Lena (256 $\times$ 256), cameraman (256 $\times$ 256), cameraman (128 $\times$ 128), and Baboon (64 $\times$ 64) images. It indicates that there is nearly 4.2%, 6.7%, 6.2%, 2.8%, and 3.2%, 7.9%, 3.8%, and 0.67% reduction in measurements on an average, while using two-measurement matrix-based CS with DCT and binDCT respectively for the above images. The measurements with CS as the only process provide acceptable image quality with reduced measurements. The results given in Table 10.2 are for no quantization case. With quantization, a further reduction in the measurements can be achieved.

The recovered images for single measurement matrix CS (with $K = 6$) and two-measurement matrix-based CS (with $S_l = 4$ and $S_h = 6$) are shown in Figure 10.6 with the corresponding PSNR values and the percentage of total number of measurements.

10.4.2 Two-Measurement Matrix-based NUS CS

The use of two-measurement matrix-based NUS strategy produces the same range of PSNR with much reduced measurements when compared to the original NUS strategy. A comparison of one-measurement matrix-based NUS CS and proposed two-measurement matrix-based NUS CS for clock (256 $\times$ 256) image is shown in Figure 10.7. The graph clearly shows that the proposed method needs much less measurements than the

Table 10.2 Comparison of single-measurement matrix CS and two-measurement matrix CS

| Image | Single-Measurement Matrix CS | | | Two-Measurement Matrix CS | | | |
	Sparsity Level per Block (K)	Total Number of Measurements	PSNR (dB)	Sparsity Level S_l	S_h	Total Number of Measurements	PSNR (dB)
Lena 256 × 256	3	12288	25.4	2	3	10884	25.2
	6	25600	28.6	4	6	24768	28.6
	10	43008	31.4	7	10	38445	31.4
	13	55246	33.1	10	13	51084	33.1
Cameraman 256 × 256	3	12288	23.6	2	3	10372	23.6
	6	25600	26.3	4	6	21768	26.3
	10	43008	28.9	7	10	36781	28.8
	13	55246	30.5	10	13	49548	30.4
Cameraman 128 × 128	3	3072	21.9	2	3	2632	21.5
	6	6400	24.8	4	6	5520	24.5
	10	10752	27.3	7	10	9322	27.2
	13	13824	28.8	10	13	12504	28.7
Baboon 64 × 64	3	768	25.2	2	3	764	25.1
	6	1600	27.7	4	6	1592	27.7
	10	2688	29.9	7	10	2454	29.6
	13	3456	31.3	10	13	3240	31

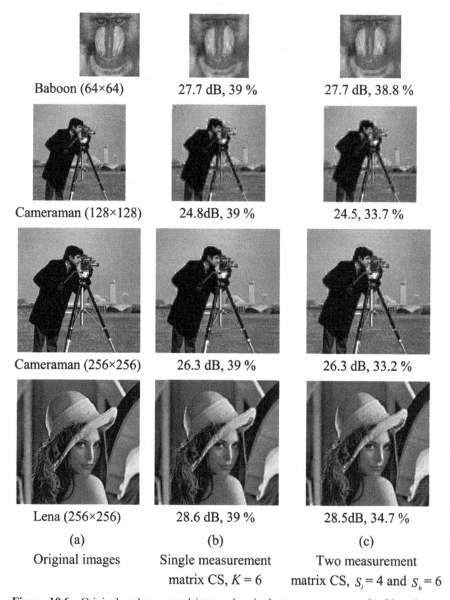

Baboon (64×64)	27.7 dB, 39 %	27.7 dB, 38.8 %
Cameraman (128×128)	24.8dB, 39 %	24.5, 33.7 %
Cameraman (256×256)	26.3 dB, 39 %	26.3 dB, 33.2 %
Lena (256×256)	28.6 dB, 39 %	28.5dB, 34.7 %
(a)	(b)	(c)
Original images	Single measurement matrix CS, $K = 6$	Two measurement matrix CS, $S_l = 4$ and $S_h = 6$

Figure 10.6 Original and recovered images by single-measurement matrix CS and two-measurement matrix CS.

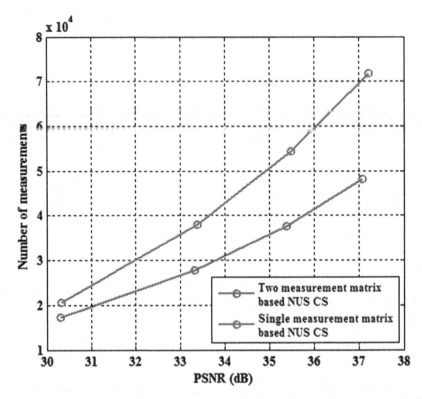

Figure 10.7   Total number of measurements required for varying image quality—NUS CS method for clock image (256 × 256).

one-measurement matrix-based NUS CS to achieve the same PSNR. It achieves nearly 26.67% reduction on an average in the number of measurements compared to single-measurement matrix-based NUS CS.

10.5 Greedy Algorithms

The signal recovery from the CS measurements of a sparse signal is done by sparse approximation. The measurements are the linear combination of K columns of the measurement matrix. The measurement vector has a K-term representation of the measurement matrix. Recovery of the original signal requires identification of the specific columns that participate in the measurement vector. The greedy algorithms use this correlation between the signal and the columns of the measurement matrix as a measure to find the elements with nonzero coefficients. There are several such algorithms in literature. The matching pursuit algorithms are concentrated in this thesis, due to its simple and fast implementation.

10.5.1 Orthogonal Matching Pursuit Algorithm

Tropp and Gilbert have proposed the OMP algorithm in which the sparse signal is recovered by choosing a highly correlated column of the measurement matrix for each iteration [6]. The indexes of the correlated columns are also accumulated in a counter consecutively. Signal estimate is found using the least-squares solution with the available column set. The contribution of the support set is removed from the residue to maintain orthogonality between the residual and the support set. Thus, the algorithm discovers a new column for each iteration. This column usually corresponds to the largest coefficient of the sparse signal. The original signal is recovered after K number of iterations. The procedure involved in OMP is provided in Table 3.1. It uses Gaussian measurement matrix to measure and recover the signals.

In some cases, OMP fails to pick up the correlated columns correctly. Hence, it needs more measurements to ensure exact recovery. Its advantage is speed and transparency. However, OMP has weaker guarantees of exact recovery.

10.5.2 Regularized Orthogonal Matching Pursuit Algorithm

Needell and Vershynin suggest ROMP, where K biggest columns of the measurement matrix are chosen and are regularized with the maximal energy concept [7]. It terminates in at most $2K$ iterations. It provides uniform guarantee of recovery. It can recover sparse signals from measurements taken using random Gaussian, Bernoulli, and partial Fourier matrices that satisfy RIP. The procedure involved in ROMP is provided in Table 10.3.

The disadvantage is that finding the least-squares solution for each iteration is tedious with more number of columns being included in the support set.

10.5.3 Ordered Orthogonal Matching Pursuit Algorithm

Baby and Pillai proposed the OOMP algorithm, an improvement to OMP. The wrongly detected columns in the support set are removed by comparing the projection coefficients and are scaled down to avoid reentry [8]. It achieves exact recovery with a reduced number of measurements when compared to

Table 10.3 Regularized OMP algorithm

Input:

- An $M \times N$ measurement matrix φ, $M \times 1$ measurement vector y.
- The sparsity level K of the ideal signal.

Procedure:

1. Initialize the residual, $r_0 = y$, index set, $\Lambda_0 = 0$ and $t = 1$.
2. Repeat the following steps K times.
3. Identify a set J of the K higher magnitude coordinates of the measurement vector $u = \varphi \times r_0$ or all of its nonzero coordinates, whichever is smaller.
4. Regularize the subsets $J_0 \subset J$ based on maximal energy $\|u|J_0|\|_2$.
5. Augment the index set, $\Lambda_t = \Lambda_{t-1} \bigcup J_0$.
6. Obtain a new signal estimate, $g = \arg \min \|u - \varphi_t z\|_2$.
7. Update the new residual, $r_0 = u - \varphi g$.
8. Reconstructed vector, $\hat{X} = g$.

OMP; however, the reduction is not much significant. It can recover sparse signals from measurements taken using random Gaussian matrix, whereas the measurements obtained using other random matrices are not tested. Moreover, the optimum values for the tolerance and reduction factors are not specified, and this plays an important role in identifying the wrongly detected columns and suppressing their reentry. The procedure involved in OOMP is provided in Table 10.4.

Table 10.4 Ordered OMP algorithm

Input:

- An $M \times N$ measurement matrix, φ, $M \times 1$ measurement vector, y.
- The sparsity level K of the ideal signal, tolerance (δ) and reduction (β) factors.

Procedure:

1. Find the index that has maximum argument, calculate the projection coefficients and residue $r_1 = y - \pi g_1$.
2. Set iteration counter $t = 2$, $k = 1$, values for threshold δ and reduction β factors. While $k \leq K$, do
3. Expand the index set $\Lambda_t = \Lambda_{t-1} \bigcup \{\lambda_t\}$ with the maximum correlated index and form the sub-matrix $\varphi_t = [\varphi_{t-1}\varphi_{\lambda_t}]$.
4. Find a new signal estimate: $g_t = \arg \min_g \|\varphi_t g - y\|_2$.
5. For $i = 1, 2, \ldots, K$, if $|g_t(i)| \leq \delta|g_{t-1}(i)|$ for some $i = j$, then

   a. Remove the corresponding index from the index set and the sub-matrix, update the residue g_t.
   b. Scale down the corresponding index $\varphi_{\Lambda_t(j)} = \beta\varphi_{\Lambda_t(j)}$.

6. Update the residue vector $a_t = \varphi_t g_t$, $r_t = y - a_t$.
7. Increment t and end while.

10.6 Enhanced Orthogonal Matching Pursuit Algorithm

The greedy algorithms considered have reduced guarantee of exact recovery with the requirement of considerably high number of measurements. Alternatively, reduction in the number of measurements for exact recovery with reduced complexity is essential for high-bandwidth applications. Therefore, the EOMP algorithm is proposed for recovery with two-column entries into the support set for each iteration, with wrong column detection and its suppression. The elements of the support set are checked to get rid of duplications in successive iterations. It provides exact recovery with a smaller number of measurements. The falsely selected columns, if any, are identified for each iteration and are removed from the support set. The approximation coefficients as in Step 6 of EOMP algorithm can also be found by $g_t = (\varphi_t^T \varphi_t)^{-1} \varphi_t^T y$. It is the projection of y on to the subspace spanned by the columns in φ_t. The projection coefficients are used to check whether a column in the present support set is falsely detected or not. To check the false detection, two parameters are included:

- Acceptance range, $\varepsilon < 1$.
- Suppression factor, $\alpha < 1$.

For every iteration, EOMP compares the projection coefficients with those obtained in the previous iteration. A column in the present support set is declared as wrongly detected if its present projection coefficient falls below ε times its value in the previous iteration. If any column fails this test; i.e., if its coefficient gets smaller, then that column will be removed from the support set. Furthermore, that column is scaled down by α so that the column is given less importance in the coming iterations. The EOMP algorithm is shown in Table 10.5.

In the proposed algorithm, two highly correlated columns are included in the index set after calculating the first estimate. If there is any duplication, it will be removed. Then, the current estimate is compared with the previous estimate, if it is up to the expected significant level, then the column identification is correct, if not the columns are taken as false identifications. The false columns are removed from the sub-matrix, and their contribution is also subtracted from the residual. They are suppressed by a factor of α. The algorithm tries to find the best approximation for the estimated input vector rather than relying on the highly correlated values.

The working of the EOMP depends mainly on the choice of the parameters ε and α. It also depends on the sparsity and size of the input vector. The data set $DS = (K, M, N)$, having an ensemble of random matrices φ of size

Table 10.5   Enhanced OMP algorithm

Input:

- An $M \times N$ measurement matrix φ.
- An $M \times 1$ measurement vector y.
- The sparsity level K of the ideal signal and the sparsity basis.
- Acceptance range, ε, and suppression factor, α.

Procedure:

1. Find $\lambda_1 = \arg\max_{j=1,2..N} |\langle y, \varphi \rangle|, \Lambda_1 = \{\lambda_1\}, \pi = \lfloor \varphi_{\lambda_1} \rfloor, g_t = (\varphi_t^t \varphi_t)^{-1} \varphi_t^t y$,
   residual, $r = y - \pi g_1$.
2. Initialize iteration counter $t = 2, k = 1$. Set values for ε and α. While $k \leq K$, do
3. Find $|\langle r_{t-1}, \varphi_j \rangle|$, take two maximum correlated indexes $\lambda_t = \{\lambda_{h1} \lambda_{h2}\}$.
4. Augment the index set $\Lambda_t = \Lambda_{t-1} \bigcup \{\lambda_t\}$ and the sub-matrix $\varphi_t = [\varphi_{t-1} \varphi_{\lambda_t}]$
   where φ_0 is an empty matrix. /Two element entry/
5. Remove the duplications in the index set and the sub-matrix /Duplication removal/
6. Find a new signal estimate: $g_t = \arg\min_g \|\varphi_t g - y\|_2$
7. For $i = 1, 2, \ldots, K$, if $|g_t(t-i)| \leq \varepsilon |g_{t-1}(t-i)|$ for some $i = j$, then /Wrong index detect/

   a. $\Lambda_t = \Lambda_t - \{\Lambda_t(j)\}, \varphi_t = \varphi_t - \{\varphi_t(j)\}$, /Wrong index removal/
   b. $g_t = \arg\min_g \|\varphi_t g - y\|_2$ /Residue updating/
   c. $\varphi_{\Lambda_t}(j) = \alpha \varphi_{\Lambda_t}(j)$ /Suppressing reentry/

8. Update the residue vector: $a_t = \varphi_t g_t, r_t = y - a_t$
9. Increment t, k and end while

$M \times N$ and an ensemble of K sparse vectors, is considered. The length of the data vector (N) is fixed as 64, since the block size of 8×8 is chosen in the processing of the image. Sparsity and the measurements are varied and the results are analyzed. The entire simulation is done using MATLAB.

10.6.1 Choice of Acceptance Range

The acceptance range, ε, is used to find the wrongly detected columns, as their entry reduces the projection coefficient values. If the current estimate value is less than ε times the previous estimate at one or more locations, then the related entries are assumed to be wrong and are removed from the support set.

To find the optimum choice for the acceptance range, the effect of the acceptance range is studied with $\alpha = 0.7$, $N = 64$, and the sparsity values of $K = 2, 6, 10$. The acceptance range is varied from 0.05 to 0.95 for all the individual cases. The percentage of recovery is calculated and is averaged over 1024 independent trials and is shown in Figures 10.8–10.10. The black, red, brown, green, and magenta curves correspond to $\varepsilon = 0.75, 0.8, 0.85, 0.9$, and 0.95, respectively. For $\varepsilon = 0.05 - 0.7$, the curve is plotted in blue color.

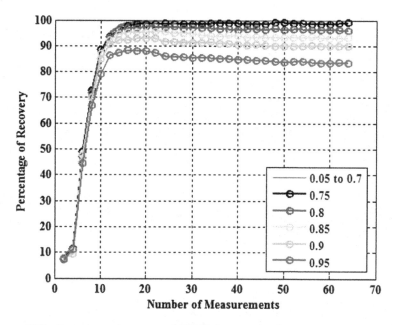

Figure 10.8   Percentage of recovery of EOMP for ε varying from 0.05 to 0.95 with $K = 2$.

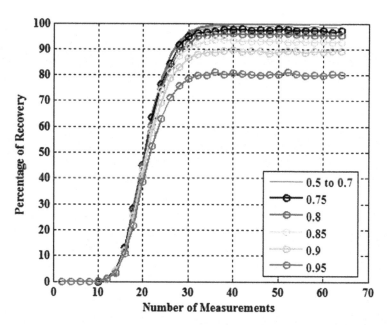

Figure 10.9   Percentage of recovery of EOMP for ε varying from 0.05 to 0.95 with $K = 6$.

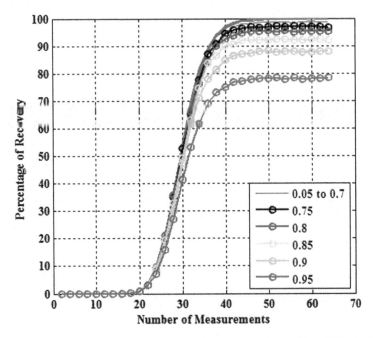

Figure 10.10   Percentage of recovery of EOMP for ε varying from 0.05 to 0.95 with $K = 10$.

In Figure 10.8, it is seen that the response for $\varepsilon = 0.05-0.7$ is not having much variation in terms of the percentage of recovery. So the lines get overlapped and are visible as a single line, whereas the percentage of recovery starts decreasing when the ε value is increased beyond 0.7.

With the increase in ε, the checking limit for the column to be identified as wrong detection is tightened. Therefore, the correct columns can also be identified as the wrongly detected ones; this leads to the performance degradation for the values of $\varepsilon > 0.7$. This is valid for the varying values of the sparsity also, which is shown in Figures 10.9 and 10.10.

In all cases, the system becomes unstable or has degraded performance for $\varepsilon > 0.7$, and as a consequence, the range above 0.7 is omitted. When $K = 2$, any value can be chosen in the range of 0.05–0.65 since there is no drastic degradation in the performance. However, when sparsity increases, $\varepsilon = 0.65$ and 0.6 has nearly 2–4% reduction in the percentage of recovery. Thus, the optimum range for ε is between 0.05 and 0.55. The performance has also been tested for varying values of sparsity. It is evident from the simulations that the specified range worked well for all sparsity levels. For forth coming analyses, acceptance range is fixed as $\varepsilon = 0.5$.

10.6.2 Choice of Suppression Factor

Suppression factor, α, is used to suppress the wrongly detected columns from entering into the support set again. It enables easy convergence toward the correct support set. The effect of suppression factor is analyzed by fixing the sparsity as 4 and the acceptable range ε as 0.5, and the α value is changed from 0.05 to 0.95. The corresponding percentage of recovery is plotted in Figure 10.11.

For each value of α, the number of measurements is varied from 0 to 64 and the percentage of recovery is observed. From Figure 10.11, it is evident that the variation in α does not affect the percentage of recovery. The change is less than 1% for all the cases; thus, all the curves are overlapped. So α value can be chosen independently from the range 0.05 to 0.95 as it has not much influence on the percentage of recovery. The percentage of success for a fixed measurement remains almost same for the entire range of variation of the α. Therefore, α can have any value between 0.05 and 0.95 as shown in Figure 10.12. The same is checked for different sparsity levels also.

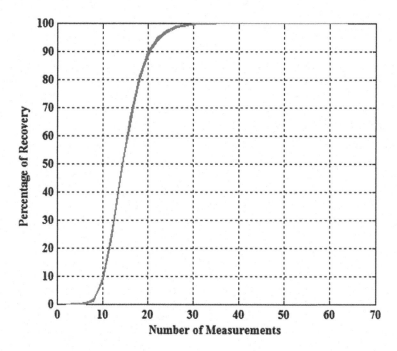

Figure 10.11 Percentage of recovery for α varied from 0.05 to 0.95 with $K = 4$, $\varepsilon = 0.5$, $N = 64$.

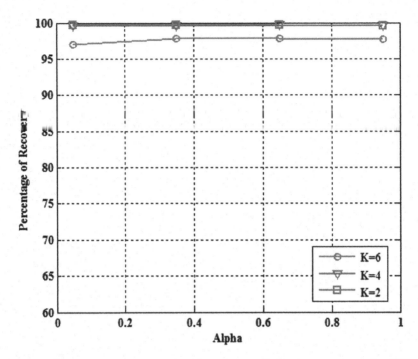

Figure 10.12 Effect of α on percentage of recovery under fixed measurements for different sparsity levels.

10.7 Performance Evaluations—Recovery Rate and Image Quality

The performance of the proposed algorithm is evaluated and compared with the OMP [6], ROMP [7], and OOMP [8]. Performance metrics such as percentage of recovery and PSNR are calculated and compared with the state-of-the-art algorithms.

Recovery rate analysis
Ensemble of input signals of size 64×1 is considered. The signals are assumed to be K sparse and are measured with varying number of measurements and recovered using the EOMP and OMP algorithms. The percentages of recovery for both cases are depicted in Figure 10.13.

The EOMP is simulated with $\varepsilon = 0.5$, $\alpha = 0.7$, and $N = 64$. The graph is drawn for $K = 2$, 4, and 6. From the graph, it is evident that EOMP has better percentage of recovery when compared to OMP. With 30 measurements

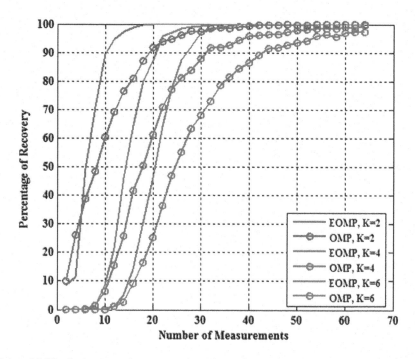

Figure 10.13 Percentage of recovery comparison of OMP and EOMP for different sparsity levels.

and sparsity $K = 6$, EOMP achieves 96% recovery, whereas OMP achieves only 70% recovery. It produces nearly 2–26% better recovery rate with the same number of measurements than the conventional OMP algorithm. The performance of EOMP is also compared with regularized OMP (ROMP) and ordered OMP (OOMP). The comparison is also done for $K = 4$ and $K = 6$. It is evident from the simulation results that EOMP outperforms both ROMP and OOMP. The algorithm performed well for other sparsity levels and different lengths of the input vector as well.

Image quality performance
The performance of the recovery algorithm for image applications is evaluated by considering an input image. The image is divided into blocks (8×8) and is converted to column vectors with specified sparsity. Then, the measurements are taken using the random binary matrix. The data are recovered with OMP, ROMP, OOMP, and EOMP algorithms. PSNR of the recovered images is tabulated in Table 10.6.

Table 10.6 Comparison of PSNR (dB) for different recovery algorithms

Image	Bridge			Cameraman			Gold Hill			Bird		
Sample Rate	20%	30%	40%	20%	30%	40%	20%	30%	40%	20%	30%	40%
OMP	15.22	21.61	24.11	15.4	20.87	23.9	16.88	23.74	26.6	18.5	28.11	31.87
ROMP	15.43	21.37	24.04	15.41	20.98	24	16.93	23.28	26.38	17.91	26.5	31.65
OOMP	15.72	21.25	23.96	15.8	21.22	23.59	16.59	23.56	26.28	18.25	25.63	30.99
EOMP	18.31	22.34	24.93	17.53	21.85	24.69	20.16	25.1	27.41	20.29	29.2	33.39

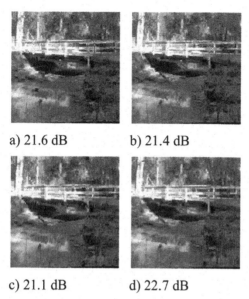

a) 21.6 dB   b) 21.4 dB

c) 21.1 dB   d) 22.7 dB

Figure 10.14   Bridge image recovered by (a) OMP, (b) ROMP, (c) OOMP, (d) EOMP (sample rate: 30%).

a) 21.2 dB   b) 21.1 dB

c) 21.0 dB   d) 21.7 dB

Figure 10.15   Cameraman image recovered by a) OMP, b) ROMP, c) OOMP, d) EOMP (sample rate: 30%).

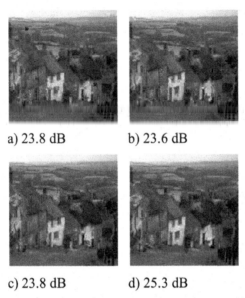

a) 23.8 dB   b) 23.6 dB

c) 23.8 dB   d) 25.3 dB

Figure 10.16   Gold hill image recovered by a) OMP, b) ROMP, c) OOMP, d) EOMP (sample rate: 30%).

a) 27.9 dB   b) 26.8 dB

c) 26.6 dB   d) 29.4 dB

Figure 10.17   Bird image recovered by a) OMP, b) ROMP, c) OOMP, d) EOMP (sample rate: 30%).

The PSNR value is averaged over 10 independent trials. The test images used are taken from the image database available in http://sipi.usc.edu/database and http://links.uwaterloo.ca. Recovery is done for different sample rates. The proposed EOMP has a better performance with almost the same level of complexity.

Table 10.6 shows a better performance for EOMP when compared to the existing algorithms. It achieves 1–3 dB increase in PSNR on an average. The images recovered by OMP, ROMP, OOMP, and EOMP with the sample rate of 30% is shown in Figure 10.14 for bridge, in Figure 10.15 for cameraman, in Figure 10.16 for gold hill, and in Figure 10.17 for bird image.

10.8 Summary

In this chapter, the two-measurement matrix algorithm and an enhanced orthogonal matching pursuit algorithm are proposed. The sampling algorithm is applied for both CS and NUS CS approaches. It leads to measurement reduction in both the approaches. However, it is more efficient for NUS CS. Simulation results confirm that the proposed two-measurement matrix algorithm performs better than the existing methods. The proposed reconstruction algorithm ensures better guarantees of recovery at reduced number of measurements. Though it picks two columns per iteration for analysis, it does not need any complex residue calculations. Simulation results show the effectiveness of the proposed algorithm in terms of percentage of recovery and PSNR.

References

[1] Sermwuthisarn, P., Auethavekiat, S., and Patanavijit, V. (2009). 'A Fast Image Recovery Using Compressive Sensing Technique with Block Based Orthogonal Matching Pursuit', In *Proceedings of International symposium on Intelligent Signal Processing and Communication Systems*, Japan, pp. 212–215.

[2] Cheng, Z., Chengyi, X., Ruixue, M., and Junbin, G. (2010). 'Compressed Sensing of Images using Nonuniform Sampling', In *Proceedings of International conference on Intelligent Computation Technology and Automation*, Shenzhen, Guangdong, Vol. 2, 483–486.

[3] Tran, T. D. (2000). 'The BinDCT: Fast Multiplierless Approximations of the DCT', *IEEE Signal Process. Letts.*, 7 (6), 141–144.

[4] Liang, J., and Tran, T. D. (2001). 'Fast Multiplierless Approximations of the DCT with the Lifting Scheme', *IEEE Trans. Signal Process.*, 49 (12), 3032–3044.

[5] Wallace, G. (1991). 'The JPEG still picture compression standard', Communs. ACM, 34 (4), 30–44.

[6] Tropp, J. A. and Gilbert, A. C. (2007) 'Signal Recovery from Random Measurements via Orthogonal Matching Pursuit', *IEEE Trans. Inform. Theor.* 53 (12), 4655–4666.

[7] Needell, D., and Vershynin, R. (2009). 'Uniform Uncertainty Principle and Signal Recovery via Regularized Orthogonal Matching Pursuit', *Founds. Comput. Math. J.* 9 (3), 317–334.

[8] Baby, D., and Pillai, S. R. B. (2012). 'Ordered Orthogonal Matching Pursuit', In *Proceedings of 2012 National Conference on Communications*, Feb, Kharagpur, India.

[9] http://sipi.usc.edu/database

[10] http://links.uwaterloo.ca

11

Real-World Applications of Compressive Sensing

11.1 Introduction

The field of compressive sensing (CS) is related to several areas in signal processing and computational mathematics, such as underdetermined linear-systems, group testing, heavy hitters, multiplexing, sparse sampling, sparse coding, and finite rate of innovation. Its broad scope and generality have enabled several innovative CS-enhanced approaches in signal processing and compression, solution of inverse problems, design of radiating systems, radar, the wall imaging, and antenna characterization. Imaging techniques having a strong affinity with CS include coded aperture and computational photography. Implementations of CS in hardware at different technology readiness levels are available.

Conventional CS reconstruction uses sparse signals (usually sampled at a rate less than the Nyquist sampling rate) for reconstruction through constrained ℓ_1 minimization. One of the earliest applications of such an approach was in reflection seismology which used sparse-reflected signals from band-limited data for tracking changes between subsurface layers. When the LASSO model came into prominence in the 1990s as a statistical method for selection of sparse models, this method was further used in computational harmonic analysis for sparse signal representation from over-complete dictionaries. One of the other applications is incoherent sampling of radar pulses. The work by Boyd et al. has applied the LASSO model for selection of sparse models toward analog-to-digital converters (the current ones use a sampling rate higher than the Nyquist rate along with the quantized Shannon representation). This would involve a parallel architecture in which the polarity of the analog signal changes at a high rate followed by digitizing the integral at the end of each time interval to obtain the converted digital signal.

413

CS has far reaching implications on compressive imaging systems and cameras. It reduces the number of measurements and hence power consumption, computational complexity, and storage space without sacrificing the spatial resolution. With the advent of a single-pixel camera (SPC) by Rice University, imaging system has seen drastic transformation. The camera is based on a single photon detector adaptable to image at wavelengths which were impossible with conventional CCD and CMOS images [1-3]. CS allows reconstruction of sparse n x n images by fewer than n^2 measurements. In SPC, each mirror in digital micromirror device (DMD) array performs one of these two tasks: either reflect light toward the sensor or reflect light away from it. Therefore, light received at sensor (photodiode) end is weighted average of many different pixels, whose combination gives a single pixel. By taking m measurements with random selection of pixels, SPC acquires a recognizable picture comparable to an n pixels picture. In combination with Bayer color filter, a SPC can be used for color images (hyperspectral camera). CS has recently gained a lot of attention due to its exploitation of signal sparsity. It is an inherent characteristic of many natural signals, which enables storage of the signal in just a few samples and recovered accurately. This section highlights the different areas of the application of CS with a major emphasis on communications and network domain [4].

Data captured by SPC can also be used for background subtraction for automatic detection and tracking of objects. The main idea is to separate foreground objects from the background ones, in a sequence of video frames. But this is quite expensive for wavelengths other than visible light. CS is being actively pursued for medical imaging, particularly in magnetic resonance imaging (MRI). Seismic data are usually incomplete, high-dimensional, and very large. Seismology exploration techniques depend on collection of data volume which is represented in five dimensions: two for sources, two for receivers, and one for time. Reduction in the number of sources and receivers which reduces the number of samples is a desirable matter considering the presence of high measurement and computational cost. Therefore, sampling technique require less number of samples simultaneously maintaining the quality of the image. CS solves this problem by combining sampling and encoding in one step, by its dimensionality reduction approach [4]. CS can also be used for biological applications due to its efficient and inexpensive sensing. Recent works show the usage of CS in comparative DNA microarray. CS has successfully been demonstrated to enhance resolution of wide-angle synthetic aperture radar. Resolution is improved by transmitting incoherent deterministic signals, eliminating the matched filter and reconstructing received signal using sparsity constraints [5]. CS is an attractive tool to acquire signals and network features in networked and communication systems.

CS has been used in communication domain for sparse channel estimation. Adoption of multiple antennas in communication system design and operation at large bandwidths, possibly in gigahertz, enables sparse representation of channels in appropriate bases. Conventional technique of training-based estimation using least-square (LS) methods may not be an optimal choice. Various recent studies have employed CS for sparse channel estimation. Compressed channel estimation (CCS) gives much better reconstruction using its nonlinear reconstruction algorithm as opposed to linear reconstruction of LS-based estimators. In addition to nonlinearity, CCS framework also provides scaling analysis. CCS-based sparse channel estimation has been shown to achieve much less reconstruction error while utilizing significantly less energy and, in some cases, less latency and bandwidth as well [6]. CS-based technique is used for speedy and accurate spectrum sensing in cognitive radio technology-based standards and systems. IEEE 802.22 is the first standard to use the concept of cognitive radio, providing an air interface for wireless communication in the TV spectrum band. Although no spectrum sensing method is explicitly defined in the standard, it has to be fast and precise. Fast Fourier sampling (FFS) is an algorithm based on CS which is used to detect wireless signals as proposed in [7]. In the emerging technology of ultra wideband (UWB) communication, CS plays a vital role by reducing the high data rate of ADC at receiver.

CS finds its applications in data gathering for large wireless sensor networks (WSNs), each consisting of thousands of sensors deployed for tasks such as infrastructure or environment monitoring. This approach of using compressive data gathering (CDG) helps in overcoming the challenges of high communication costs and uneven energy consumption by sending "m" weighted sums of all sensor readings to a sink which recovers data from these measurements. CS can be utilized for inexpensive compression at encoder, making every bit even more precious as it carries more information. CS is again utilized as a channel coding scheme [9] for enabling correct recovery of the compressed data after passing through erasure channels. Such compressive sensing erasure coding (CSEC) techniques are not a replacement of channel coding schemes; rather, they are used at the application layer, for added robustness to channel impairments and in low power systems due to their computational simplicity. CS provides an attractive solution as CS-based encoded images provide an inherent resilience to random channel errors. Since the samples transmitted have no proper structure, as a result, every sample is equally important and the quality of reconstruction depends on the number of correctly received samples only. To enhance the error resilience of images, joint source channel coding (JSCC) using CS is also an active area of research.

Another application domain for CS is traffic volume anomaly detection. In the backbone of large-scale networks, origin-to-destination (OD) traffic flows experience abrupt unusual changes known as traffic volume anomalies, which can result in congestion and limit the extent to which end user quality-of-service requirements are met. A major challenge in network data-mining applications is when the full information about the underlying processes, such as sensor networks or large online database, cannot be obtained in practice due to physical limitations such as low bandwidth or memory, storage, or computing power. The spectral methods used for volume anomaly detection can be directly applied to the CS data with guarantee of performance. Distributed source coding is a compression technique in WSNs in which one signal is transmitted fully and rest of the signals is compressed based on their spatial correlation with main signal. DSC performs poorly when sudden changes occur in sensor readings, as these changes reflect in correlation parameters and main signal fails to provide requisite base information for correct recovery of side signals.

11.2 Compressive Sensing for Real-Time Energy-Efficient ECG Compression on Wireless Body Sensor Nodes

Wireless body sensor networks (WBSNs) hold the promise to be a key enabling information and communication technology for next-generation patient-centric telecardiology or mobile cardiology solutions. Through enabling continuous remote cardiac monitoring, they have the potential to achieve improved personalization and quality of care; increased ability of prevention and early diagnosis; and enhanced patient autonomy, mobility, and safety. However, state-of-the-art WBSN-enabled ECG monitors still fall short of the required functionality, miniaturization, and energy efficiency. Among others, improvement in energy efficiency is possible through embedded ECG compression, for reducing airtime over energy hungry wireless links. In this section, the potential of the emerging compressed sensing (CS) signal acquisition compression paradigm is quantified for low-complexity energy-efficient ECG compression on the state-of-the-art Shimmer WBSN mote. The results show that CS represents a competitive alternative to state-of-the-art DWT-based ECG compression solutions in the context of WBSN-based ECG-monitoring systems. Its lower complexity and CPU execution time enable it to ultimately outperform DWT-based ECG compression in terms of overall energy efficiency. Achievement of CS-based ECG compression is accordingly shown as a achieve 37.1% extension in node lifetime relative to its DWT-based counterpart for "good" reconstruction quality.

The increasing prevalent cardiac diseases require escalating levels of supervision and medical management, which are contributing to skyrocketing healthcare costs and, more importantly, are unsustainable for traditional healthcare infrastructures. WBSN technologies promise offer of large-scale and cost-effective solutions to this problem. These solutions help the patients with wearable, miniaturized, and wireless sensors which are able to measure and report cardiac signals to telehealth providers wirelessly. They enable the required personalized, real-time, and long-term ambulatory monitoring of chronic patients with the caretaker or hospital. Its seamless integration with the patient's medical record and its coordination with nursing/medical support while the resting ECG monitoring is standard practice in hospitals, its ambulatory counterpart is still facing many technical challenges. For instance, the three-lead ECG is still nowadays recorded on a rather bulky and obtrusive commercial data logging (Holter) device for observing normal daily activities of a patient. The important limitations of these systems are limited autonomy, bulkiness, and limited wireless connectivity. Recently, the realization of wireless-enabled low-power ECG monitors for ambulatory use has received significant industrial and academic interest. The most important highlights of these research and development efforts are as follows: (1) Toumaz's Sensium Life Pebble TZ203082 [12], an ultra-small and ultra-low-power monitor for heart rate, physical activity, and skin temperature measurements with a reported autonomy of five days on a hearing aid battery; (2) Intel's Shimmer [11], a small wireless wearable sensor platform, is able to record and wirelessly transmit three-lead ECG data as well as accelerometer, gyroscope, and galvanic skin response information; and (3) IMEC's wireless single-lead bipolar ECG patch for ambulatory monitoring claiming over ten days of monitoring on a 160-mAh Li-ion battery (for undisclosed use conditions). The clinical relevance of the first system is still being validated, as Toumaz aims to achieve more than the system's so far established accurate measurement of heart rate. The second system, which is based on commercial off-the-shelf components such as the TI MSP430 microcontroller and the CC2420 radio chipset, operates on a Li-ion battery that provides about 1 Wh of energy. According to the measurements, it is able to support a maximum of 6-and-a-half day single-lead raw ECG sensing and storage on local memory. This autonomy figure is reduced by 25%, when the raw ECG data are wirelessly streamed using the ultra-low-power CC2420 in a perfect point-to-point link with no wireless protocol overhead.

This autonomy figure dramatically decreases mostly under-realistic ambulatory monitoring. IMEC ultra-low-power wireless biopotential sensor node achieves its enhanced autonomy due to a proprietary customized ultra-low-power analog read-out ASIC, signal acquisition and amplification. More importantly, it has dedicated signal processors to preprocess and compress the sensed data using state-of-the-art techniques, for reducing the airtime over power-hungry wireless links. The achievement of truly WBSN-enabled ambulatory monitoring systems requires more breakthroughs not only in terms of ultra-low-power read-out electronics and radios, but also increasingly in terms of ultra-low-power-dedicated digital processors and associated embedded feature extraction and data compression algorithms. This has been acknowledged on the basis of the premises mentioned.

A thorough system-level performance comparison between the state-of-the-art thresholding-based DWT algorithm and a CS-based algorithm in the context of ECG data compression on the embedded Shimmer wireless mote has been performed and is shown in Figure 11.1. It consists of three processing stages: A linear transformation is first applied to the original ECG signal, followed by an optional "sparsification" stage, and a final encoding stage which outputs the compressed signal to be wirelessly transmitted. As highlighted in Figure 11.1, the fundamental difference between the two approaches consists of the former algorithm explicitly exploiting the sparsity of the ECG signal by computing its sparse expansion and *adaptively* encoding the coefficients of this expansion. The latter algorithm *non-adaptively* acquires a few random measurements of the ECG signal and only implicitly relies on ECG signal sparsity to guarantee accurate reconstruction. This section further describes in detail the two algorithms and introduces the corresponding data models.

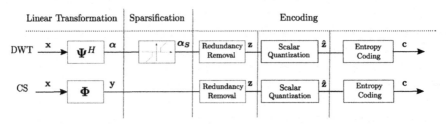

Figure 11.1 Block diagram of the two ECG compression schemes implemented on the Shimmer wireless mote.

11.2.1 CS-based Compression Algorithm

Sensing and processing information have traditionally relied on the Shannon sampling theorem, one of the central tenets of digital signal processing. This theorem states that, given a signal of bandwidth Ω, it is sufficient to sample it at 2Ω samples per second (i.e., the Nyquist rate) to ensure faithful representation and reconstruction. However, this traditional ADC paradigm has been challenged later. First, there are many situations where Ω is so large that constraints put on sampling architectures are simply intolerable. Second, even for relatively low signal bandwidths such as our target wearable ECG application, given the established sparsity of the ECG signal, (above) Nyquist-rate sampling produces a large amount of redundant digital samples, which are costly to wirelessly transmit, and severely limit lifetime of the sensor nodes. If one sets course to design energy-efficient embedded ECG sensors, it is desirable to reduce the number of acquired ECG samples by taking advantage of the sparsity, or the reduced "information rate" of the ECG signal. CS is a methodology that has been recently proposed to address this problem [12–14]. It is also particularly well suited for low-power implementations as it dramatically reduces the need for resource (both processing and storage)-intensive digital signal-processing operations. As aforementioned, the original ECG signal **x** has a sparse approximation; i.e., it can be represented by a linear superposition of S elements of an orthonormal wavelet basis, $x \approx \sum_{k=1}^{S} \alpha_k \psi_k^2$, with $S << N$. Conventionally, ECG samples can be collected at the Nyquist rate forming **x** and *then* compressing it using nonlinear digital compression techniques. CS offers a striking alternative by showing roughly S samples can be collected using simple *analog* measurement waveforms, thus sensing/sampling and compressing at the same time. Moreover, by merging sampling and compression steps, CS removes a large part of the digital architecture. This so-called analog CS, where the compression occurs in the analog sensor read-out electronics prior to ADC, is our ultimate goal. Its demonstration still requires extensive work on the analog sensor read-out electronics. Consequently, in this work, it is approached through "digital CS," where the linear CS compression is applied after the ADC [15].

11.3 Real-Time Compressive Sensing MRI Reconstruction Using GPU Computing and Split Bregman Methods [16]

CS has been shown to enable dramatic acceleration of MRI acquisition in some applications. Being an iterative reconstruction technique, CS MRI reconstruction can be more time-consuming than traditional inverse Fourier reconstruction. CS MRI reconstruction is accelerated by factors of up to 27 by using a split Bregman solver combined with a graphics processing unit (GPU) computing platform. The increase in speed found is similar to that measured for matrix multiplication on this platform, suggesting efficient parallelizing of the split Bregman methods efficiently. The combination of the rapid convergence of the split Bregman algorithm and the massively parallel strategy of GPU computing can enable real-time CS reconstruction of acquisition data matrices of dimension 4096^2 or more, depending on available GPU VRAM. Reconstruction of two-dimensional data matrices of dimension 1024^2 and smaller took ~ 0.3 s or less, showing that this platform also provides very fast iterative reconstruction for small-to moderate-sized images.

CS in MRI [17] has the potential for significant improvement of both the speed of acquisition and the quality of MR images, but requires an iterative reconstruction that is more computationally intensive than the traditional inverse Fourier reconstruction. CS accelerates MR acquisitions by reducing the amount of data requiring acquisition. Reconstruction of this partial data set is then accomplished by iteratively constraining the resulting image to be sparse in some domain while enforcing consistency of the measured subset of Fourier data. One practical barrier to the routine adoption of CS MRI is the delay between acquisition and reconstruction of images. CS solvers work almost entirely with vector and image arithmetic, making them an excellent candidate for acceleration through using GPUs for parallelization. Here, it is illustrated that how GPUs can be used to achieve significant increases in speed of CS reconstructions of large MRI data sets. GPU computing means using the GPU to perform general-purpose scientific and engineering computation. High-end video cards can contain hundreds of separate floating-point units, allowing for massively parallel computations at a fraction of the cost of CPU-based supercomputers, as measured on a per gigaFLOP basis (one gigaFLOP is equivalent to one billion floating-point operations per second). GPU computation works best in a single-instruction, multiple-data (SIMD) situations, such as the solution of large systems of linear equations. The power of GPU computing is already being realized in several advanced medical image reconstruction applications [18–21].

The optimization problem chosen for exploration is the reconstruction of partial MRI data sets using CS. MRI is proving to be a fertile application for CS as many MR images can be highly compressed by transform coding with little loss of information. Conventional MRI data are acquired according to the Fourier sampling pattern required to satisfy the Nyquist criterion while producing an image of a given field of view and spatial resolution. Only a random subset of the full-Nyquist Fourier sampling scheme is acquired for sampling the Fourier domain in a compressive manner. Reconstruction of the conventional fully sampled MRI data requires simply an inverse Fourier transform,

$$u = F^{-1}b, \tag{11.1}$$

but the inverse problem becomes underdetermined when part of the Fourier data is omitted, so approximate methods must be used. One highly successful method in particular has been to formulate the CS MRI reconstruction as a sparse recovery problem, in which an image is found that is consistent with the acquired Fourier data while having the sparsest representation in a chosen basis (e.g., gradient and wavelet). The typical formulation of the reconstruction of a complex image u from a partial Fourier data set b is then

$$u = \arg\min_{x} \|x\|_1 + \lambda \|Ax - b\|_2^2 \tag{11.2}$$

where A is a measurement operator that transforms the sparse representation x to the image domain and then performs the subsampled Fourier measurement, and the $l1$ and $l2$ norms are, respectively,

$$\|x\|_1 = \sum_i |x_i|$$

$$\|x\|_2^2 = \sum_i \bar{x}_i x_i \tag{11.3}$$

where the bar denotes complex conjugation. With the addition of the $l1$ norm, the problem is more difficult to solve, and iterative techniques such as interior point methods, iterative soft thresholding, and gradient projection are typically employed.

11.3.1 Software

The open-source split Bregman code of Goldstein and Osher [22] was chosen as the starting point for the GPU-based CS solver. The solver was originally written in Matlab. This solver was chosen for its rapid convergence and lack of array reduction steps, which hinders parallelization. The original code was modified to work with Jacket 1.8.0 (AccelerEyes, Atlanta, GA) and Matlab R2010b (Mathworks, Natick, MA). CUDA Toolkit 4.0 and CUDA developer driver 270.41.19 were used for all computations. Algorithm 1 briefly outlines the procedure for running the split Bregman reconstruction on the GPU. Note that the split Bregman algorithm runs for a fixed number of iterations, so there is no variation in run time due to different descent trajectories as with a tolerance-based stopping criterion. Furthermore, the choice of image reconstruction has no bearing on the results, since the fixed number of iterations ensures functionalities same number of operations on any input data set. At the beginning of the reconstruction, the Fourier data in main memory must be transferred to the GPU with Jacket's *gsingle* and *gdouble* Matlab commands.

Next, temporary storage is allocated on the GPU using Jacket's *gzero* and *gones* commands. All subsequent arithmetic operations, including the Fourier transform, are carried out on the GPU using function overloading. Function overloading simplifies code syntax by enabling a single function to encapsulate different functionality for different types of arguments. The specific behavior is typically chosen by the compiler or at run time. In our case, Matlab automatically calls the Jacket library when an operation is requested on a matrix that lies in GPU memory, while identical operations on a matrix in main memory are carried out with Matlab's built-in functions. After the last loop iteration on the GPU, the solution is transferred back to main memory with overloaded versions of Matlab's *double* and *single* commands; all temporary storage is automatically freed.

11.3.2 Experiments

Two numerical experiments were performed. The first was a pure matrix multiplication, designed to measure practical peak floating-point performance of the CPU and GPU as realized by Jacket 1.8.0. With a view to remove dependencies on the multithreading performance of Matlab R2010b and provide easier comparison of the CS reconstructions, the CPU experiment was run both with and without multithreading enabled. The second experiment is CS MRI reconstruction of a $T1$-weighted breast image subjected to a 50% undersampling in Fourier (spatial frequency) space. Total variation was used as the sparsity constraint and was defined as the sum of the magnitudes of pixels in the gradient image. CS MRI reconstruction was performed for powers of two image sizes ranging from 32^2 to 8192^2 (up to 4096^2 only for double precision, due to memory limitations). This covers the range of realistic MR acquisition matrix sizes for 2D scans with allowance for specialty techniques at very-low- or very-high-resolutions or future developments in imaging capabilities. The largest matrix sizes can also be indicative of the performance of three-dimensional reconstruction problems. For example, a 256^3 data set is the same size as a 4096^2 one. The CPU-based reconstructions were performed with and without multithreading enabled.

The entire reconstruction was done eleven times, with the first iteration discarded and the following ten iterations averaged. This avoids biasing the results with startup costs associated with both Jacket and CUDA. Jacket uses just-in-time (JIT) compilation to improve the performance of repeated function calls, so the first call to the Jacket library is slowed by this compilation step. The CUDA driver, upon initial invocation, optimizes the low-level GPU code for the particular hardware being used. These two processes increase code performance across multiple runs but reduce it for the initial function calls. For timing experiments, the startup penalty is a confounding factor. Accurate timing thus requires that the reconstruction be "warmed up" with a similar problem before timing the full-scale computation. Here, the simplest approach of discarding the first iteration is used. In principle though, the warm-up problem can be much smaller in data size as long as it uses the same set of Jacket functions needed in the full-size reconstruction.

The significant acceleration of GPU computation is shown in this section. CS MRI reconstruction of all but the smallest of the tested image sizes. Most of the theoretical performance of the GPU can be realized over the combination of Matlab- and Jacket-processing package through requirement of minimal code development. The speedup realized by the GPU for the smallest images was progressively hampered by communication overhead, while the largest images suffered from the limited GPU RAM. The optimal image dimensions, however, seem to be beneficially close to those of the high-resolution MRI data; so for a rapid CS MRI reconstruction, GPU computing coupled with the Osher split Bregman and Goldstein algorithm appears to be a well-suited platform. Future improvements to these methods include algorithm modifications to allow unified reconstruction of multiple two-dimensional or a single three-dimensional Fourier data set on the GPU with a single call to the reconstructor, thus reducing the communication penalty. The split Bregman algorithm parallelizes extremely well and so additional cores on the GPU card could allow higher acceleration. Finally, the GPU allows larger data sets to be reconstructed more efficiently.

11.4 Compressive Sensing in Video Surveillance

A compressive sensing method combined with decomposition of a matrix formed with image frames of a surveillance video into low-rank and sparse matrices is proposed to segment the background and extract moving objects in a surveillance video. The video is acquired through compressive measurements, which are used for the reconstruction of the video by a low-rank and sparse decomposition of matrix. The low-rank component represents the background, while the sparse component is used for identifying moving objects in the surveillance video. The decomposition is performed by an augmented Lagrangian alternating direction method. Experiments are carried out to demonstrate that moving objects can be reliably extracted with a small amount of measurements.

In a network of cameras for surveillance, a massive number of cameras are deployed, some with wireless connections. The cameras transmit surveillance videos to a processing center where the videos are processed and analyzed. Surveillance video processing is able to detect anomalies and moving objects in a scene automatically and quickly. Detection of moving objects is traditionally achieved by background subtraction methods which segment background and moving objects in a sequence of surveillance video frames. The mixture of Gaussians technique assumes that each pixel having a distribution that is a sum of Gaussians and the background and foreground is modeled by the size of the Gaussians. In low-rank and sparse decomposition, the moving objects are identified by a sparse component and the background is modeled by a low-rank matrix. These traditional background subtraction techniques require all pixels of a surveillance video to be captured, transmitted, and analyzed.

Bandwidth is the major challenge in the network of cameras. Since traditional background subtraction requires acquisition of all pixels of video, an enormous volume of data is transported in the network due to a large number of cameras. At the same time, most of the data are uninteresting due to inactivity. There is a high risk of the network being overwhelmed by the mostly uninteresting data to prevent timely detection of anomalies and moving objects. Therefore, it is highly desirable to have a network of cameras in which each camera transmits a small amount of data with enough information for reliable detection and tracking of moving objects or anomalies. CS allows achievement of this goal. In CS, the surveillance cameras make compressive measurements of video and transmit measurements in the network. Since the number of measurements is much smaller than the total number of pixels, transmission of measurements, instead of pixels, helps prevention of network congestion. Furthermore, the lower data rate of compressed measurements helps wireless cameras to reduce power consumption.

When a surveillance video is acquired through compressive measurements, the pixel values of the video frames are unknown, and consequently, direct application of the traditional background application techniques is not feasible. A straightforward approach is to recover the video from the compressive measurements and then, after the pixel values are estimated, to apply one of the known background subtraction techniques. But such an approach is undesirable for two reasons. First, a generic video reconstruction algorithm does not take advantage of special characteristics of surveillance video in which a well-defined, relatively static background exists. The existence of a background provides prior information that helps reduction in the number of measurements. Secondly, in the straightforward approach, additional processing is needed to perform background subtraction after the video is recovered from the measurements.

A method is proposed in this section for segmentation of background by using a low-rank and sparse decomposition of matrix. In this method, the compressive measurements from a surveillance camera are used to reconstruct video which is assumed to comprise a low-rank and sparse component. As in [17], the low-rank component is the background, and the sparse component identifies moving objects. Therefore, the background subtraction becomes part of the reconstruction, with no additional processing being needed after reconstruction. Furthermore, the reconstruction takes advantage of the knowledge that there exists a background in the video, which helps to reduce the number of measurements required. The proposed method is inspired by the work of [17] and extends it to the measurement domain, rather than the pixel domain, for use in conjunction with CS. This method is motivated where a matrix equation is solved with the assumption that the solution is a sum of a low-rank matrix and a sparse matrix for 4D-CT reconstruction. CS has been used in background subtraction previously, but the method requires the pixel values of the background to be known a priori, such as acquired from a training process. The method of this paper may be considered to be the training process in which the compressive measurements are used to obtain the background.

Low-rank and sparse decomposition constitute an effective method for processing surveillance video when it is combined with CS. This is a good reconstruction method of the surveillance video as it takes advantage of the well-defined low-rank and sparse components in the surveillance video signal. The background subtraction and moving object extraction come from the process of reconstruction at no additional cost. The processing is done after the acquisition of a large number of frames (using compressive measurements), and therefore, it is not done in real time. It is possible to extend this concept to "online," real-time processing by adaptively updating the low-rank component. The framework of our method is shown in Figure 11.2. The video is treated as black and white, having only the luminance component.

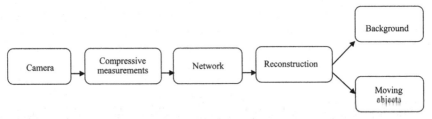

Figure 11.2   Compressive video sensing framework.

11.5 An Application of Compressive Sensing for Image Fusion

Wireless sensor networking is a technology that promises unprecedented capability for monitoring and manipulating the physical world via a network of densely distributed wireless sensor nodes, which can sense the physical environment in a variety of modalities, including image, radar, acoustic, video, seismic, thermal, and infrared. In WSNs, how to fuse multiple-sensed information is very challenging. Image fusion is an important issue in the field of digital image processing. Traditional image-fusion algorithms are always difficult for meeting the practical demands of real-time and low bit-rate transmission in WSNs because of their huge amount of calculation. In recent years, CS has inspired significant interests because of its compressive capability. It provides great inspiration to balance the relationship between the quality of fused images and the computation complexity. The focus is on the pixel-level fusion problem of infrared and visible images of the same scene. Literature [9] claims decomposition of the fused image composed by the images through two multi-resolution basis functions in succession shows better quality than the image fused in a single multi-resolution domain. As wavelet function and other multi-resolution tools are often used as sparse basis in CS, inspiration is provided for application of the idea of blending two multi-resolution functions to CS image fusion [18].

11.5.1 Image Fusion in CS Domain

With the development of the CS theory, CS has many practical applications in recent years. It is also an attractive scheme for image fusion. Literature in some areas has reported researches on image fusion in CS domain [11–14]. The core idea of these is summarized in Figure 11.1, and it shows that the core idea of applying CS to image fusion is to fuse the measurements of the two input images in CS domain, and then, the composite measurements can be used to reconstruct the fused image by a nonlinear procedure. Wavelet transform is a widely used sparse transform and a traditional image multi-resolution analysis tool. It is often used for image sparse decomposition. So, it is more common to use wavelet as a sparse basis in CS-based image fusion. However, wavelet transform does not have the superiority of anisotropy on the presentation of two-dimensional signals. So, edges of the images fused by wavelet-based algorithm tend to be blurred, motivating the exploration of a new way to combine the advantages of different multi-resolution analysis tools in image-fusion process.

11.5.2 Introduction to Multi-Resolution Analysis Tools

In the pixel-level image fusion based on transform domain, the commonly used multi-resolution analysis tools are wavelet transform, pyramid transform, contourlet transform, and so on. In this section, two multi-resolution analysis methods, wavelet transform and NSCT (non-subsampled contourlet transform), are selected for comparative analysis. It can be seen that the two basis functions have their own features and their advantages are complementary.

11.5.2.1 Wavelet transform

Wavelet transform is a multi-resolution analysis tool which has extensive use. It can decompose the signals into different scales with different levels of resolution by dilating a prototype function, that is, to decompose the signals into shifted and scaled versions of the mother wavelet [15]. Any details of the signals can be focused adaptively by wavelet transform. So it is called "digital microscope." It also shows good performance in two-dimensional signal processing such as enhancement, image denoising, and fusion. However, since the 2D wavelet transform has only limited numbers of direction, it cannot express the high-dimensional signals that have optimal line singularity. But line singularity is a typical performance of the edges in natural images. Wavelet transforms show its deficiency in the processing of edge signals.

11.5.2.2 Non-subsampled contourlet transform

Non-subsampled contourlet transform is proposed based on contourlet transform. It has not only the frequency characteristics of multi-resolution, but also the feature of being anisotropic, with capacity to have a good grasp of the geometry of the images. The basic idea of NSCT is to use the non-subsampled pyramid decomposition for decomposing the image into multiple scales.

The signals of each scale are decomposed into different directional sub-bands later, through the non-sampled directional filter bank. The number of sub-bands can be any power of two in each scale. NSCT has no downsampling process in the two-step decomposition, so it has the feature of translation invariant [16]. Since NSCT has the directional characteristics, its advantage of image-edge processing is obvious.

11.5.2.3 The idea of blended basis function

The above analysis on the characteristics of wavelet transform and NSCT helps understanding of the complementary advantages of the two algorithms. Literature [9] proposes a novel algorithm that combines two multi-resolution analysis functions for fusing the image. The method provides results that are than the traditional multi-resolution-based image fusion. The process of decomposing signals by two basis functions successively is called a blended basis functions representation. Considering that blended basis functions have given promising results in multi-resolution-based image fusion, it is proposed that the application of CS image fusion is based on blended basis functions.

11.5.3 Applying Blended Basis Functions to CS Image Fusion

In the multi-resolution analysis of image signals, the low-frequency components are not as sparse as the high-frequency components. So, it is proposed that fusion of the two kinds of components separately. First, NSCT is employed for decomposing the image into multi-scales. Then, the high-frequency NSCT components are sparsely represented by wavelet basis, while the low-frequency parts can be fused in the NSCT domain directly. Since the high-frequency NSCT coefficients have the sparse features, their sparsity is enhanced, after being sparsely represented again by wavelet transform.

The algorithm steps are listed below:

1. Decompose the two input images by NSCT and divide the coefficients into high-frequency parts and low-frequency parts according to their layers.
2. Fuse the low-frequency components of the two images according to the low-frequency fusion rule in NSCT domain directly.
3. Sparsely represent the high-frequency components by wavelet basis.
4. Obtain the compressed measurements matrix with the sampling rate M rate.
5. Fuse the measurements of the high-frequency components according to the high-frequency fusion rule in CS domain.

6. Reconstruct the fused high-frequency components via OMP algorithm and apply inverse wavelet transform on them.
7. The fused image is obtained by inverse NSCT transform.

11.5.4 Image Fusion in the Compressive Domain

In this section, an image-fusion algorithm is formulated that uses compressive measurements for the fusion of multiple images into a single representation. Recent theoretical results show if the signal is sparse or nearly sparse in some basis, then with high probability, the measurements essentially encode the salient information in the signal. Further, the unknown signal can be estimated from these compressive measurements to within a controllable mean-squared error [1, 2]. In this sense, similar fusion schemes are applied to that used in the wavelet domain in the compressive domain, so the difference is that image fusion is performed on the compressive measurements rather than on the wavelet coefficients.

Algorithm 1: Compressive image-fusion algorithm

1. Take the compressive measurements Yi, $i = 1, \ldots, I$ for the ith input image using the double-star-shaped sampling pattern.
2. Calculate the fused measurements using the formula: $YF = YM$ *with* $M = arg$ max $i = 1, \ldots, I$ $(|Yi|)$.
3. Reconstruct the fused image from the composite measurements YF via the total variation optimization method [1].

11.5.5 Advantages of CS-based Image Fusion

CS-based image fusion has a number of advantages over conventional image-fusion algorithms. It offers computational and storage savings by using a CS technique. Compressive measurements are progressive in the sense that larger numbers of measurements will lead to higher quality reconstructed images. Image fusion can be performed without acquiring the observed signals. Additionally, the recently proposed compressive imaging system [7], which relies on a single photon detector, enables imaging at new wavelengths inaccessible or prohibitively expensive using current focal plane imaging arrays. This new development of the imaging system has made it promising to exploit the CS-based image fusion in a practical use. This will significantly reduce the hardware cost and meanwhile expand image fusion in modern military and civilian imaging applications in a cheaper and more efficient way. However, the compressive measurements lose spatial information due to the

CS measurement process. Therefore, traditional image-fusion rules operating on local knowledge cannot be applied to compressive image fusion.

11.6 Sparse MRI: The Application of Compressive Sensing for Rapid MR Imaging

The sparsity which is implicit in MR images is exploited to significantly undersample k-space. Some MR images such as angiograms are already sparse in the pixel representation; other, more complicated images have a sparse representation in some transform domain—for example, in terms of spatial finite differences or their wavelet coefficients. According to the recently developed mathematical theory of CS, images with a sparse representation can be recovered from randomly undersampled k-space data, provided an appropriate nonlinear recovery scheme is used. Intuitively, artifacts due to random undersampling add as noise-like interference. Significant coefficients stand out above the interference in the sparse transform domain. A nonlinear thresholding scheme can recover the sparse coefficients, effectively recovering the image itself. In this section, practical incoherent undersampling schemes are developed and analyzed by means of their aliasing interference. Incoherence is introduced by pseudorandom variable–density undersampling of phase encodes. The reconstruction is performed by minimizing the $\ell1$ norm of a transformed image, subjects to data fidelity constraints. Examples demonstrate improved spatial resolution and accelerated acquisition for multi-slice fast spin–echo brain imaging and 3D contrast enhanced angiography.

The sparsity of MR images

The transform sparsity of MR images can be demonstrated by applying a sparsifying transform to a fully sampled image and reconstructing an approximation to the image from a subset of the largest transform coefficients. The sparsity of the image is the percentage of transform coefficients sufficient for diagnostic quality reconstruction. Of course, the term "diagnostic quality" is subjective. Nevertheless, it is possible for specific applications to get an empirical sparsity estimate by performing a clinical trial and evaluating reconstructions of many images quantitatively or qualitatively. To illustrate this, experiment on two representative MR images was performed: an angiogram of a leg and a brain image. The images were transformed by each transform of interest and reconstructed from several subsets of the largest transform coefficients. The results are depicted in Figure 11.3.

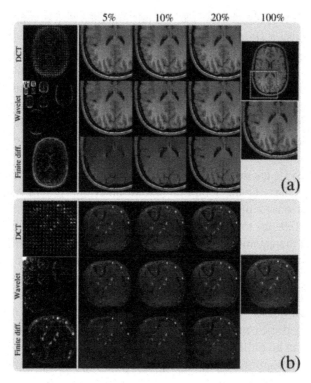

Figure 11.3 Transform domain sparsity of images.

The left column images show the magnitude of the transform coefficients; they illustrate the sparsity of transform coefficients is more than the images themselves. The DCT and the wavelet transforms have similarly good performance with a slight advantage for the wavelet transform for both brain and angiogram images at reconstructions involving 5–10% of the coefficients. The finite-difference transform does not sparsify the brain image well. Nevertheless, finite differences do sparsify angiograms because they primarily detect the boundaries of the blood vessels, which occupy less than 5% of the spatial domain.

11.7 Shortwave Infrared Cameras

In general, cameras used for imaging in short- and mid-wave infrared spectra are significantly more expensive than their counterparts in visible imaging. As a result, high-resolution imaging in that spectrum remains beyond the reach of most consumers. Over the last decade, CS has emerged as a potential means to realize inexpensive shortwave infrared cameras. One approach for doing this is the SPC where a single detector acquires coded measurements of a high-resolution image. A computational reconstruction algorithm is then used for recovering the image from these coded measurements. Unfortunately, the measurement rate of a SPC is insufficient to enable imaging at high spatial and temporal resolutions [19]. A focal plane array-based compressive sensing (FPA-CS) architecture that achieves high spatial and temporal resolutions is presented.

An array of SPCs uses that sense in parallel to increase the measurement rate and also to achieve spatio-temporal resolution of the camera. Shortwave infrared cameras based upon CS are available. These cameras have light sensitivity from 0.9 μm to 1.7 μm, which are wavelengths invisible to the human eye. The theory and practice of CS to enable high-resolution SWIR imaging from low-resolution sensor arrays is used. CS relies on the ability to obtain arbitrary linear measurements of the scene; this requires a fundamental redesign of the architecture used for imaging the scene. The SPC is an example of such architecture [20]. The SPC uses a DMD as a spatial light modulator and acquires coded measurements of an image onto a single photodetector. An SWIR SPC can be built by employing a photodetector sensitive to SWIR along with the appropriate choice of optical accessories. The sum of any arbitrary subset of pixels can be obtained by the programmable nature of the DMD which enables photo detection. Then, the high-resolution image can be computationally reconstructed from a small number of such measurements.

SPC is incapable of producing high-resolution images at video rate. The measurement rate of a SPC is determined by the operating speed of its DMD which seldom goes beyond 20 kHz. At this measurement rate, conventional Nyquist-based sampling with a single pixel can barely support a 20-fps video at a spatial resolution of 32×32 pixels. A measurement rate of tens of millions of measurements per second is needed for sensing at a megapixel resolution and video rate using a CS-based SPC.

11.7.1 Applications of SWIR Imaging

A large number of applications that are difficult or impossible to perform using the visible spectrum become much simpler due to the characteristics of the SWIR spectrum. SWIR imaging is currently used in a host of applications including automotive, electronic board inspection, product inspection, solar cell inspection, identification and sorting, anti-counterfeiting, surveillance, process quality control, and much more [21]. Some unique properties of SWIR that enable many of these applications include (a) improved penetration through scattering media including tissue, fog, and haze (b) seeing through many commonly used packaging materials which are transparent in SWIR while opaque in visible spectrum, and (c) observing defects and gauging quality of fruits and vegetables. Finally, for night-time surveillance application, the nightglow of the sky provides sufficient SWIR illumination even on a moonless night; this enables long-distance SWIR imaging without the need for extra illumination sources that could compromise reconnaissance.

11.7.2 Spatio-Temporal Resolution (STR)

11.7.2.1 Nyquist cameras STR

The STR of a camera is limited by the product of the number of pixels and the maximum frame rate. For example, a 1-megapixel sensor operating at 30 fps provides a measurement rate, which is denoted as Mr, equals to 30×10^6 samples per second. Traditional cameras rely on the principle of Nyquist sampling; thus, for such cameras, STR = Mr. Pixel count and frame rate of visible sensors have significantly improved. It is now common to obtain sensors that can achieve megapixel resolution at 30 fps. Unfortunately, sensors outside the visible spectrum either provide a much lower spatial resolution or are quite expensive.

11.7.2.2 Compressive cameras STR

Let us consider a CS-based camera operating at measurement rate, M_r samples per second, that can provide a high-resolution video with αM_r pixels per seconds. The effective STR can be written as STR = αM_r, where $\alpha \geq 1$ represents the compression factor by which the sampling rate gets reduced due to the CS framework. In a SPC, the measurement rate is typically limited by the maximum rate at which synchronization can be achieved between the DMD modulator and the sensor. While a photodetector can be operated at very high rates (even GHz), commercially available DMD seldom operates faster than 10–20 kHz. Hence, it is not possible to achieve synchronization between any of the current high-resolution spatial light modulators and a photodetector at greater than f_{DMD} = 20 kHz. This directly imposes a limit on the STR of compressive cameras based on single-pixel sensors; i.e., STR $\geq \alpha M r = \alpha f_{DMD}$ samples per second.

11.7.2.3 Increasing the measurement rate

The previous section brings out the need to increase the measurement rate for increasing the STR of CS-based imaging systems. Given that the operating speed of the DMD poses strictly limits the number of frames that can be obtained in unit time, one approach is to increase the measurement rate by reading multiple measurements in parallel. As an example, a compressive imaging system, in which a $K \times K$ pixel image sensor array is used for acquiring multiplexed measurements in synchronization with a DMD at an operational rate f_{DMD} Hz, provides a measurement rate of $M_r = K^2 f_{DMD}$ samples per second—a K^2 times improvement over the SPC. This increased measurement rate enables the acquisition of videos at higher spatial and temporal resolution. In the next section, SWIR prototype is discussed that uses a 64×64 focal plane array sensor along with a DMD operating at $f_{DMD} = 480$ Hz to achieve measurement rates in millions of samples per second.

11.8 Holography

CS can be used for improving image reconstruction in holography by increasing the number of voxels one can infer from a single hologram.

11.8.1 Compressive Fresnel Holography

Let us consider the one-dimensional free-space propagation formula in the Fresnel approximation, which relates the complex values of a propagating wave, measured in a plane perpendicular to the direction of propagation and separated by a distance z.

$$U_z(x) = A \exp\left\{\frac{j\pi}{\lambda z}x^2\right\} \int U_{in}(\xi) \exp\left\{\frac{j\pi}{\lambda z}\xi^2\right\} \exp\left\{\frac{-j2\pi}{\lambda z}x\xi\right\} d\xi.$$

$$(11.4)$$

where λ is the wavelength of the light wave and A is a multiplicative constant. If the latter equation can be written in discrete form (sampling uniformly on a Cartesian grid) by defining $x = n\Delta_d$ and $\xi = m\Delta_o$ changing the integral to summation,

$$U_z(n) = A \exp\left\{\frac{j\pi\Delta_d^2}{\lambda z}n^2\right\} \cdot \sum_m U_{in}(m)\exp\left\{\frac{j\pi\Delta_o^2}{\lambda z}m^2\right\}$$

$$\exp\left\{\frac{-j2\pi\Delta_d\Delta_o}{\lambda z}mn\right\} d\xi = A\, Q_z F_{1/\lambda z} Q_z u_{in} \qquad (11.5)$$

where the last equality is the matrix–vector form of the discrete Fresnel transform, u_{in} is a one-dimensional vector representing the object, Q_z is a diagonal matrix whose elements are the discrete quadratic-phase elements of the Fresnel transform, and $F_{1/\lambda z}$ is a scaled version of the discrete Fourier transform. Invalidity of the discrete approximation (2) for small propagation distances z is seen.

Application of CS scheme on Fresnel digital holograms is attractive for two reasons. First is the availability of the entire complex field. This is in contrast to most imaging applications where variations in the squared modulus of the propagating optical field are detected. This, in turn, implies that the sensing operator has strictly positive entries. Strictly positive sensing operators are unsuitable for CS applications, since they are relatively large in number. Second, the Fourier and Fresnel transforms are closely related. As z increases, Q_z becomes the identity canonical basis, and the Fresnel transform becomes the Fourier transform (yielding the Fraunhofer approximation). The quadratic-phase factor in (1) shows the Fresnel approximation behaving like the Fourier transform in a small area by the origin. This infers the need for more samples to be taken around the origin for taking full advantage of the incoherence properties of the Fourier–wavelet transform. Inline digital holograms are classically reconstructed using linear operators to model diffraction. It has long been recognized that such reconstruction operators do not invert the hologram formation operator. Classical linear reconstructions yield images with artifacts such as distortions near the field of view boundaries or twin images. When objects located at different depths are reconstructed from a hologram, in focus and out of focus images of all objects superimpose upon each other. Additional processing, such as maximum of focus detection, is thus inevitable for any successful use of the reconstructed volume [22, 23].

11.8.2 Compressed Sensing with Off-Axis Frequency-Shifting Holography

This section reveals an experimental microscopy acquisition scheme successfully combining CS and digital holography in off-axis and frequency-shifting conditions. A genuine CS-based imaging scheme for sparse-gradient images, acquisition of a diffraction map of the optical field with holographic microscopy, and recovery of the signal from as little as 7% of random measurements are proposed. The experimental results help inference of the ability of CS to lead an elegant and effective method for the reconstruction of images opening the door for new microscopy applications [24]. A novel microscopy imaging framework combines an iterative image reconstruction and digital holography for performing quadrature-resolved random measurements of an optical field in a diffraction plane. The CS approach enables optimal image reconstruction while being robust to high noise levels. The proposed technique is expected to improve many microscopy applications, allowing the acquisition of high-dimensional data with reduced acquisition time increasing imaging throughput and opening the door to sample friendly acquisition protocols.

11.8.3 Off-Axis Compressed Holographic Microscopy in Low-Light Conditions

An acquisition protocol relying on a single exposure of a randomly under sampled diffraction map of the optical field, recorded in high heterodyne gain regime, is proposed. The image acquisition scheme is based on CS, a theory establishing the possibility of near-exact recovery of an unknown sparse signal from a small number of non-structured measurements. Image reconstruction is further enhanced through introduction of an off-axis spatial support constraint to the image estimation algorithm. Accurate experimental recovery of holographic images of a resolution target in low-light conditions is reported with a frame exposure of 5 µs, scaling down measurements to 9% of random pixels within the array detector. Off-axis holography is well-suited to dim light imaging. Shot-noise sensitivity in high optical gain regime can be achieved with few simple setup conditions [25]. Holographic measurements are made in dual domains, where each pixel exhibits spatially dispersed (i.e., multiplexed) information from the object. The measurement domain and the image domain are "incoherent," which is a requirement for using CS sampling protocols. In particular, CS approaches using frequency-based measurements can be applied to holography sampling the diffraction field in amplitude and phase. In biological imaging, images are typically compressible or sparse in some domain due to the homogeneity, compactness, and regularity of the structures of interest. Such property can be easily formulated as mathematical constraints on specific image features. CS can be viewed as a data acquisition theory for sampling and reconstructing signals with very few measurements. Instead of sampling the entire data domain and then compressing it for taking advantage of redundancies, CS enables compressed data acquisition from randomly distributed measurements. Image reconstruction relies on an optimization scheme enforcing some specific sparsity constraints on the image. CS can be used for improving image reconstruction in holography by increasing the number of voxels one can infer from a single hologram and canceling artifacts. CS is also used for image retrieval from undersampled measurements in millimeter-wave holography and off-axis frequency-shifting holography.

11.8.4 Millimeter-Wave Compressive Holography

An active millimeter-wave holographic imaging system that uses compressive measurements for three-dimensional (3D) tomographic object estimation is described. The system records a two-dimensional (2D) digitized Gabor hologram by translating a single-pixel incoherent receiver. Two approaches for compressive measurement are undertaken: nonlinear inversion of a 2D Gabor hologram for 3D object estimation and nonlinear inversion of a randomly subsampled Gabor hologram for 3D object estimation. The object estimation algorithm minimizes the convex quadratic problem using total variation (TV) regularization for 3D object estimation. Object reconstruction is compared using linear back-propagation and TV minimization. Simulated and experimental reconstructions are presented from both the compressive measurement strategies. In contrast to back-propagation, which estimates the 3D electromagnetic field, TV minimization estimates the 3D object that produces the field. Despite under sampling, range resolution is consistent with the extent of the 3D object band volume [26].

11.9 Network Tomography

Network tomography is the study of internal characteristics of a network using information derived from end point data. The word tomography is used for linking the field to other processes that infer the internal characteristics of an object from external observation. It is done in MRI or positron emission tomography (even though the term tomography strictly refers to imaging by slicing). The field is a recent development in electrical engineering and computer science, initiated in 1996. Network tomography advocates the possibility of mapping the path data through the Internet by examining information from "edge nodes," the computers in which the data are originated and from which they are requested.

The field is useful for engineers who attempt development of highly efficient computer networks. Data derived from network tomography studies can be used for increasing the quality of service by limiting link packet loss and increasing routing optimization [27].

Monitoring of link properties (delay, loss rates, etc.) within the Internet has been stimulated by the demand for network management tasks such as fault and congestion detection or traffic management. This would help network engineers and Internet service providers (ISP) to keep track of network utilization and performance. The need for accurate and fast network monitoring method has increased further in recent years due to the complexity of new services (such as videoconferencing, Internet telephony, and online games) that require high-level quality-of-service (QoS) guarantees. In 1996, the term "network tomography" was coined for encompassing these classes of approaches that seek inference of internal link parameters and identification of the link congestion status. Current network tomography methods can be broadly categorized as follows:

- *Node-oriented:* These methods are based on cooperation among network nodes on an end-to-end route using control packets, for example, active probing tools such as ping or trace route, measure, and report attributes of the round-trip path (from sender to receiver and back) based on separate probe packets [28]. The challenges for such node-oriented methods arise from the fact that many service providers do not own the entire network and hence do not have access to the Internal nodes [29].
- *Edge-oriented:* In networks with a defined boundary, availability of access is available to all nodes at the edge (and not to any in the interior) is assumed. A boundary node sends probes to all (or a subset) other boundary nodes for measuring packet attributes on the path between network end-to-end points. Clearly, these edge-based methods do not require exchanging special control messages between interior nodes.

A novel approach to estimate link delay in a network is based on the idea of binary CS, which has received significant attention in the past few years. Upper bound is used on delay recovery of the links inside the network using an end-to-end probe sending method. Sending probes between nodes on the boundary of a network comes with the cost of increasing traffic inside the network. Thus, the design of routing matrix for a given network for minimizing the number of injected probes while the network remains 1-identifiable is showed. It is the simplest identifiability problem, and as shown here, it is a significant challenge worth study. A possible future research avenue is to extend the work here for networks which are $k > 1$ identifiable.

11.10 Applications of Compressed Sensing in Communication Networks

CS has attracted much interest among the research community and found wide-ranging applications in astronomy, biology, communications, image and video processing, medicine, and radar since its inception in 2006. CS has also found successful applications in communication networks and also in the detection and estimation of wireless signals, source coding, multi-access channels, data collection in sensor networks, and network monitoring. In many cases, CS has shown the ability to bring performance gains on the order of 10X.

CS can be applied in various layers of communications networks. CS can be used in detecting and estimating sparse physical signals such as UWB signals, MIMO signals, and wideband cognitive radio signals at the physical layer. CS can also be used as erasure code. At the MAC layer, CS can be used for implementing multi-access channels. At the network layer, CS can be used for data collection in WSNs, where the sensor y signals are usually sparse in certain representations. CS can be used to monitor the network itself, where network performance metrics are sparse in some transform domains at the application level. In many cases, CS was shown to bring performance gains on the order of 10X.

11.10.1 Applications of CS in the Physical Layer

CS can be used for detecting and estimating sparse physical signals, such as MIMO signals, wideband cognitive radio signals, and UWB signals. The details are provided below.

A. MIMO signals

Channel state information (CSI) is essential for coherent communication over multi-antenna (MIMO) channels. Convention holds that the MIMO channel exhibits rich multi-path behavior and the number of degrees of freedom as proportional to the dimension of the signal space. However, in practice, the impulse responses of MIMO channel actually are dominated by a relatively small number of dominant paths. This is especially true with large bandwidth, long signaling duration, or large number of antennas. Because of this sparsity in the multi-path signals, CS can be used to improve the performance in channel estimation [30].

B. Wideband cognitive radio signals

Dynamic spectrum access (DSA) is an emerging approach for solving today's radio spectrum scarcity problem. Key to DSA is the cognitive radio (CR) that can sense the environment and adjust its transmitting behavior accordingly by not causing interference to other primary users of the frequency. Thus, spectrum sensing is a critical function of CR, whereas wideband spectrum sensing faces hard challenges. There are two major approaches to do wideband spectrum sensing. A bank of tunable narrowband filters is used for searching narrowbands one by one. The challenge for this approach lies in the require ment of a large number of filters requiring use, leading to high hardware cost and complexity. Then, a single RF front end and DSP are used for searching the narrowbands. The challenge in this approach lies in the requirement of very high sampling rate and processing speed for wideband signals. CS can be used to overcome the challenges mentioned above. Today, a small portion of the wireless spectrum is heavily used, while the rest is partially or rarely used fcc2002. Thus, the spectrum signal is sparse and CS is applicable.

An analog approach:

In this approach, CS is directly performed on the analog signal, which has the advantage of saving the ADC resources, especially in cases where the sampling rate is high. A parallel bank of filters is used for acquiring measurements y_i. With a view to reduce the number of filters required, which is equal to the number of measurements M and can be potentially large, each filter samples time-windowed segments of signal. Let NF denotes the number of filters required, and NS denotes the number of segments each filter acquires. As long as NFNS = M, the measurement is sufficient. The performance for an OFDM-based CR system with 256 subcarriers where only 10 carriers are simultaneously active shows that a CS system with 8–10 filters can perform spectrum sensing at 20/256 of the Nyquist rate.

C. Ultra-wideband (UWB) signals

UWB communication is a promising technology for low-power, high-bandwidth wireless communications. In UWB, an ultrashort pulse, in the order of nanoseconds, is used as the elementary signal for carrying information. The advantages of UWB are as follows: (1) The implementation of the transmitter is simple because of the use of base-band signaling; (2) UWB has little impact on other narrowband signals on the same frequency range, since its power spreads out on the broad frequency range. However, one of the challenges for UWB is that it requires extremely high sampling rate (several GHz) to digitize UWB signals based on the Nyquist rate, leading to high cost in hardware. Since UWB signals are sparse in the time domain, CS can be applied which provides an effective solution to this problem by requiring much lower sampling rate.

11.10.2 CS Applications as Erasure Code and in the MAC Layer

A. CS as erasure code

Many physical phenomena are compressible or sparse in some domains. For example, virtual images are sparse in the wavelet domain, while sound signals are sparse in the frequency domain. The conventional approach is to use source coding for compressing the signal first and for using erasure coding for protection against the missing data caused by the noisy wireless channel. Let x denotes the n-dimensional signal. Let S and E denote the $m \times n$ source coding matrix and $l \times m$ erasure coding matrix, respectively. When the signal is k-sparse in some domain, m is close to k. If the expected probability of missing data is p, then $l = m/(1 - p)$. The transmitted signal is ESx, and the received signal is CESx, where the linear operator C models the channel. C is a sub-matrix of the identity matrix Il with e rows deleted, e being the number of erasures. If $e \leq l - m$, the decoding at the receiver is successful; otherwise, the data cannot be decoded and are discarded. CS can be used as an effective erasure coding method. In [30], CS was applied in WSNs as erasure code. At the source l, measurements of the signal are generated by random projections $y = \phi x$ and sent out, where ϕ is a $l \times n$-dimensional random matrix. The received signal is $y' = Cy$, with e measurements erased. Suppose each measurement carries its serial number, where the erasure occurred is known and therefore the matrix C. At the receiver, the standard CS procedure is carried out.

$$x = \arg \min \|\Psi x\|_1 \text{ subject to } y' = C\Phi x \qquad (11.6)$$

where Ψ is the representation, under which x is sparse. Since erasure occurs randomly, $C\Phi$ is still a random matrix. CS performs data compression and erasure coding in one stroke. Information about the signal is spread out among l measurements, of which m measurements are expected to be correctly received, with m on the same order of k, the sparsity of the data.

CS has compression performance similar to the conventional erasure coding methods but with two outstanding advantages: (1) CS allows graceful degradation of the reconstruction error when the amount of missing data exceeds the designed redundancy, whereas the conventional coding methods do not. Specifically, if $e \leq l - m$, conventional decoding cannot recover the data at all. However, if RIP holds, CS can still recover partial data, with an error no larger than that of the approximate signal, which keeps the largest $l - e$ elements of the sparse signal and sets the rest of the elements to zero. (2) In terms of energy consumed in the processing, performance of CS erasure coding is 2.5 times better than performing local source coding and 3 times better than sending raw samples.

B. CS in on–off random access channels

In [31], a connection was made between CS and on–off random access channels. In an on–off random multiple access channel, there are N users communicating simultaneously to a single receiver through a channel with n degrees of freedom. Each user transmits with probability λ. Typically, $\lambda N < n << N$. User i is assigned as code word and n-dimensional vector φ_i. The signal at the receiver from user i is $\varphi_i x_i$, where x_i is a nonzero complex scalar if the user is active and zero otherwise. The total signal at the receiver is given by

$$y = \sum_{i=1}^{N} \varphi_i x_i + \omega = \Phi x + \omega y$$

where ω is the noise, $x = [x_1, x_2, \ldots, x_N]$, and $\phi = [\varphi_1, \varphi_2, \ldots \varphi_N]$ is the codebook. The active user set is defined by

$$\Omega = \{ i : x_i \neq 0 \}$$

The goal of the receiver is to estimate Ω. Since $|\Omega| << n$, there is a sparse signal detection problem. A formulation in terms of CS is as follows:

$$\hat{x} = \arg \min_{x} \mu \|x\|_1 + \|y - \Phi x\|_2^2$$

where $\mu > 0$ is an algorithm parameter that weights the importance of sparsity in $\hat{x}$. It was shown in [31] that the CS-based algorithms perform better than single-user detection in terms of the number of measurements required to recover the signal and have some near-far resistance. At high signal-to-noise ratio (SNR), CS-based algorithms perform worse than the optimal maximum-likelihood detection. However, CS-based algorithms are computationally efficient, whereas the optimal maximum-likelihood detection is not computationally feasible.

11.10.3 Applications of CS in the Network Layer

As mentioned before, most natural phenomena are sparse in some domains, while CS could be effective in WSN that monitors such phenomena. CS was used for gathering data in a single-hop WSN. Sensors transmit random projections of their data simultaneously in a phase-synchronized channel.

The base station receives the summation of the randomly projected data, which constitutes a CS measurement. The l_1-norm minimization is performed at the base station for recovering the sensory data. Three approaches for reducing the measurement costs are as follows: (1) using the joint sparsity in data for reducing the number of measurements, (2) using sparse random projection for reducing number of measurements, and (3) using data routing for reducing the cost of data transport.

11.11 Summary

Magnetic resonance imaging (MRI) is an important application of CS which accelerates MR acquisitions by reducing the amount of the required data. Real-time Compressive Sensing MRI Reconstruction Using GPU Computing and Split Bregman Methods. GPU computation significantly accelerated CS MRI reconstruction of all but the smallest of the tested image sizes. Most part of the theoretical performance of the GPU can be realized over the combination of Matlab and Jacket-processing package by requiring minimal code development. Surveillance video processing is able to detect anomalies and moving objects in a scene automatically and quickly. The video is acquired by compressive measurements, and the measurements are used to reconstruct the video by a low-rank and sparse decomposition of matrix. CS has proposed some novel solutions in many practical applications. Focusing on the pixel-level multi-source image-fusion problem in WSNs, an algorithm of CS image fusion is proposed on the basis of multi-resolution analysis. The advantages of CS-based image fusion are also discussed. MRI is an essential medical imaging tool with an inherently slow data acquisition process. Applying CS to MRI offers potentially significant scan time reduction, with benefits for patients and in the area of healthcare economics. A focal plane array-compressive sensing (FPA-CS) is presented with a new imaging architecture for parallel compressive measurement acquisition that can provide quality videos at high spatial and temporal resolutions in SWIR. CS can be used for improvement of image reconstruction in holography by increasing the number of voxels that can be inferred from a single hologram. CS is also used for image retrieval from undersampled measurements in millimeter-wave holography and off-axis frequency-shifting holography. CS has applications in the detection and estimation of wireless signals, source coding, multi-access channels, data collection in sensor networks, and network monitoring. Finally, architecture for compressive imaging without a lens is also discussed.

References

[1] Duarte, M., Davenport, M., Takhar, D., Laska, J., Sun, T., Kelly, K., and Baraniuk, R. (2008). "Single-pixel Imaging via Compressive Sampling", *IEEE Signal Process. Mag.* 25 (2), 83–91.

[2] Wakin, M., Laska, J., Duarte, M., Baron, D., Sarvotham, S., Takhar, D., Kelly, K., and Baraniuk, R. (2006). "An Architecture for Compressive Imaging", *In 2006 IEEE International Conference on Image Processing*, 1273–1276.

[3] Takhar, D., Laska, J., Wakin, M., Duarte, M., Baron, D., Sarvotham, S., Kelly, K., and Baraniuk, R. (2006). "A New Compressive Imaging Camera Architecture Using Optical-Domain Compression", *IS&T/SPIE Computational Imaging IV*, 6065.

[4] Saad, Q., et al. (2013). "Compressive Sensing: From Theory to Applications, a Survey", *J. Commun. Networks* 15.5, 443–456.

[5] Moses, R., Potter, L., and Cetin, M. (2004). "Wide Angle SAR Imaging", *In Proc. SPIE*, 5427, 164–175, Citeseer.

[6] Bajwa, W., Haupt, J., Sayeed, A. and Nowak, R. (2010). "Compressed Channel Sensing: A New Approach to Estimating Sparse Multipath Channels", *Proce. IEEE*, 98 (6), 1058–1076.

[7] Tachwali, Y., Barnes, W., Basma, F., and Refai, H. (2010). "The Feasibility of a Fast Fourier Sampling Technique for Wireless Microphone Detection in IEEE 802.22 Air Interface", *In INFOCOM IEEE Conference on Computer Communications Workshops*, 2010, 1–5, IEEE.

[8] Luo, C., Wu, F., Sun, J., and Chen, C. (2009). "Compressive Data Gathering for Large Scale Wireless Sensor Networks", In *Proceedings of the 15th annual international conference on Mobile computing and networking*, 145–156, ACM.

[9] Charbiwala, Z., Chakraborty, S., Zahedi, S., Kim, Y., Srivastava, M., He, T., and Bisdikian, C. (2010). "Compressive Oversampling for Robust Data Transmission in Sensor Networks", In *INFOCOM, 2010 Proceedings IEEE*, 1–9, IEEE.

[10] Toumaz Technology. (2009). Available: http://www.toumaz.com/public/news.php?id = 92

[11] Shimmer Research. (2008). Available: http://shimmer-research.com

[12] Candes, E., Romberg, J., and Tao, T. (2006). "Robust Uncertainty Principles: Exact Signal Reconstruction from Highly Incomplete Frequency Information", *IEEE Trans. Inform. Theory*, 52 (2), 489–509, Feb.

[13] Donoho, D. L. (2006). "Compressed Sensing", *IEEE Trans. on Inform. Theory*, 52, (4), 1289–1306, Apr.

[14] Candes, E. and Tao, T. (2006). "Near Optimal Signal Recovery from Random Projections: Universal Encoding Strategies", *IEEE Trans. Inform. Theory*, Vol. 52, 12, 5406–5425, Dec.

[15] Hossein, M., et al. (2011). "Compressed Sensing for Real-Time Energy-Efficient ECG Compression on Wireless Body Sensor Nodes", *IEEE Trans. Biomed. Eng.* 58.9, 2456–2466.

[16] Smith, David S., et al. (2012). "Real-Time Compressive Sensing MRI Reconstruction using GPU Computing and Split Bregman Methods", *Int. J. Biomed. Imag.* 2012.

[17] Candes, E. J., Li, X., Ma, Y., and Wright, J. (2009). "Robust Principal Component Analysis?" *J. ACM*, 58 (1), 1–37.

[18] Tong, Ying, et al. (2014). "Compressive Sensing Image-Fusion Algorithm in Wireless Sensor Networks Based on Blended Basis Functions", *EURASIP J. Wireless Commun. Network.* 2014.1, 1–6.

[19] Chen, Huaijin, et al. (2015). "FPA-CS: Focal Plane Array-based Compressive Imaging in Short-wave Infrared", *arXiv preprint arXiv: 1504.04085*.

[20] Duarte, M. F., Davenport, M. A., Takhar, D., Laska, J. N., Sun, T., Kelly, K. F., and Baraniuk, R. G. (2008). Single-Pixel Imaging via Compressive Sampling. IEEE Signal Processing Magazine, 25 (2), 83–91, Mar.

[21] Gupta, M., Agrawal, A., Veeraraghavan, A., and Narasimhan, S. (2010). Flexible Voxels for Motion-Aware Videography. In European Conference on Computer Vision, Crete, Greece, Sep.

[22] Rivenson, Y., Stern, A., Javidi, B. (2010). "Compressive Fresnel Holography". *J. Display Technol.* 6 (10), 506–509. doi:10.1109/jdt.2010. 2042276.

[23] Denis Loic; Lorenz, Dirk; Thibaut, Eric; Fournier, Corinne; Trede, Dennis (2009). "Inline Hologram Reconstruction with Sparsity Constraints". *Opt. Lett.* 34 (22), 3475–3477. doi:10.1364/ol.34.003475.

[24] Marim, M. M., Atlan, M., Angelini, E., Olivo-Marin, J. C. (2010). "Compressed Sensing with Off-axis Frequency-Shifting Holography". *Optics Lett.* 35 (6), 871–873. doi:10.1364/ol.35.000871.

[25] Gross, M., and Atlan, M. (2007). Digital Holography with Ultimate Sensitivity. *Optics Lett.* 32, 909–911.

[26] Cull, C. F., Wikner, D. A., Mait, J. N., Mattheiss, M., and Brady, D. J. (2010). "Millimeter-Wave Compressive Holography", *Appl. Opt.* 49, E67–E82.

[27] Firooz, M. H., and Sumit Roy. (2010). "Network Tomography via Compressed Sensing", *Global Telecommunications Conference (GLOBE-COM 2010), 2010 IEEE*. IEEE.

[28] Richard, S. W. (1994). *TCP/IP illustrated*. (New York: Addison-Welsey Publishing Company).

[29] Bejerano, Y., and Rastogi, R. (2003). "Robust Monitoring of Link Delays and Faults in IP Networks", in *Twenty-Second Annual Joint Conference of the IEEE Computer and Communications Societies (INFOCOM'03)*, Vol. 1, 134–144,

[30] Huang, H. et al. (2013). "Applications of Compressed Sensing in Communications Networks", *arXiv preprint arXiv:1305.3002*.

[31] Fletcher, A. K., Rangan, S. and Goyal, V.K. (2009). "On-off Random Access Channels: a Compressed Sensing Framework", *arXiv preprint*, Mar.

Index

About the Authors

Dr. S. Radha, Professor & Head, Department of ECE, has 26 years of teaching and 16 years of research experience in the area of Wireless Networks. She has graduated from Madurai Kamraj University, in Electronics and Communication Engineering during the year 1989. She has obtained her Master degree in Applied Electronics with First Rank from Government College of Technology, Coimbatore and Ph.D. degree from College of Engineering, Gunidy, Anna University, Chennai. She also worked as a visiting researcher at Carnegie Mellon University for a period of six month in the area of Wireless Sensor Networks.

She has 95 publications in International and National Journals and conferences in the area of Mobile Ad hoc Network and Wireless Sensor Networks. 11 scholars are graduated their Ph.D. under her guidance and presently 7 research scholars are doing research in the area of sensor networks, IoT and cognitive radios.

She is the recipient of IETE – S K Mitra Memorial Award in October 2006 from IETE Council of INDIA, Best paper awards in various conferences and CTS – SSN Best Faculty Award – 2007 & 2009 for the outstanding performance for the academic years 2006–07 and 2008–09.

She has received research funding for sponsored projects as Principal investigator and co-investigator from various funding agencies such as Indira Gandhi Atomic Centre for Research (IGCAR), Kalpakkam, National Institute of Ocean Technology (NIOT), Chennai, INTEL Bangalore, DST and SSN. She is a Senior member of IEEE, Fellow of IETE, and Life member of ISTE.

Dr. R. Hemalatha, has received her B.E. degree in Electronics and Communication Engineering from PGP College of Engineering and Technology, and M.E. degree in Communication Systems from SSN College of Engineering (with College Gold medal). She obtained her Ph.D. degree from Anna University for her research work on Image transmission in WMSN. Currently she is an Associate Professor in the department of ECE, SSN College of Engineering.

She has around 12 years of teaching experience. She has research publications in International Journals, National and International level conferences. Her research interests include Wireless Sensor Networks, Energy Harvesting Techniques, Compressive Sampling applications and Image Processing. She is a life member of ISTE.

Aasha Nandhini Sukumaran is a Senior Research Fellow at SSN College of Engineering, India. She received the B.E. degree in Electronics and Communication Engineering from Rajalakshmi Engineering College, India, in 2010 and the M.E. degree in Communication Systems from SSN College of Engineering, India, in 2012. She is currently pursuing her Ph.D. from Anna University, India. She has research publications in International Journals, National conferences and International conferences. Her research interests include security issues and compressive sensing in wireless sensor networks. She is a member of IEEE Communication Society (IEEE ComSoc).

Lightning Source UK Ltd.
Milton Keynes UK
UKOW06n1147270117
293027UK00002B/44/P